Work Study and Ergonomics for Mechanical Engineering

Arvind Kaushal

Rekha Ahirwar

Published by

BONFRING®
Intellectual Integrity

Work Study and Ergonomics for Mechanical Engineering

9 789386 638885 >

ISBN 978-93-86638-88-5

Authors

Arvind Kaushal

Rekha Ahirwar

Bonfring

309, 2nd Floor, 5th Street Extension, Gandhipuram,

Coimbatore-641 012.

Tamilnadu, India.

E-mail: info@bonfring.org

Website: www.bonfring.org

Phone: 0422 4213231

Preface

In present scenario, it is a time of modern technology that is depends upon the growth of industrial sector since all the sectors are in competitive environment, each and every industrial have a notion to achieve better position in competitative world. But the question is arises here how can we achieved it, answer is very simple i.e. Enhancement of the productivity Further Productivity enhancement can be achieved by using the principles of industrial engineering such as Work Study and Ergonomics. This book established the basic foundation of work-study & ergonomics for modern manufacturing technology in order to improve the work performance of operator/work men thereby productivity of manufacturing process is positively affected. this book should be provided correct path for the students who will be become a parts of mechanical manufacturing in future , industrial operator, research scholars' & technician for valuable reference in the field of manufacturing . The overall topics covers are especially designed and prepared for undergraduate and post-graduate students & applied in mechanical engineering, production engineering, industrial engineering syllabus. the subject work study and ergonomics is recognize by different titles some of the synonyms may be work design and ergonomics, work study and ergonomics, industrial ergonomics, occupational health and ergonomics etc. This book is divided into fifteen parts/chapters, **chapters 1** is an introduction of work study & ergonomics ,this chapters also deals with the relational concepts of work-study & productivity. **Chapter 2** - Described about the Purpose & Basic approach of work-study, Method study and procedure, various type of charts & diagram applied in work-study . Therbligs, memo & micro motion study, critical examination, Principle of motion economy. **Chapter 3** – deals to Work – Measurements, Purpose of Work- Measurement, Procedure of Work Measurement , Time - study method , Time study equipments , Rating , type of Rating , Allowances, type of Allowances, No of cycle to be timed, Economic considerations, Breaking a job into elements, Work sampling, Theory of sampling, Advantages and Disadvantages of Work Sampling , Standard time. **Chapter 4** - In this chapter we focused on the Incentive schemes, type of Incentive schemes, Predetermined motion time systems (PMTS), Method time measurement (MTM), MOST (Maynard Operation Sequence Technique), Work Factor System, Standard data system. **Chapter 5** – Described about the human factors engineering , Definition of human factor engineering, Effect of ergonomics in the workplace, Sitting posture for operator , display system, detail study Ergonomics, Man-machine-system & their type, Automation and Control, Coding of information, Data processing. **Chapter 6** – This chapter included Anthropometry, Anthropometric variables, Anthropometric principles,

Applications of Anthropometry, Display systems, type of Display systems, Safety Signs and Colour code, Factorial and Graphic display, Pie Charts, Bar Charts, and various others charts, Factors influencing visual display effectiveness, characteristics of display. **Chapter 7** – This chapter deals with introduction to Working Condition, Color coding system for shop safety, Working space and seating, Accidental prevention at work-place, Fire precautions & Fire colour code. **Chapter 8** – deals with concept of Plant layout, Objective of plant layout, Classification of layout, Material handling technique, Principle of material handling system, The purpose of material handling, Type of material-handling equipment, selection of material-handling equipment, Site Selection, The Site Selection Process. **Chapter 9** – This chapter concentrated upon Quality & Reliability, aspect of quality, Quality definitions, Quality control, Quality Control Procedure, Quality Improvement tools, Customer orientation, Customer orientation steps, External& Internal Customer, Life cycle approach to quality costs-Prevention, Quality Philosophies, Quality Control tools, Control Charts, PDCA-CYCLE, Process capability concepts, Reliability, Bath tub curve, Design for reliability, Availability, Maintainability, Redundancy and their types , Types of System Reliability. **Chapter 10** - This chapter deals with introduction to Inventory Management, Type of Inventory, Inventory costs, Principles of Inventory Management, Demand forecasting, EOQ, ABC Analysis, VED Analysis SDE analyses, MNG ANALYSES, MRP, Master production schedule, The product structure records. **Chapter 11**- This chapter deals with introduction to PERT/CPM, Scheduling, Principle of scheduling, Job shop scheduling, Flow-shop scheduling, Job Shop Sequencing, Line balancing. **Chapter 12** – This chapter deals with introduction to Production planning and control, Principle of lean manufacturing, Agile manufacturing system. **Chapter13** - This chapter deals with introduction to Product design and development, Product function, Product design Concept, Economics of a new design, Product Cost analyses, The Product Lifecycle, aspects of the products, Profit and competitiveness, Break-Even Analysis. **Chapter 14** - This chapter deals with Capacity Planning, Process of Capacity Planning, Capacity planning technique, Factors Affecting Capacity Planning, Aggregate planning, Demand Forecasting Demand patterns, Factor Influences the Demand, Principle of forecasting, Forecasting Methods , Casual Forecasting , Objectives & Selection of Demand Forecasting , Forecasting models. **Chapter 15** - This chapter deals with introduction to Operations Research, contribution of operation research in manufacturing production, objectives of operation research, important aspect of operation research, Linear Programming.

Acknowledgement

Any kind of **manuscript** or **book** preparation is very complicated and difficult task which could not be achieved by a single person. At very initial it is a constant supports & motivation of God for giving me the strength and thought for this task, my parents and others family members, teachers, colleagues, friends, as well as discussion and question form under graduates students throughout the teaching session except it many others individuals have also been contributed to the preparation of this book. I also extend my sincere thanks to Professor Dr.R.S. Rajpoot who is Director of UIT (University Institute of Technology) **Rajiv Gandhi Proudyogiki Vishwavidylaya Bhopal (the technical university of MP** for their constant supports and encouragement. Last but not the least I am thanks to all teachers who have taught me thereby I have been able to achieve this state of study.

Finally this work is wholly dedicated to GOD

Chapter	Contents	Page No

CHAPTER 1

1.0. Introduction

Work study is nothing but, it is a management tool applied to improving the work performance of workman who is associated in manufacturing industries by applying the various kinds of sub tools for example – method study or motion study , work- measurement , study of many types of charts in order to get visualizing the correct & accurate analyses the worker performance.

Work study consists a concept of economics, engineering, ergonomics, behavioural sciences, computer science, statistics and management science. Work-study also has a great significance in resource management or we can say that it provided us productivity improvement criteria. How productivity is improved, what is correct procedure to applying & implementing the process? Work study & ergonomics is a conceptual thought of our management gurus that point out how performance improved with minimum investment of available resources , which is directly proportional to output or in other word proportional to profit earned by the manufacturing organization, it also helps in deciding the wages & incentive of workmen's .

Work study can reduced the probability of accidents throughout the manufacturing operation by reducing the no of possible operation and distance travelled by goods as well as raw material , process charts , process analysis , and critical examination of work also reduces the major chances of accidents in an manufacturing organization.

Ergonomics is also a branch of science including engineering , anatomy and physiology, psychology, engineering and statistics in which workplace is design in such a manner that the operator or worker can perform their task comfortably , stress free & without any dangerous or harm Ergonomics aims to improve workspaces design and environments to minimize risk of injury or harm.

Ergonomics included some other very important aspect for detecting the suitability of work place & working environment such as – **Anthropometry** related to body sizes, shapes , populations and variations, **Biomechanics** related to muscles, levers, forces, strength , **Environmental physics** related to noise, light, heat, cold, radiation, vibration , **body systems** related to hearing, vision, sensations, **Applied psychology** related to skill, learning, errors, differences, **Social psychology** related to groups, communication, learning, behaviours.

1.1. What is Work Study and Ergonomics

Work-study is a composition of various tools & techniques used in manufacturing industries , for improving the productivity of manufacturing system that is productivity ,means minimum wastage of available resources like- men , machine , material , time , space , capital and so many others factors. In broad sense work-study is a composition of method study & work measurement.

Productivity is defined as the ratio of input to output, all the manufacturing units wants to improve the productivity i.e. minimum wastage of available resources , the resources must be utilized in such a manner that it can be converted fully in useful way . The minimum wastage of available resources means improving the profit the organization.

$$\textbf{Productivity} = \frac{\textbf{Output}}{\textbf{Input}}$$

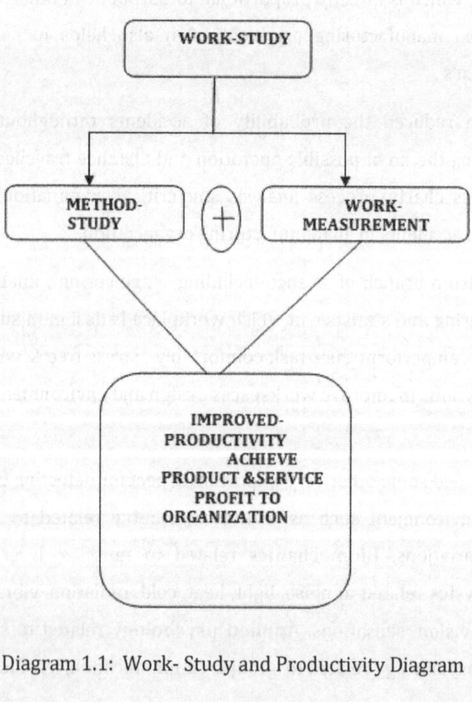

Diagram 1.1: Work- Study and Productivity Diagram

Classification of Productivity

In general Productivity is of following types:

- Labour productivity.
- Capital productivity.
- Material productivity.

Labour productivity :- Labour productivity is defined as how many units are produced per hours by the worker or operator in manufacturing

Capital productivity :- Capital productivity is defined as how much unit capital input is done for producing the output by the worker or operator in industries.

Material productivity :- Material productivity is defined as how many units are produced per material input throughout the manufacturing process.

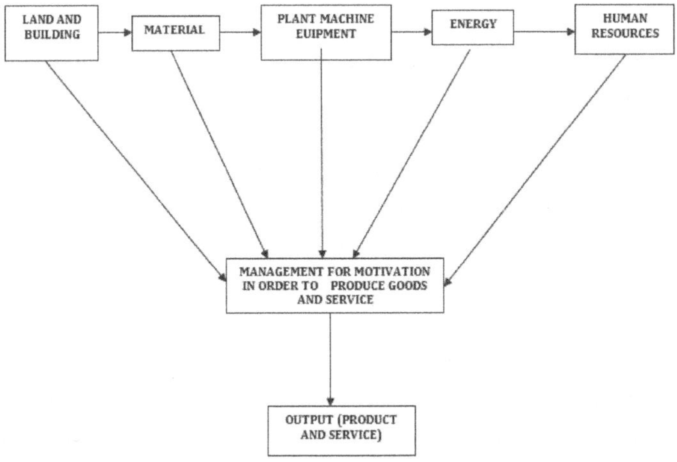

Diagram 1.2: Productivity Diagram

Following are Very Important Advantage of Improved Productivity

1. Increase the wages to employees.
2. Distribution of products systematically between the stock-holders.
3. Selling of products as possible as lower cost.
4. Calculating the labour productivity index with respect to others companies. How much annually labour performance is?
5. Improve the standard of living and reduce inflation.

1.2. Ergonomics

Ergonomics is related to the study of working environment , safety , precaution , worker health , in order to get optimal performance of the operator or worker throughout the manufacturing or production of articles . concept of the ergonomics is same as that of work-study i.e. improve the productivity in term of profit to the manufacturing organization because good environmental working condition of manufacturing shop , improve the interest of operator and efficiency and hence finally we have got the improved productivity on the other hand bad working condition decreases the performance & efficiency of operator .thus the no of factor is considered in designing the workplace or workshop layout , the study of science related to the concern factor called Ergonomics . it is relationship between man and his working environment for improving the optimal performance of human operator . it also considered the physical & mental balance of the operator or worker.

Following are some important definition of ergonomics that has-been given by famous management guru's , these are given below .

(i) Ergonomics is the science which deals with the relationship between man and his working conditions or operational environment .

(ii) Ergonomics is the process of designing or arrangement work places, product, and system so that they fit the peoples who use .

(iii) Ergonomics is the application of human bio-logical science in association with engineering science to the worker and his working environment to obtain maximum productivity .

(iv) Engineering is a multi-disciplinary science comprising subject like anatomy, psychology, physiology, sociology, engineering, anthropology, physics and medicine.

(v) Ergonomics is the study of human abilities and characteristics, which affects the design of equipments , system and jobs .

1.3. Effect & Significance of Work-Study on Enterprises

In any kind of enterprises, the basic & very important prestigious role play by the manager in an enterprises, because it faces so many challenges in operation at work-place. An intelligent enterprises manager always try to investigate the basic problem associated with the production how it can be controlled for improvement and then correction is carried out , or may be eradicated if necessary in such a manner that does not affect the overall output. a manager must have suitable or sufficient skill , knowledge , experience , have always been able to make adequate advancement for enterprises. The prime value of work study lies in the fact that, by carrying out its systematic procedures, a manager can achieve results as good as or better than the less systematic genius was able to achieve in the past

Work-Study Consists of Two Segments

1. The systematic investigation of the problem should consider related to production.
2. Resolving the problem in order to achieving the correct & feasible solution including their development.

Systematic investigation consuming to much time but it is very necessary to understand the problem . this work is carefully carried out by the management regarding the job.There are so many activity is regularly happened day to day problem like- human and material in an enterprises for that our managers and supervisors are very aware to it but they could not afford to devote a long time to study a single activity in work-place which is very impossible.But for correct implementation of procedure, required accurate information that is achieved by continuous observation & systematic study of work-place where in the the problematic activity is going on in order to get a actually existing data in an operating area. To fulfill the required purpose management appointed a person who can carried out the responsibility very seriously and engaging fulltime with the related activity, the responsible person is chosen from the senior working staff member or may be appointed from management team , thus work-study is an accurate management & supervision of the activity to predicts the reality about the activity means what is actually being happened in course of activity. Work-study is a clause of various tools & techniques of management to affirm the growth of any kind of manufacturing or production industries. Following are some important sub-tools which had following significance .

(i) Any plant or operating unit associated with some kind of production or manufacturing of articles, for that any specific method, tools, and equipments is used to facilitations and completing the task. This can be completed with minimum effort by applying the work-study and hence enhancing the overall productivity of the system.

(ii) Work-study helps us to analyses the method, if necessary , improving the method or developing the new method thereby it can improve the efficiency of production in systematic manner.

(iii) It is an effective tool for proper production , planning & control .

(iv) This can helps in control the situation of work-place in terms of safety , working environment , precaution regarding hazardous operations and developing a new advanced method for doing the job effectively & safely .

(v) Generally it has always be connected to work-study with the manufacturing industries or plant , but this tool can also be implemented in other areas like office , stores , laboratories , Service industries , restaurants , other firms etc.

(vi) This method is very simple, less time consuming, minimum capital investing procedure, and easy for applying & implementation .

(vii) It is excellent weapon for management to overcome in-efficiency in any kind of manufacturing organization for short out the problem & their correct and feasible solution .

In many time it is found that so many supervisors and managers have been generally failed to get a positive feed- back by implementing the work-study, because they have been unable to apply themselves continuously to such activity due to the lake of training . the work study can be applied everywhere for improving the productivity, intolerance of waste in any form, whether of material , time , effort or human ability.

Question for Discussion

1. What is the role of ergonomics in industrial engineering ?
2. Defme productivity. What productivity measure should be adopted as a basis of optimizing the working of an organisation? Discuss.
3. Bring out the importance of productivity measures in any organization. Also in detail enumerate the different productivity enhancement tools used by the organizations in the present competitive scenario.
4. How could you measure improvement in productivity?
5. How use of work study leads to higher productivity in manufacturingunit? Explain briefly.
6. Explain different productivity Ratios. What are the different factors affect as productivity.
7. What are the responsibilities of Industrialists Environment and Ecology.
8. What is Productivity? Give various ways to improve productivity and also Explain at least one way, with the help of suitable example.
9. Describe the ways to increase the industrial productivity.
10. Define productivity. How the productivity of land, machine, man and material can be increased.
11. What is Productivity? Discuss the factors affecting the productivity of the manufacturing organization.
12. Discuss the following:
    1. Work Study and Management.
    2. Work Study and the Worker.

13. State the relationship between productivity and work study. Explain various tools for increasing productivity.

References

1. J.E. Thurman, "Higher productivity and a better place to work: Action manual", (Geneva, ILO, 1988.
2. W.T. Singleton, "Introduction to ergonomics", (Geneva, World Health Organization), 1972.
3. Prokopenko Joseph, "Productivity management, a practical handbook", (Geneva, ILO), 1987.
4. Higher productivity through better management. Management Development Programme technical paper Man Dev/56 (Geneva, ILO), 1990.
5. Lawler Alan, "Productivity improvement manual", (Aldershot, Hampshire, Gower), 1985.
6. J. Baranson, "Industrial Technology for Developing Economies", New York: Praeger, 1969.
7. H.B. Maynard, "Industrial Engineering Handbook", New York. Mc Graw-Hill. Second Edition, (1963).
8. I.L.O, "Introduction to Work Study", (Geneva International Labour Office), 1957.
9. Psychology and Productivity by Paul Whitney (Author), Robert B. Ochsman (Author).
10. RN. Sen, "Application of ergonomics to industrially developing countries", Ergonomics, Vol.27, Pp. 1021-1032, 1984.
11. Dan Macleod, "The ergonomics edge improving safety quality & productivity".
12. Ravi Shankar, 'Industrial engineering and management',.
13. Jack Greene, 'Industrial engineering',.
14. Alan Hedge, 'Ergonomics work place design for health & wellness',.

CHAPTER 2

METHOD STUDY

2.0. Purpose of Work-Study

The basic purpose of work-study is to improving the overall efficiency of enterprises or productivity that's includes , enhancing the productivity of manufacturing or production , because the basic resources of any manufacturing for creating any kimd of articles are **man , machine , material , space , time , capital investment** etc.

All those are the input for production that should be minimize as possible as to achieve the desired output with feasible profit so therefore the productivity may also be considered as efficiency of system. Or in other word we can say that productivity is the utilization of available resources in such a manner that fully utilized with minimum wastage.

$$\text{Productivity} = \text{Output} \,/\, \text{Input}$$

Following are very important available input for creating a desired output for a manufacturing system.

1. **Men's / worker**: Worker may be a man or women that perform the task or operation, he or she must be very well trained according to the proposed method of doing the job. And also perform others task if necessary such as maintenance or administrative and secretarial job
2. **Machine**: Machine tools are another very important requirement of operational activity in a plant without which nothing can be produce others included, transportion and handling, heating or air conditioning, office equipment, computer terminals etc.
3. **Material:** Before producing any kind of articles, there is a requirement of raw material because the raw material is converted into finished product that should be standardized, easily available at low cost, if the required standardized raw material is not available then their auxiliary material may be used to fulfill the desired purpose.
4. **Space or land:** Space or land is on prime consideration for establishing the plant because it should be well connected to transporting medium, for receiving all kind of accessories that required in plant as well as for transporting the finished goods. Finally location should be convenient in all aspect.
5. **Power sources:** All kind of power sources must be available near about the plant, for example electricity, gas, oil, or coal etc.

6. **Time:** Time is also a very important factor in association to the above.

Except all those input resources, some others are also important from subject point of view, the combination of all resources helps in calculating the productivity of enterprises.

Minimum wastage of resources may improve the productivity or enterprises got an optimal profit.

In order to get a very close concept regarding the work-study , it may also be defined as

Work –study is a complete study of manufacturing an articles involving a method study / motion study, time study, work-measurement that has been used for investigating all the factors regarding the manufacturing thereby we can analyzing the effect upon efficiency, economic growth, productivity and also profit, actually it is the complete overview of the title we can also make a deep & minute analyses of others factors whose effect will be on prime consideration .

Basically it has been consists of two broad problems for analyses

1. In which manner a job should be completed / how can it be perform ?

This question is concentrated to method study / motion study , work- simplification

2. How much time should be taken for completing the job, study is done can it be minimize ?

It has been associated with the , time study and work-measurement

2.1. Basic Approach for Work-Study

The ray diagram 2.1 given below shows the basic steps of work-study.

Objective of Work-Study

1. Due to study, we can achieve a best method of performing each & every operation regarding the job.
2. Improve productivity by reducing the associated wastage of available resources, increases rate of production, also favourable for worker and operator due to less fatigue.
3. For set-up a standard time for a particular job, that is being taken by qualified / experienced operator or worker.
4. It has also been helpful in fixation of wages & incentive for worker.

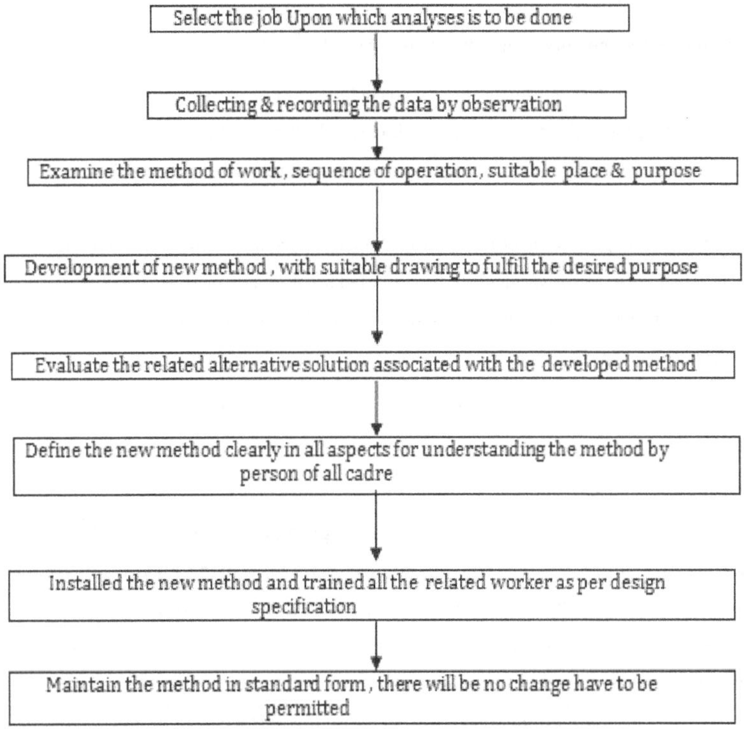

Digarm 2.1: Basic Approach for Work-Study

Application

There are various field of application of work-study, generally it has been understand by so many authors that it is not only apply for industrial or production related problems, but it can also be applied in following field.

1. The very first field of application is industries, production operation, research and development.
2. In proper/systematic management of offices, stores, ware-housing.
3. For selecting the exact material handling technique through the manufacturing process.
4. For achieving correct work--method design.
5. In building & others related constructions problem.
6. It may also apply in transportation, hospital, army, agriculture etc.

Advantages

Followings are some important advantages of work-study.

1. Improved production rate, uniformity is achieved throughout the production.
2. Due to improved production rate, we have got higher production efficiency.
3. The manufacturing cost of article creation is reduced due to the proper & systematic utilization of available resources .
4. It has also helpful in selecting the wages for workers, as per their performance.
5. Satisfactory & better service can be provided for the customers.
6. For creating favorable and better working condition for operator or worker

2.2. Method Study

The term work-study is an integration of various tools & technique, such as method study or motion study, work-measurement and so many others, in this portion we will discussed about the method study.

Method study is nothing but related to way of performing the job and it is defined as systematic procedure and critical examination of way of doing the job as per the existing & proposed method to fulfill the purpose of development of existing method in such a manner which will be more effective & economical , easy to implementation ,safe operable to operator thereby we can reducing the production cost of an article.

The basic purpose of method study is to facilitate the performance of work while meeting pre-determined standards using the least amount of time, effort and resources. The best way to conduct method study is in collaboration with the people actually performing the work to be studied. The people doing the work usually have the best insight into how methods may be analyzed and improved. Conduction of method study is perform with the help of questionnaires which consists of no question related to method comes on the front line as a problem that will give us a correct guiding path to innovation. Implementation of method study not only saving the available resources but also eliminating unnecessary activities that is time consuming thereby it also reduced the capital investment .

Conduction of method study involving the following stages for any appropriate job or task.

Stage – 1

In this stage we will considering the detail study of the job regarding the various operation , study of job performing method , how many & which kind of machine is required , including various type material handling technique like AGVS , LIFT , FORK ,TRUCK, etc as well as how many skilled worker/operator are required for doing the job . Recording all the facts regarding the job by direct observation and collect all the necessary information in respect of job .

Stage – 2

In this stage critical examination is performed, this examination included , the way of performing the job , how many tools & equipment is required as per present method. Critical examination consists following questions

Element	Description
Purpose	• What is the requirement of the present operation ? • Does this give the required output as per the management ? • If this operation may be eradicated or remove without affecting the purpose ?
Place	• Study about the suitable place for doing the operation without affecting the Performance ?
Sequence	• In this we will study about the correct sequence of operation, correct timing the operation ? • Can it be perform at same time or before time or after a specific operation ? • Defined what will the suitable, effective & economical time be for Performing ?
Method of performance	• Study about which method is used for doing the task , in which manners the operation can be performed ? • How many & which kind of machine tools is required ? • Who is best person for performing the job ? • Can the present method be easier,safest, or effective ?

Stage – 3

In this stage overall analyses & concentration is made to develop a new method for doing the job which will be most practical , easy , economic and effective method. in this section following question must be considered

1. **Elimination or removing the operation :** It included complete details of each & every operation associated with the job , study should be made in such a way that any operation may be removed or eliminate without affecting the performance the job operation .
2. **Combining or parallel operation :** In this clause, we considered the provision of combination of two or more operation can be perform without affecting the performance of task.
3. **Simplification :** In this section , concentration is made upon the simplification & re-arrangement of sequence of operation thereby we can reduced the material handling operation, unnecessary storage , also reduced internal transportation of un–finished

goods . Simplification also leads to adopted that method which must be very safe, economical and easy to implementation. It has been consists following guiding points.

A. Required material , tools and equipments should be placed in such a manner that can be used easily in working area.
B. Both hands should be used in such a way that utilized fully.
C. Gravity feed hopper , special type of jigs & fixture should be used.

Stage – 4

In this section we have to evaluate different alternatives to developing a new improved method comparing the cost-effectiveness of the selected new method with the current method of performance.

Stage – 5

In this section the new method should be installed , new method must be clear in all respect that has been approved by management, supervisors and workers. Worker must be trained according to new method .

Stage – 6

Once a method is installed, used practically in work-place must be save in standard form could not be changed in any aspects if so management permission is necessary when management realize the positive feedback .

2.3. Charting Technique Applied in Method Study

2.3.1. Flow Process Charts

flow process charts is a very important charts in method study , it is defined as the graphical representation of complete set of activity of manufacturing an articles from initiation to finished product . The basic observation is made upon sequence of operation, transportation of raw material/semi-finished product , inspection , delay as well as storage. This gives the complete information regarding the distance travelled throughout the manufacturing process & how much time is being consumed in various activities. This charts also helps in improving the efficiency , proper selection of layout etc . The American Society of Mechanical Engineers has established as standard the five symbols listed below. This set of symbols is a modification of Gilbreths which is widely used in an particular operation . the brief description of all the symbols are given below .

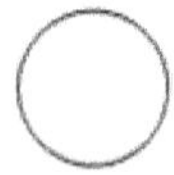

(i) Operation : An operation occurs when an object is intentionally changed in one or more of its characteristics. An operation represents a major step in the process and usually occurs at a machine or work station.This activity is represented by **circles.**

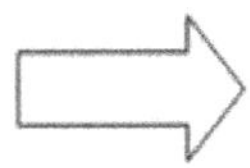

(ii) Transportation : A transportation occurs when an object is moved From one place to another, except when the movement is an integral part of an operation or an inspection. This activity is represented by **closed arrow.**

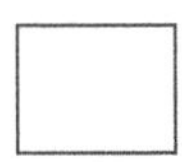

(iii)Inspection : An inspection occurs when an object is examined for identification or is compared with a standard as to quantity or quality. This is generally represented by **square.** "D"

(iv) Delay: A delay occurs when the immediate performance of the next planned action does not take place . delay is represented by capital letter **" D "**

(v) Storage : Storage occurs when an object is kept under control such that its withdrawal requires authorization. The given activity is represented by **inverted triangle.**

The Flow Process is Generally Three Types

- **Flow process chart for worker or operator :** In this process charts , the operator activity is recorded throughout the process , means what's the operator does .
- **Flow process chart for material :** In this process charts , we examine the handling of material at various stages in manufacturing the product .
- **Flow process chart for Equipment :** In this process charts we study about , how the equipment is used for completing the operation.

Job Title – Inspection of Component

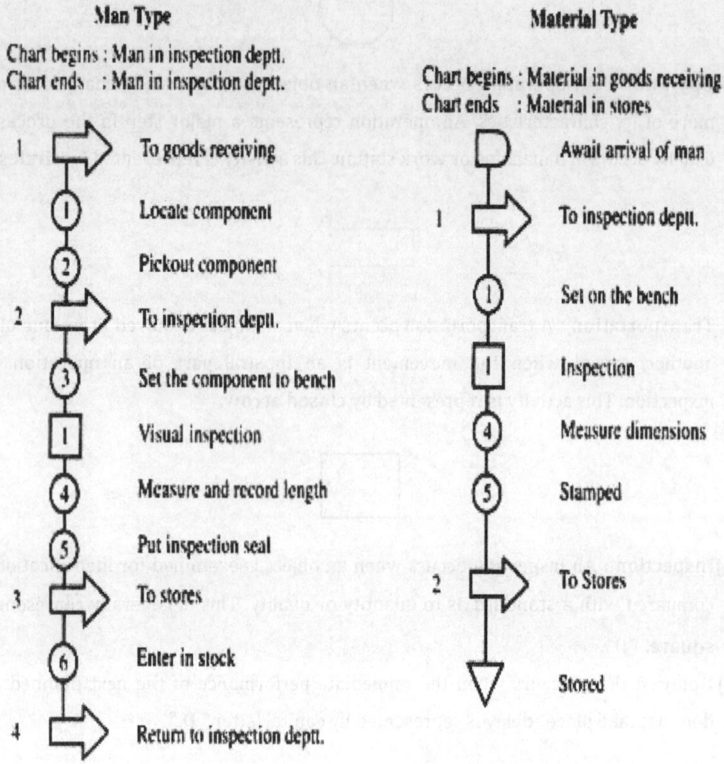

Diagram 2.2: Flow Process Charts (Man –Material Type)

Job Title : Gear Manufacturing

Shop-Machine Shop

SUMMARY			
Activity	Symbol	Present	Proposed
Operation	O		5
Transport	⇒		8
Inspection	□		2
Delay	D		5
Storage	∇		2
Distance		116 m	

ELEMENT DESCRIPTION	SYMBOL					DISTANCE (m)	TIME
	O	⇒	□	D	▽		
50 mmΦ c 1004 bar from stores					●		
Sent to Cutting Machine		●				15	
Cut to Size	●						2.5
Sent to Lathe		●				35	
Delay or wait				●			1.8
Facing Drilling and Reaming	●						
To Lathe		●				5	
Facing of other side, turn to size	●						1.8
To gear hobbing machine		●				4	
Wait				●			
Machine the Gear	●						4.5
To inspection		●				10	
Wait				●			
Inspection for size			●				
To heat treatment deptt.		●				12	0.5
Wait				●			
Hardening	●						
To inspection		●				15	
Wait				●			
Inspection for hardness			●				
To spare part stores		●				2	
Storage for reissue					●		

Diagram 2.3: Flow Process Chart Material Type

Flow Process Chart (Equipment Type)

For Inspecting the Turned Shaft

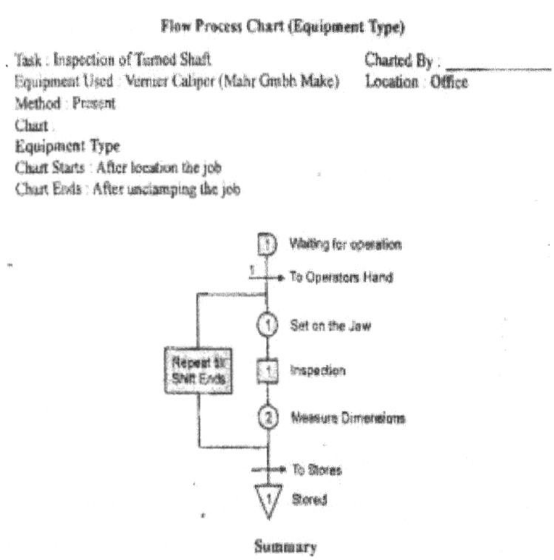

Diagram 2.4: Flow Process Chart (Equipment Type) For Inspecting the Turned Shaft

2.3.2. *String Diagram*

It is also a very important type of flow diagram generally used in method study. string diagram is the simplest & oldest technique for recording and examine the technique for various activity in limited working area / operational area and records the movement of workers / operators , movement of equipment used , movement of raw material or semi-finished products during the particular operation for specified period of time in shop area thereby we can calculated the distance moved also frequency .

The string diagram is a plan scale model on which a thread is used to trace and measure the path of workers , material or equipment during a specified sequence of event in this diagram a string or thread is used to measure distance , the threads may be a simple or colored . Because of this it is necessary that the string diagram be drawn correctly to scale, whereas the ordinary flow diagram will probably be drawn only approximately to scale , A string diagram can be used to plot the movements of materials , and this is sometimes done , especially when a work study person wants to find out easily just how far the materials travel . The work study person proceeds to follow the worker being investigated as he or she moves from point to point in doing the job.

The study person notes methodically each point to which the worker moves a , if the journeys are fairly long , the times of arrival and departure. It will save a good deal of writing if the observer codes the various machines , stores and other points of call by numbers , letters or other means . String diagram is generally used in following condition .

(i) When group of operator working together .

(ii) Single operator,operated two or more machines .

(iii) To and fro motion of sub-assembly to another's .

(iv) Operator moves one work-place to another's .

(v) Effects on various layouts value .

Construction of String Diagram

1. Developed a scale layout of shop area or working area and highlights various features like machines , equipments , work-bench , and stores . all these are indicated on drawing sheets with suitable scale .
2. Kept the drawing sheet on drawing board and strike out the pins on highlighted area between which the material flow or operators movement or equipment movement is going on .

3. Then string or thread is to be stretched form highlighted starting point to the last end point . The string used should be of different colour for clear identification of each activity.

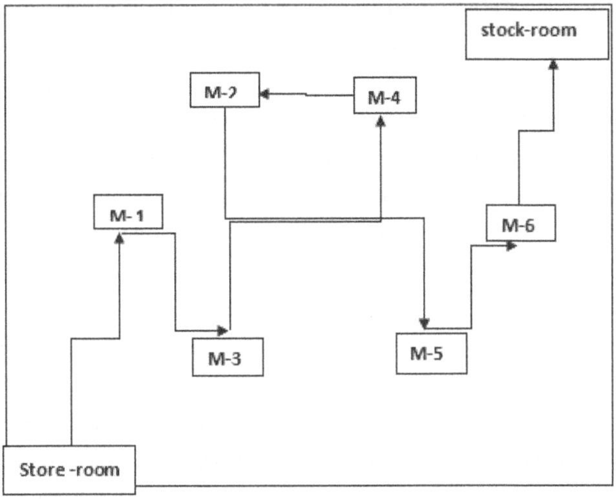

Diagram 2.5: Represented the String Diagram of Operation on product that's passes through six- machines , we can calculated the distance moved & also frequency in work-shop area

The Manufacturing Of A Metal Box

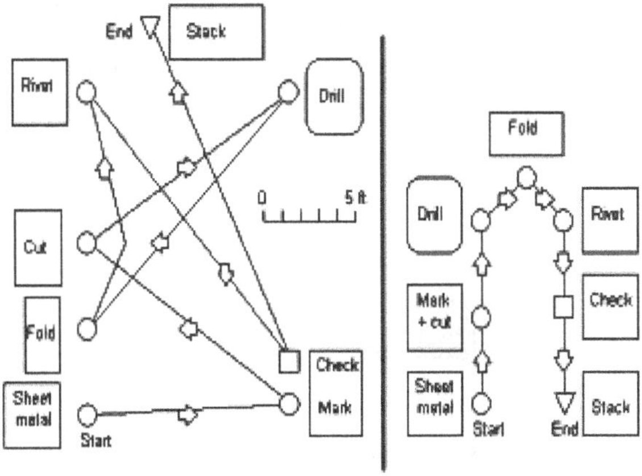

Diagram 2.6: String Diagram For Manufacturing Of A Metal Box

2.3.3. *Multiple Activity Charts*

Multiple activity charts is usually applied where we want to establish a interrelation-ship between various activity during the production of an articles , such kind of charts are generated when number of workers /operator or group of operator , engaging two of more machine or equipments simultaneously and their activities are recorded on common time scale , thus representing a inter-relationship at a each moments of time , such kind of charts is called a multiple activity charts . it is very important in team operation work or mass production or in brief we can defined as a multiple activity chart is a chart on which the activities of more than one subject (worker, machine or item of equipment) are each recorded on a common time scale to show their interrelationship.By using separate vertical columns, or bars, to represent the activities of different operatives or machines against a common time scale, the chart shows very clearly periods of idleness on the part of any of the subjects during the process. A study of the chart often makes it possible to rearrange these activities so that such ineffective time is reduced , for example man – machine charts is a multiple activity charts in which an operator is attending the two machine on common time scale , thus we have got an inter-relationship between operator and machine. Diagram below representing the multiple activity charts .

Multiple Activity Charts

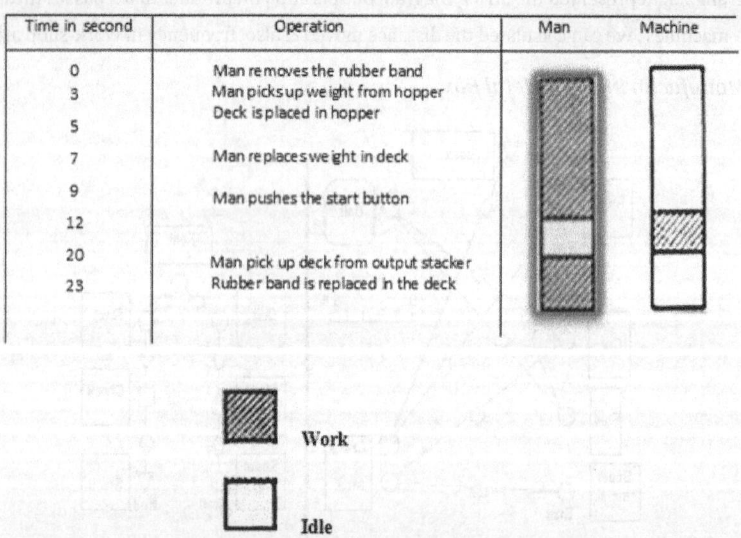

Diagram 2.7: Man Machine Multiple Activity Chart

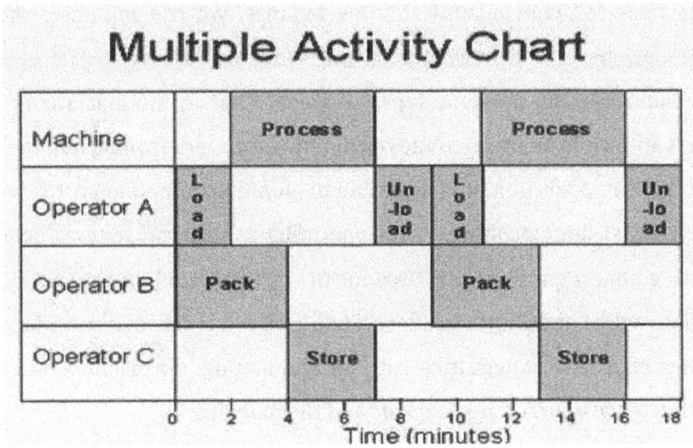

Diagram 2.8: Multiple Activity Charts

The multiple activity charts is extremely useful in organizing teams of operatives on mass-production work, and also on maintenance work when expensive plant cannot be allowed to remain idle longer than is absolutely necessary. It can also be used to determine the number of machines which an operative or operatives should be able to look after. After constructing a charts work-cycle is rearrange such that , that minimize men & machine idle time also simplify the operation combine or eliminate some elements .

Goal of Multiple Activity Charts

(i) To detect the idle time enforced on men and machines and thereby rearranging the work cycle in order to reduce the same.

(ii) To establish the number of machines that may be conveniently managed by an operator.

(iii) To assess the number of workmen required to perform a job involving an organized team work.

(iv) To reallocate the activities among the workers in order to achieve optimal work distribution.

(v) To combine or eliminate some of the work elements if observed unnecessary.

2.3.4. Man Machine Charts

A man-machine chart graphically represents the relationship between the manual work performed by one or more operators and one or more machines& their activity has-been recorded on common time scale involved in a manufacturing process .when shop has a no. Of operator and machines, the analyses can be performed to find out how many machine can each operator canbe operated conveniently. In this manner, we can minimize the idle time of worker aswellas machines and fully utilized the available resources. The operator and the machine work intermittently on some types of work. That is, the machine is idle while the operator loads it and while he or she removes the finished work from it, and the worker is idle while the machine is in operation. It is desirable to eliminate idle time for the worker, but it is equally important that the machine be kept operating as near capacity as possible. The first step is eliminating unnecessary waiting time for the operator and for the machine is to record exactly when each works and what each does. Such a record is called Man & Machine chart. The man and machine chart often suggesting path for eliminating idle machine time and promotes a better balancing of the work of the operator and the machine.

Any Operations Consist of three Main Steps

1. GET READY, such as putting material in the machine.
2. DO (doing the work), for operation like grinding , or drilling etc.
3. Waiting for an operation to be completed or "Idle.

For Example

1. An operator attending one machine.
2. An operator attending two- machine.
3. An operator attending three- machine.

The graphical arrangement & their conclusion are given below

1. An Operator Attending One Machine

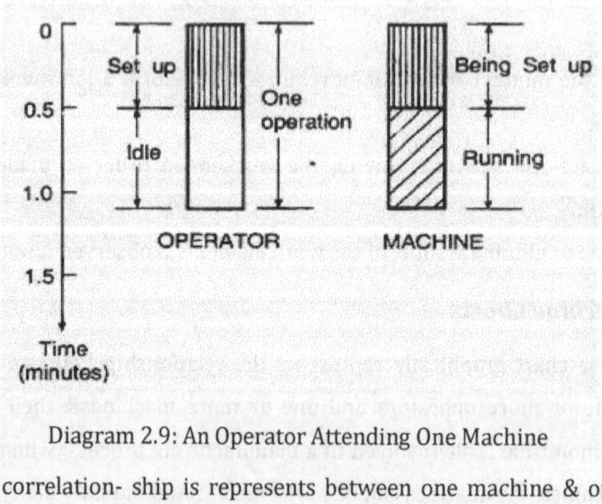

Diagram 2.9: An Operator Attending One Machine

In this charts correlation- ship is represents between one machine & one operator, it is clearly show that the operator was idle for 0.75 minutes on the same time machine is running.

2. An Operator Attending Two- Machine

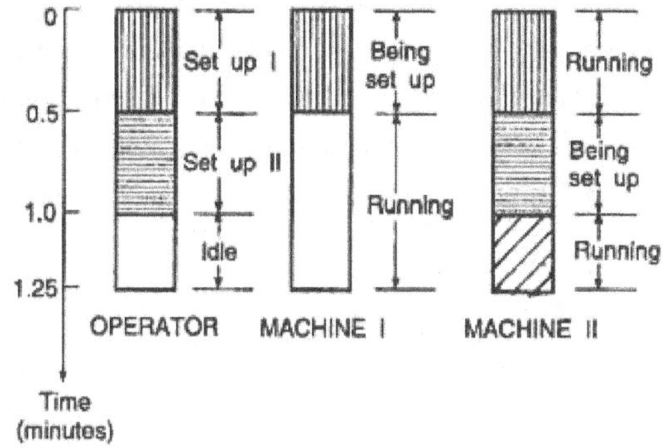

Diagram 2.10: An Operator Attending Two Machine

In this case a single operator operates a two machine against common time scale, representing that the operator is ideal for 0.25 minutes

3. An Operator Attending Three- Machine

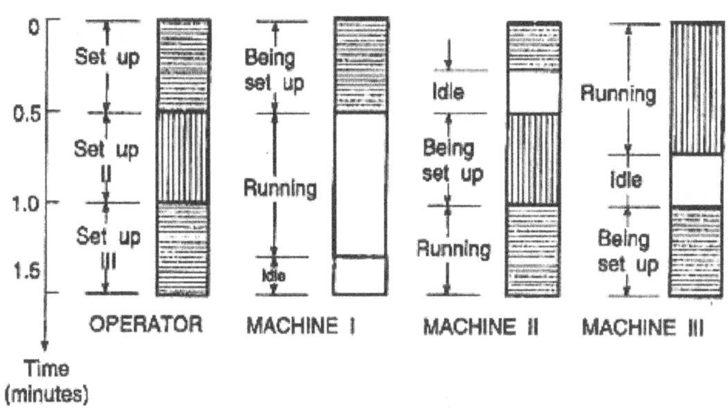

Diagram 2.11: An Operator Attending Three Machine

In this position a single operator operates three machines on common time scale, concluded that each machine is idle for 0.35 minutes.

Notation

1. The first vertical line indicating a common time scale in **minuits.**
2. The **1st** vertical coloumn representing the **position of operator**.
3. The**2nd ,** **3rd**& 4th vertical bar representing the **position of machine.**

2.3.5. *Simo-Charts*

(simultaneous motion charts) : Simo-charts is a left & right hand operation charts indicating the recording of very minute movement of left hand & right hand or both hands simultaneously , such kind of charts not only recording the hands motion but also recorded the movement of others limbs of the body against the common time scale. The scale reading is represented in **"WINKS" (1/2000 of minit)** .this charts is generally used in micro motion study

Procedure for Simo-Charts

(i) Taking a camera for recording a repetitive work performed by the operator.

(ii) This short film is then viewed &analyze.

(iii) Some time special projector is also applied for clear recording of the film.

(iv) The recorded data is then converted into therbligs.

(v) Simo-charts is then drawn for further analyses &study.

(vi) The basic concept of simo-charts is of motion –economy.

(vii) For new method, new simo-chart is constructed also considering the respective advantages. But Simo-charts gives an accurate analyses but it is very costly method.

The Simo-charts summary for any particular operation may be represented as,

Simo- Charts Summary

Operation.................................. Date...

Name of operator...................... Film No-...................................

Componentname....................... Operation No............................

Applied Method.......................

Table 2.1: Simo- Charts

Left-hand description	Symbol	Time (WINKS)	Symbol	Right-hand description
GRASP	G	0 10	G	GRASP HAMMER
TO JOB	TL	20	TL	TO JOB
POSITION	P	30 40	AD	IDLE
HOLD	H	50	U	USE

SIMO Chart for Filing Process of Work Piece

S.No.	Left hand description	Therblig	Time	Therblig	Right hand
1.	Searching and lifting	SH,H	0.2		
2.			0.4	U	Opening the vice
3.	Clamping workpiece	PP	0.8	PP	clamping work piece in the vice piece in the vice.
4.			1.0	TL	Take the file
5.	Do the hand filling operation.	U	2.0	U	Do the hand filing Operation.
6.			2.2	TL	Taking the micrometer
7.	Check the dimension	I	3.0	I	Check the dimension
8.			3.2	U	Open the vice
9.	Remove the work piece	TL		3.4	

2.3.6. Flow Diagram

In any production shop, repair shop or any other department. There is a movement of men or material from one work station to other work station . The frequent movement of men & material represented by flow diagram. These movement may be minimized then we have got a cost &effort saving throughout the process , if path of material is simple and not frequent thus we can recorded the achieved movement . the flow diagram is a diagram which is drawn to scale . the position of various machineries , material handling equipments , gang ways is indicated first or drawn first . after that the flow of men & machine path are indicated in the diagram . Different movement are represented by different colour , processes symbol are used to identify the various activity at different work-stations. Diagram shows a flow diagram , in which raw -material is collected from store thereafter machining operation is performed & then finished components stock in a store .

Job Title - Job Material from Store Machined and Finished Components Stock in a Store

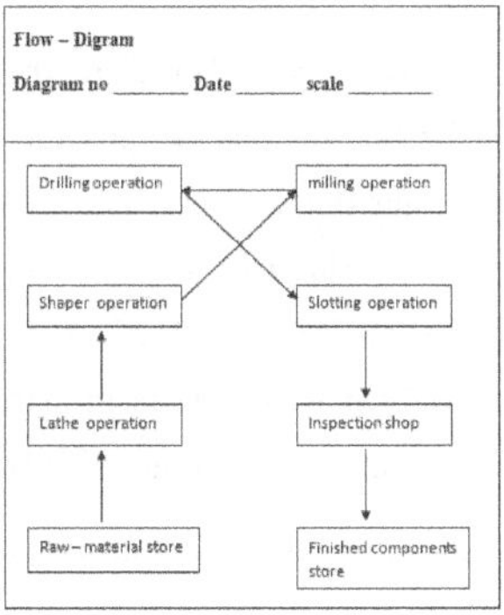

Diagram 2.12: Flow Diagram for Existing Method

Flow Diagram of the Manufacturing of A Bicycle Pedal Axle

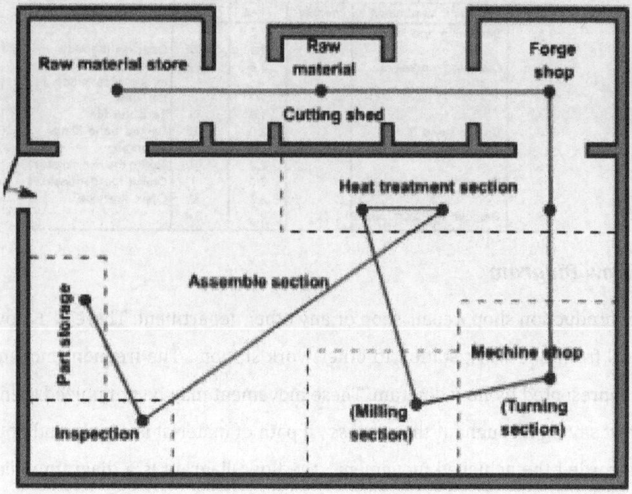

Purpose of flow- diagram.

1. To remove unwanted flow of material.
2. To remove back-tracking.
3. To control the traffic throughout the flow line.
4. For improving the plant- layout.

2.3.7. Two- Hand Charts

It is a motion study whose function is to analyzed in performing an activity during the particular operation .

The aim of this investigation is to eliminate / eradicte the unwanted motion to minimum as possible as and to arrange the best of motions in a possible sequence. Two handed process chart is also known as Left and Right Hand process chart. For example assembly of small instrument such as pressure gauge , carbrettor , time piece and fuel pump etc. Where both hands are used for assemble the various component of articles , in order to achieve a correct sequence of motion left hand / right hand charts is prepared . thereby we can achieve a defective arrangement of tool , incorrect motion sequence and defective motion

Following Two hands process charts is preparing for a nut- bolt assembly of any machine component is given as,

Left Hand Description	Symbol (L.H.)	Symbol (R.H.)	Right Hand Description
1 Move to bolt	⇒	⇒	Move to first washer.
2. Grasp the bolt	O	O	Pickup/grasp the washer.
3. Move to position	⇒	⇒	To position
4. Hold	∇	O	Assembled to bolt.
5.		⇒	Move to second washer.
6.		O	Pickup/grasp the washer.
7.		⇒	To position
8.		O	Assembled to bolt.
9.		⇒	Move to nut.
10.		O	Pick up the nut.
11. Move to box	⇒	⇒	To position.
12. Aside to box	O	O	Assembled to bolt

Diagram 2.13: Two Hands Process Charts

Job : Assembly of washer and nut to a bolt
Chart begins : Both hands free before assembly Assembly Bolt Washer Nut
Chart ends : Both hands free after assembly
Chart : Existing method/Proposed method Date : Operator
Operator : Chart No :

Left hand | Right hand

Sl. No.	Description of the activities	Symbols O ⇒ ∇	Sl. No.	Description of the activities	Symbols O ⇒ ∇
1.	To the bolt tray		1.	To the washer tray	
2.	Picks up one bolt		2.	Picks up one washer	
3.	Returns to original position		3.	Returns to the initial position	
4.	Holding the bolt		4.	Assembles washer over bolt	
5.	Idle		5.	To the nut tray	
6.	Idle		6.	Picks up one nut	
7.	Idle		7.	Returns to initial position	
8.	Idle		8.	Assemble nut to the bolt	
9.	To the assembly tray		9.	Idle	
10.	Puts the bolt in the tray		10.	Idle	
11.	Returns to the original position		11.	Idle	

Diagram 2.14: Two Hands Process Charts

Construction of two hand charts: following step should be taken for constructing two-hand process charts.

1. Analyze the repetitive operation for few times.
2. Observe and recorded the motion of one hand, indicating as a symbol.
3. Recording should be done carefully in such a manner that they can be easily understand.
4. Each and every activity should be performedcarefully with great precaution.
5. Activity remainng the operator motion must be neglected or avoided, consider only operator motion.

2.3.8. Outline Process Charts

a outline process charts is usually applied for recording about the overall view of process in respect of operation and their sequence , to full fill the required purpose only symbol of operation and inspection is indicated. Or in very brief An outline process chart is a process chart giving an overall picture by recording in sequence only the main operations and inspection

Example -1 Here we put an example for outline process charts , in which riveting of two – metal plates is being performed thus the operation and their sequence is indicated as

Job Title – Rivetingof Two Plates

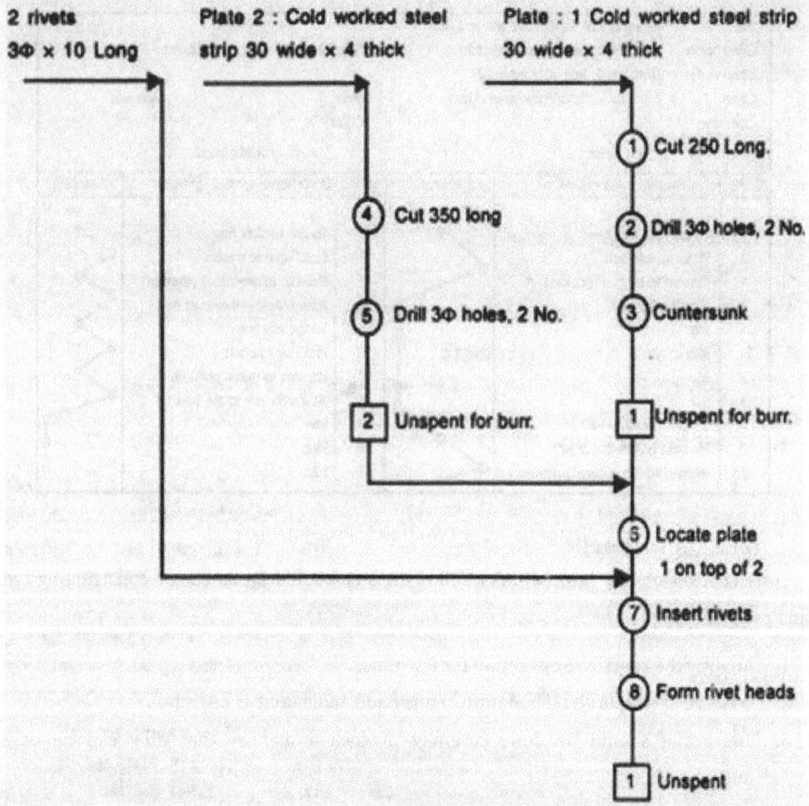

Diagram 2.15: Outline Process Charts

Example 2 In this example, an outline process charts has been constructed for manufacturing an electric motor.

Job Title – Out Line Process Charts for Electric Motor Manufacture

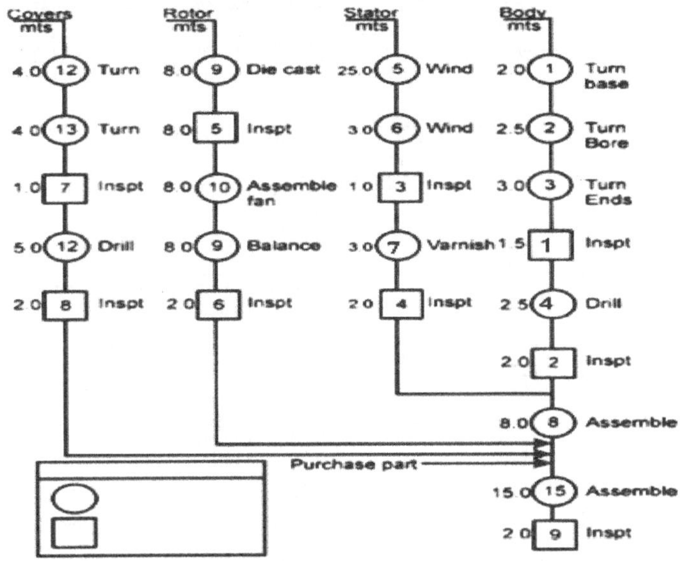

Diagram 2.16: Out Line Process Charts for Electric Motor Manufacturing

2.4. Therbligs

Therbligs are 18 kinds of operational elemental motions that is performed by the operator when he/she is doing the task.

These elements are very important in order to improving the productivity of manufacturing process . therbligs are proposed & invented by the Frank and Lillian Gilbreth in year between 1908 and1924.

Therbligs comprise a system for analyzing the motions involved in performing a task. The identification of individual motions, as well as moments of delay in the process, was designed to find unnecessary or inefficient motions and to utilize or eliminate even split-seconds of wasted time.

Therbligs is an industrial tool to cut down the unnecessary movement thereby we can reduced the time It is a reversal of the name " **Gilbreth",** with 'th' treated as one letter.

Table 2.2: Therbligs Symbols & their Description

S. no	Therbligs	Symbol	Colour	Description
1	search		black	**Search (Sh):** attempting to find an object using the eyes and hands.
2	Find		Grey	**Find (F):** A momentary mental reaction at the end of the Search cycle. Seldom used.
3	Select	→	Light grey	**Select (St):** choosing among several objects in a group.
4	Grasp		Lake red	**Grasp (G):** grasping an object with the active hand.
5	Hold		Gold ochre	**Hold (H):** holding an object.
6	Transport loaded		Green	**Transport loaded (TL):**moving an object using a hand motion.
7	Transport empty		Olive green	**Transport empty (TE):** receiving an object with an empty hand.
8	Position	9	Blue	**Position (P):** positioning and/or orienting an object in the defined location.
9	Assemble	#	Violet heavy	**Assemble (A):** joining two parts together
10	Use		Purple	**Use (U):** manipulating a tool in the intended way during the course working
11	Inspect		Burnt orange	**Inspect (I):** determining the quality or the characteristics of an object using the eyes and/or other senses
	Preposition		Sky blue	**Preposition (PP):** positioning and/or orienting an object for the next operation and relative to an approximation location.
13	Release load		Carmine red	**Release load (RL):** releasing control of an object.
14	Unavoidable delay		Yellow ochre	**Unavoidable delay (UD):** waiting due to factors beyond the worker's control and included in the work cycle.
15	Avoidable delay		Lemon yellow	**Avoidable delay (AD):** waiting within the worker's control which causes idleness that is not included in the regular work cycle.
16	Plan		Brown	**Plan (Pn):** deciding on a course of action.
17	Rest	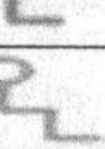	Orange	
18	Disassemble	≠	Violet – light	**Disassemble (DA):** separating multiple components that were joined.

1. **Search :-**The Search motion starts when operator eyes or hand start to seek the object required it may be tool / equipments / devices / raw material . The Gilbreths stated that in a search, "the time and attention required varies with the number of dimensions in which the search is performed." A one-dimensional search might be locating a piece

of paper on a desktop. A two- dimensional search might be finding a light switch on a wall and the three -dimensional search would be locating ahanging pull-chain for a light or fan.

2. **Find :-**find is a process in therbligs is to looking forward an object . but some times it is neglected .
3. **Select :-** This Therblig may be considered a part of Search for indicating / locating an object from a group of similar objects.Another aspect that the Gilbreths included under Select was in picking objects that required acertain quantity.
4. **Grasp :-**In simplest terms, Grasp is when the worker's hand grabs the object. In this Therblig, the time taken is directly proportional with the ease of thegrasp.The Gilbreths also recognized that when Grasp was a static position, such as holding a block of wood while a screw was being inserted, it should be eliminated by using a jig or foot-activatedclamp or other holding device.An important element in saving time was whether the initial grasping of an object would bethe proper grasp for the Use or Assemble function. In this respect, Grasp has a close relationshipwith the Position and Pre-Position Therbligs.
5. **Hold :-** This Therblig was "...the retention of an object after it has been grasped, [with] nomovement of the object taking place." To clarify, we can call Hold a Grasp, of an object, occurring in one hand, while the other hand performs a Use or Assemble function.
6. **Transport Loaded :-** This Therblig begins after Grasp where the hand is doing "work" by moving the weight of an object, and ends when just before the Release Load, Use or Assemble Therbligs. The main objective of this Therblig is to reduce the distance and subsequent time involved for transport .
7. **Transport Empty :-**This is the motion of moving the unloaded hand from the point of Release Load to the next function within the sequence. It can also be considered the hand motions involved between Select andGrasp, where the eye identifies the object and the hand moves towards it to grasp. This Therbligis a non-productive one, and as such, should be kept to a minimum.
8. **Position :-**This motion is the act of placing the object in the proper orientation for Use.This Therblig, like Hold, is one, which can completely be eliminated by the design of thework place. If a tool or part is placed such that it is stored in the proper position for Grasp, theobject doesn't have to be reoriented.

9. **Assemble :-** This Therblig starts when two or more parts are placed together (a peg into a hole) and ends wheneither the assembled object is Transport Loaded or when the hand reaches for another part(Transport Empty).
10. **Use :-** Use is when an object is being operated asit was intended, and typically denotes a tool. For example, we would like to make a hole with the help of lathe machine , to achieve the desired purpose the drill is kept in 3- jaw or 4- jaw chuck and tightening it , but we Use the drill to bore holes. Operation of controlson a machine would also be considered Use.
11. **Disassemble :-** The Disassemble Therblig should always be examined for possible elimination. If an itemmust be partially disassembled for Inspection or for addition of other parts, the basic concept s such kind of operations is to reviewe / eliminate unnecessary steps as well as unusable parts .
12. **Inspect :-** This Therblig involves the act of comparing the object with a predetermined standard. This action can be employed one or all human senses, depending up on the object and the desirable attributes beingchecked. The inspection may be for amount , dimension or **quality**. The motion starts whenthe item is first picked up or viewed and ends when it is either released or used in assembly.
13. **Pre-Position :-** This is the motion of replacing an item in the proper orientation for its next Use . If the time to Pre-Positiona tool (after Use) is longer than Select, Grasp and Position motions, then one should consider improvements in the way Pre-Position is done.
14. **Release Load :-** This motion involves releasing the object when it reaches its destination. The actual time takenwill be fractions of a second and would vary with such things as if it were being Pre-Positioned orif the release was merely down a hole, into a gravity chute depending upon the position of load releasing .
15. **Unavoidable Delay :-** This Therblig is measured from the point where a hand is inactive to the point where it becomesactive again, with another Therblig. These delays were defined by Gilbreth as being out of thecontrol of the particular worker being studied. The Gilbreths felt that these time periods should be used for rest.
16. **Avoidable Delay :-** This counterpart to Unavoidable Delay involves inactive time the worker encounters over whichhe/she has control. For example, if the person is Selectingparts for Assembly, exclusively with one hand, while the other remains idle, the work is not only slowed, butthe hand being used is becoming more fatigued.

17. **Plan :-** This Therblig is a mental function, which may occur before Assemble or prior to Inspection. The extent of the use of Plan variesgreatly with the type of job performed. However, in routine jobs, the time spent in the PlanTherblig should be kept to a minimum through arrangement of parts and tools, Plan time, by the worker, can also be eliminated by the system

Amongs these elements only few are used for simplicity. Some important symbols are given below that has-been generally used in operation process.

Table 2.3: Common Therbligs Symbols used in Manufacturing Process

Symbols	Activity	Description
	Operation	Indicates the main steps in a process, method or procedure. Usually the part, material or product concerned is modified or changed during the operation
	Inspection	Representing the inspection for quality /and check quantity
	Transport	Indicates the movement of workers, materials or equipment from place to place
	Temporary storage and dealy	Indicates a delay in the sequence of events: for example, work waiting between consecutive operations, or any object laid aside temporarily without record until required
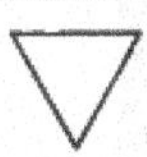	Permanent storage	Indicates a controlled storage in which material is received into or issued from a store under some form of authorization, or an item is retained for reference purposes
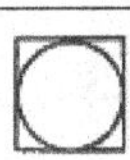	Operation and inspection	The circle within the square represents a combined operation and inspection. This occurs when two or more activity isperformed at the same time, like—operation &inspection, loading of tool and work piece.

2.5. Cycle Graph and Chro-Nocycle Graph

Both graph are very important from subject point of view, the basic significance of both the graph is to identify the path of motion of an operator hands , finger and other limb of the body . In case of cycle graph , a small electric bulb is installed to the finger , hand , or other limb of the body of the operator then path of motion is to be recorded with the help of camera & photographed is made for analyses , the light of electric bulb is trace the path as per the motion of operator limbs. This analyses gives a chance for minimizing the motion of an operator & calculating the economic path , which will befeasible and favorable for an operation.

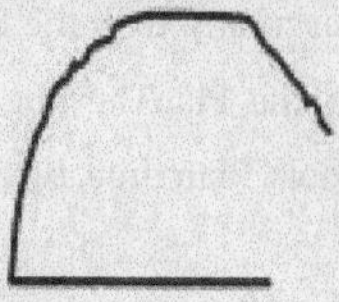

Cycle Graph Cycle Graph Representing Change of Tools

2.6. Chronocycle Graph

It is a special type of cycle graph in which ,a small electric bulb is installed to the finger , hand , or other limb of the body of the operator but an interrupter is connected in the electric circuits due to that the bulb is glow quickly and reduced the shine slowly till the off position thereby we have got the path in the shape of dotted line . Direction of dots indicating the direction of motion. the space between the dots will represent the speed of hands or finger or other limbs of the body , the no of dots will give the total time taken by that particular motion .this technique of recording the motion &evaluating the time for a particular motion is called chronocycle-graph.

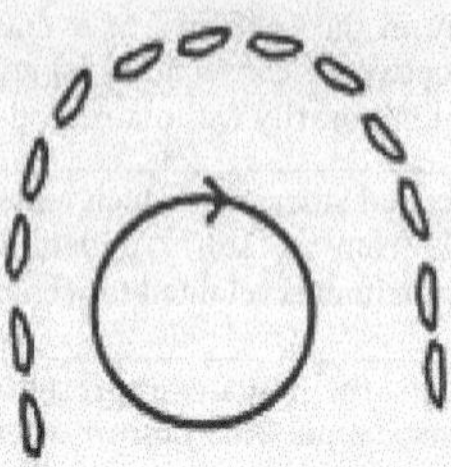

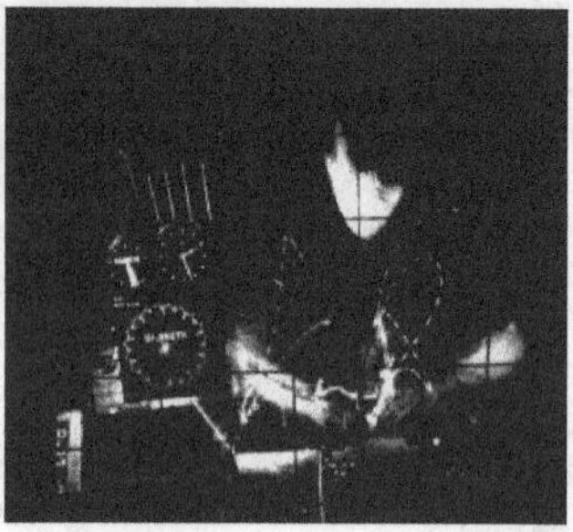

Chronocycle Graph Chrono Cycle Graph of An Operator Representing Diamond Variations

Procedure of Preparation for Cycle / Chronocycle Graph

1. Set the frequency of flashes - depending on the type of job.
2. Attach bulbs to hands.
3. Decide the exposure time.
4. Take a time exposure for the period of the job cycle whilst the operator performs the job - the location would be a darkened room in former times.
5. Take a second, instantaneous exposure on the same film, if it is required to superimpose the scene on the chronocyclgraph.

Application of Cycle / Chronocycle Graph

1. Developing a better work place - chronocyclegraphs reveal obstructions and bad locations.
2. Analysis of a complex movement.
3. An aid to training.
4. Comparison of two methods.
5. Publicity and advertising.
6. Design of new equipment.

Advanntages of Cycle Graph and Chronocycle Graph

1. All **types** of movements i.e. complex, very fast or unrestricted movements can be recorded.
2. It can be used for training purposes when the superiority of one motion pattern is shown as better than other.
3. It can be used for method improvement.
4. These graphs provide an aid in explaining the pattern of motion which is used in performing the operation.

2.7. Micro-Motion & Memo-Motion Study

In certain types of operation, the large movement can be recoded very easily , but in association the large movement some smallest movement is also involved which has a very short period of time , these respective time is very difficult to measure for any operation and could not be neglected because it repeated again and again, thus motion are recorded on picture film with the help of camera . a very small time accuracy up to 0.0005 minit can be measured with this technique . thus such kind of analyses of motion with the help of picture camera is called micro-motion study.

Advantages of Micro Motion- Study

1. Permit greater detailing than eye observation
2. Provide greater accuracy than pencil, paper and watch techniques
3. More convenient & gives positive result.
4. Help in the development of the work studytechnique.

Memo-motion study : -It is very special type of micro-motion study , in which motion is filmed at very slower speed .The time intervals usually lie between 0.5 sec. and 4 sec. In memo

– motion study, A camera is placed with a view over the whole working area to take pictures at the rate of one or two per second instead of the usual rate of 24 frames per second.

The result is that the activities of ten or 20 minutes may be compressed into one minute and a very rapid survey of the general pattern of movements may be obtained, from which the larger movements giving rise to wasted effort can be detected and steps taken to eliminate them. This method of analysis is very economical than that of micro-motion study.

Procedure for Memo-Motion Study

1. Making a detailed study of operation which is being filmed.
2. Developed a film.
3. Analyses the film.
4. Draw a multiple activity charts.
5. Apply the principle of motion economy for suitable improvement

Advantages of Memo-Motion Study

1. It is very economical than micro-motion-study.
2. It calculated complete view of motion aswellas timing for each motion.
3. Analyses can be made in minimum time than that of micro-motion study.

Micro- motion study is to be recorded the movements of people and materials in a given workplace, and the type of handling or storage facilities used. A multitude of useful data can then be viewed quickly by the work study specialist in order to assist him or her in drawing up charts, analyzing the operation and developing a new method.

2.8. Critical Examination

Critical examination is nothing but it is a questioning technique for creating any particular article or job in an manufacturing industries . questioners is made in such a manner therby we can improve a present method of doing a job it also involve way of operation , tools & equipments required , the procedure of questioning called critical examination .

In critical examination there are two section of question are considered for detailed study.

1. Primary questions.
2. Secondary questions.

Primary questions :- primary question have been involve following necessary step.

Table 2.4: Primary Questions for Critical Examination

Steps	Response
Purpose	• What is actually purpose of operation? • Without this operation job can be created? • If any kind of elimination &additionfulfill the desired purpose?
Place	• Which place is very suitable for performing the job &why?
Sequence	• When this operation can be performed economically? Means indicating the time of doing the operation. wheter two or more operation can be performed at same time without affecting the efficiency of operation.
Person	• Which place is very suitable for performing the job &why?
Means	• By which means, the operation can be performed? which kind of machine tools , equipments , tools are required as per the operation thereby we can perform economically , easily , safely for both machine tools aswellas operator .

Considering the above question , we will also have to concentrating some other questions , like elimination which responds to removing un-necessary parts of the job or respective operation, combining means whether there may be any possibility of two or more operation can be combined together , rearrangement is the activity of achieving the sequence operation for better & effective results. For example – in any particular manufacturing process, the sequence of operation should adopt as given below, to achieve minimum time investment throughout the process.

1 **Assembly** → 2 **Inspection** → 3 **Storage** → 4 **Dispatching**

Secondaryquestions: -The secondary questions cover the second stage of the questioning technique, during which the answers to the primary questions are subjected to further query to determine whether possible alternatives of place, sequence, persons and/or means are practicable and preferable as a means of improvement upon the existing method .

Table 2.5: Secondary Questions for Critical Examination

Steps	Significance
PURPOSE	• What is done? • Why is it done? • What else might be done? • What should be done?
PLACE	• Where is it done? • Why is it done there? • Where else might it be done? • Where should it be done?
SEQUENCE	• When is it done? • Why is it done then? • When might it be done? • When should it be done?
PERSON	• Who does it? • Why does that person do it? • Who else might do it? • Who should do it?
MEANS	• How is it done? • Why is it done that way? • How else might it be done?

Combining the two primary questions with the two secondary questions under each of the headings "purpose, place", etc., yields the following list, which sets out the questioning technique in full.

Combining the two primary questions with the two secondary questions under each of the headings "purpose, place", etc., yields the following list, which sets out the questioning technique in full.

These questions, in the above sequence, must be asked systematically and analyses should be performed in critical examination.

2.9. Principle of Motion Economy

In any kind of manufacturing / production of anarticles , we will always try to be improve the productivity , There may be various type of factor upon which concentration is to be made . Among those expenditure is one , in method-study we have been discussed about the method of doing the job & always full concentration has been made on development of present method of doing the task so therefore we can adopt the improve method hence expenditure may be minimize .

But other principle may also be used for economic point of view these principle are called principle of motion economy.

In very clear and close vision the principles of motion economy form a set of rules and suggestions to improve the manual work in manufacturing and reduce fatigue and unnecessary movements by the worker, which can lead to the reduction in the work related trauma. these principle consists of three basics principle which is given below .

1. Principle related to use of " **Motion of human body**
2. Principle related to the ' **work-place**'
3. Principle related to the design of " **tool &equipments**"

2.9.1. Pinciple Releated to Use of " Motion of Human Body

1. The two hands should begin as well as complete their motions at the same time.
2. The two hands should not be idle at the same time except during rest periods.
3. Motions of the arms should be made in opposite and symmetrical directions, and should be made simultaneously.
4. Hand motion should be such that, operator feeling least fatigue & should be most economics for organization. For least fatigue following procedure must be carefully implemented.

Table 2.6: Motion of Human Body Limbs

Type of motion	Parts of motion
First motion	Finger motion
Second motion	Finger and wrist motion
Third motion	Finger, wrist and lower arm motion
Fourth motion	Finger, wrist, lower arm and upper arm motion
Fifth motion	Finger, wrist, lower arm, upper arm and motion

5. Hand and body motions should be such that it can perform the work satisfactorily.
6. Momentum of body motion should be employed to assist the worker whenever possible, and it should be reduced to a minimum.
7. Smooth continuous curved motions of the hands are preferable to straight-line motions involving sudden and sharp changes in direction.
8. Always follow Ballistic movements because it is faster, easier, and more accurate than "controlled" movements.
9. Work should be arranged to permit easy and natural rhythm wherever possible. Like opening & closing of button of shirts .
10. Eye fixations should be as few and as close together as possible .
11. Hand motion should be avoided when ever it can perform by feet or other parts of body.

2.9.2. Principles Related to the ' Work-Place'

1. Definite and fixed station should be provided for all tools / equipments and materials in order to easy reach for the operator.
2. Tools / equipments / devices and materials should be pre-positioned and very nearer to the operator in order to reduced searching & wastage of time.
3. Gravity feed, bins and containers should be used to transporting the materials, semi-finished products / finished products / devices.
4. Tools, materials and controls should be located within the maximum working area and as near to the worker as possible.
5. Materials and tools should be arranged to permit the best sequence of motions.
6. The color of workplace should contrast or illuminated properly the work place by proper lighting there by reduced eye fatigue.
7. Provision should be made for adequate lighting, and a chair of the type and height to permit good posture should be provided.
8. The height of the workplace and seat should be arranged for suitable standing and sitting.

9. A chair of the type and height to permit good posture should be provided for every worker.

2.9.3. *Principles Related to the Design of "Tool & Equipments"*

1. The hands should be relieved of all work that can be done more advantageously by a jig, a fixture, or a foot-operated device.
2. Two or more tools should be combined wherever possible.
3. Tools and materials should be pr-positioned whenever possible.
4. Handles such as those on cranks and large screwdrivers should be designed so as to permit as much of the surface of the hand as possible to come in contact with the handle.
5. Levers, crossbars and hand-wheels should be so placed that the operator can use them
6. Levers, crossbars and hand-wheels should be so placed that the operator can use them with the least change in body position and the greatest "mechanical advantage."

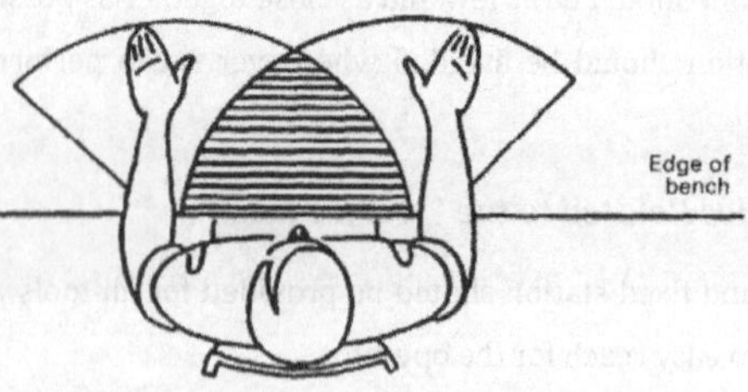

A Normal Working area Finger, Wrist and Elbow Movement

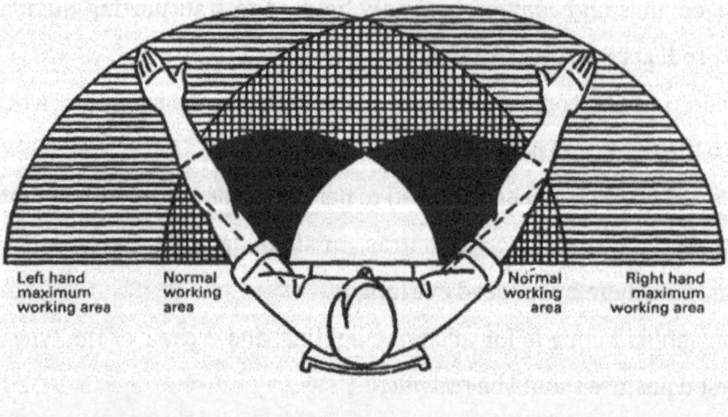

Maximum Working Area for Shoulders Movement

Diagram 2.17: Minimum and maximum working area for an operator

Question for Discussion

1. Define productivity ? Explain objectives of work study.
2. Explain steps in conducting method study.
3. What are the objectives, significance, and applications of Work Study? Discuss briefly.
4. Discuss the criteria in selection of the task for conducting time study,
    a) When such study has already been carried out earlier.
    b) When no time study has been conducted earlier.
5. Describe various recording techniques in Method Study.
6. Discuss the various principles of motion economy
7. Differentiate between cyclegraph and.chronocyclegraph.
8. Write down the various types of symbols used in therbligs
9. Describe various stages of method study. Also make a left and right hand operation chart for filling ink in a fountain pen.
10. Define Method Study and explain technique with its various steps ?
11. Differentiate between Micro & Macro motion study. Explain in brief various techniques for micro motion study.
12. Define Method Study. Discuss the objectives of method study.
13. What is method study? Explain in brief various charts and diagrams utilized in method study.
14. What is method study? Write its procedures (steps) and explain any two steps in detail.
15. What is therblig? State the importance of micro motion study.
16. List the various Therbligswith their symbols. Explain applications of micro motion study.
17. Explain principle of motion economy and work place layout.
18. Explain the principles of motion economy.
19. How does the multiple activity charts help to reduce the ideal time of men and machine?
20. Briefly discuss multiple activity chart with their objectives.
21. Explain outline process chart and flow process chart.
22. Explain the procedure of work study. State the applications of work study in the industries
23. Draw man-machine chart for the following condition and find working time and % utilization of man and machine: No. of operator= 1, No. of machines= 2 (M1& M2)
    (a) Time for clamping the work and setting the tool on M1& M2 each = 0.2 minute
    (b) Machining time on M1= 0.4 minute
    (c) Machining time on M2= 0.6 minute
    (d) Groove cutting and parting off on M1 =0.3 minute

(e) Parting off on M2 = 0.1 minute

24. What is human machine system?
25. Explain the procedure for making flow process chart.
26. What do you understand by Man-Machine system? Explain its characteristics.
27. Discuss the principle of motion economy used in-method study
28. Elaborate the different types of methods study principles and motioneconomy.
29. Write short notes on the following:
    a) String diagram.
    b) Micrornotion and macromotion study.
    c) SIMO charts.
    d) Flow process chart.
    e) String Diagram.
    f) Two handed process char.
30. Explain man - machine system and its characteristics. How can you say if the fit is poor or good ? Explain
31. What is man-machine system? Discuss its working. Give its general characteristics and features.
32. Explain about String diagram & its uses.

References

1. L.A. Sylvestbr, "The Handbook of Advanced Time Motion Study", New Y9rk. Funk and Wagnalls, 1950.
2. G. Shaw Airne, "The Purpose and Practice of Motion Study. London. Columbine Press. Second Edition.1960.
3. Idein: "How Many Time Study Readings to Take", Factory Management and Maintenance, Vol. 113, No. 2, Pp. 132-1 33, 1955.
4. G. Nadler, "Motion and Time Study", New York. McGraw-Hill, 1955.
5. R.L. Moanow, "Motion Economy and Work Measurement", New York. Ronald Press Company. Second Edition, 1957.
6. M.B. Mundel, "Motion and Time Study: Principles and Practice", Englewood Cliffs, NJ. Prentice Hail, Inc. Third Edition, 1960.
7. S.M. LowRy, H.B. Maynard and G.J. Stegemerten, "Time and Motion Study", New York. McGrawHill", Third Edition, 1940.
8. W. Gomberg, "A Trade Union Analysis of Time Study", Engle wood Cliffs, N.J. Prentice Hall. Second Edition, 1955.

9. Idem: How to Chart Data. New York. McGraw-Hill. 1960.
10. Niebel and W. Benjamin, "Motion and time study", (Homewood, Illinois, Richard Irwin), 1972.
11. M.E. Mundel, "Motion and time study: Principles and practice", (Englewood Cliffs, New Jersey, and Hemel Hempstead, Hertfordshire, Prentice-Hall, 6th éd.,), 1985.
12. Barnes and M. Ralph, "Motion and time study: Design and measurement of work", (New York and London, John Wiley, 7th éd.,), 1980.
13. R.M. Barnes, "Motion and Time Study Design and Measurement of Work", 7th edn, John Wiley and Sons, NY, 1980.
14. W. Antis, "Stopwatch time study", Industrial Engineering Handbook, 3rd edn, W.B. Maynard, raw-Hill, New York, 1971.
15. Adler and S Paul, "Time-and-Motion Regained", Harvard Business Review, Pp. 97-108, 1993.
16. Antara Shantilal Rathod, Rajat Ganesh Jadhav, Amey Balasaheb Babar, "An Overview of Method Study and Study of Different Recording Techniques", International Journal of Science and Research (IJSR).

9. Idem Kaur to Chart [illegible], New York McGraw Hill, 1960.

10. Niebel and [illegible], "Motion and time study", Homewood, Illinois, Richard Irwin, 1972.

11. Marvin E. Mundel, "Motion and time study: Principles and practice", Englewood Cliffs, New Jersey and [illegible], Prentice Hall, 6th ed., 1985.

12. Barnes and M. Ralph, "Motion and time study: Design and measurement of work", (New York and London: John Wiley, 7th edn), 1980.

13. Ralph Barnes, "Motion and Time Study Design and Measurement of Work", 7th edn, John Wiley and Sons (NY) 1980.

14. W. Willis, "Stopwatch time study", Industrial Engineering Handbook, 3rd edn, W.B. Maynard, McGraw-Hill New York, 1971.

15. Abruzzi and [illegible], "Time and Motion Reviewed", Harvard Business Review, [illegible], 97-103, 1990.

16. Anjani Mandal, Rahul, Rajat [illegible] Industry, Vijay [illegible], "An Overview of Method Study and Study of Current Prevailing Techniques", International Journal of Science and Research (IJSR).

CHAPTER 3

WORK MEASUREMENTS

3.0. Work-Measurement

Work - measurement is defined as the time required for completing an unit task of work. How long it takes to perform a task depends on the worker's pace, operating conditions, and work method. Work measurement is very important procedure for determining the time, for any particular part of complete work. it also helps in to understand the clear vision of the present applied method . if any kind of improvement in applied method is required then it will be given a correct path aswellas altering or training the operator as per developed method and proper planning is implemented according to the work . It is also a productivity enhancement tool for a manufacturing organization / unit or in very brief we can say that

Work measurement is a systematic procedure for the analysis of work and determination of times required to perform key tasks in processes

Or

According toBritish Standard Institution "The application of techniques designed to establish the time for a qualified worker to carry out a specified job at a defined level of performance". This time is called standard or allowed time.

Work – Measurement is representing a co-relations between **time & price** of an articles , for an example when ever customer want to purchase any articles , they generally search a similar articles of various manufacturing companies & selecting that product which have lower price , a simple question arises here that is why ? The price variation is identified in various similar products fulfill the similar purpose &functioning. Thecorrect answer of the question is how much time has been consumed in completing the unit task of work. Thus we have concluded that, If large amount of time isconsumed in completing the task thereby labour cost is in increasing order hence the pricing of an article also increases so therefore it is required that we have to estimate / calculate the correct time for completing the work . it also provided a platform for fixation of wages , salary and incentives . Some important terms that is generally used in work-measurement are given below.

Normal time:-Normal time is the expected time required to perform some work activity at a normal pace, under normal operating conditions, and using a prescribed method. By a normal pace, we mean a pace that can be consistently performed by the average employee without fatigue under normal operating conditions, the normal time is given by the following equation

$$\text{Normal time} = \text{Observed time} \times \text{Performance rating factor} = OT \times PRF$$

Fair Day's Work

Amount of work that can be produced by a qualified worker/employee when working at normal pace and effectively utilizing his time and where work is not restricted by process limitations.

Allowances: - Allowances are very important segments of time in manufacturing process. That includes time for labor fatigue and personal needs, equipment breakdowns, rest periods, information delays, and so on.

Standard time:-Standard time is normal time adjusted for allowances. It is computed using the following equation:

$$\text{Standard time} = \text{Normal time } (1 + \text{Allowance factor})$$

3.1. Purpose of Work- Measurement

The basic purpose of work measurement and methods that organizations used to perform time studies, calculate standard times, and estimate the proportion of time spent performing different types of tasks.

Following are the Important Objective of Work – Measurement

(i) For proper estimation of work-force and equipment capacity.

(ii) For complete establishment of budgets.

(iii) Determining what new work procedures must be applied & their respective cost.

(iv) Evaluating time and cost trade-offs among process design alternatives.

(v) For fixation of wage-incentive scheme for labour.

(vi) Monitoring and Evaluating employee performance and productivity.

(vii) Gives accurate information for scheduling and sequencing.

The uses of work Measurement:- Revealing existing causes of ineffective time through study, important though it is, is perhaps less important in the long term than the setting of sound time standards, since these will continue to apply as long as the work to which they refer continues to be done and will show up any ineffective time or additional work which may occur once they have been established. In the process of setting standards it may be necessary to use work measurement:

(i) To compare the efficiency of alternative methods. Other conditions being equal, the method which takes the least time will be the best method.

(ii) To balance the work of members of teams, in association with multiple activity charts, so that, as nearly as possible, each member has a task taking an equal time to perform.

(iii) To determine, in association with worker and machine multiple activity charts, the number of machines an operative can run .

(iv) To provide the basis for production planning and control for the choice of a better layout and for process planning, and for establishing just-in-time inventory control systems.

(v) To provide information that can enable estimates to be made for tenders, selling prices and delivery dates.

(vi) To set standards of machine utilization and labour performance which can be used for any of the above purposes and as a basis for incentive schemes?

(vii) To provide information for labour-cost control and to enable standard.

It is thus clear that work measurement provides the basic information necessary for all the activities of organizing and controlling the work of an enterprise in which the time element plays a part. Its uses in connection with these activities will be more clearly seen when we have shown how the standard time is obtained.

3.2. Procedure of Work Measurement

Since work measurement is an art of investigating and eliminating ineffective time and to set standard times for carrying out the work. Following are the important steps that must be followed by organization for conducting the work-measurement.

Table 3.1: Procedure of Work Measurement

STEPS	ACTIVITY
SELECT	The work to be studied
RECORD	All the relevant data relating to the circumstances in which the work is being done, the methods and the elements of activity in them.
EXAMINE	The recorded data and the detailed breakdown critically to ensure that the most effective method and motions are being used and those unproductive and foreign elements are separated from productive elements.
MEASURE	The quantity of work involved in each element, in terms of time, using the appropriate work-measurement technique.
COMPILE	The standard time for the operation, which in the case of stop-watch time study will include time allowances to cover relaxation, personal needs, etc.
DEFINE	Precisely the series of activities and method of operation for which the time has been compiled and issue the time as standard for the activities and methods specified.

The Work -measurement technique is performed by the following method.

1. Time – study.
2. Work- sampling.
3. Pre-determined Motion Time System.
4. Analytical Estimating.

3.2.1. *Time - Study Method*

Time study is a basic method of work- Measurement, and it is defined as "the process by which we are observing and recording the time required for completing the each operation throughout the manufacturing of an article. The very close and exact definition may be given as

> Time study is a work measurement technique for recording the times of performing a certain specific job or its elements carried out under specified conditions, and for analyzing the data so as to obtain the time necessary for an operator to carry it out at a defined rate of performance.

Whenever the operation is performed it consists of following three types of steps.

Manual operation: - It consists of manual time having a handling of tools/equipments , handling of machines , material etc .

Mental time :- for starting , processing , completion of any operation depends upon the worker look out about the operation , how , when , where & in which manner the operation task will be completed , this will required a mental time for thinking .

Machining time: it consists the share of work included by every machine.

Procedure for time study: For conducting the time study we must have to consider following important steps

1. Define the task or objective to be studied.
2. Decide how many times to measure the task. Time study is a sampling process. It is important to decide the number of cycles or samples needed and the required level of confidence in the estimated time standards to set standard time
3. Divide the task into precise elements.
4. Time and record each element time a worker would take to complete the task element. Rate the performance of worker. Let the experienced analyst record the time.

(i) Complete the average cycle time for each task element as given below:

$$\text{Average Cycle Time} = \frac{[\text{Sum of the times recorded to perform each element}]}{\text{Number of cycles observed}}$$

(ii) Determine the performance rating and normal time for each task element as given below:

Normal time= [Average observed time] x [Performance rating factor]

(iii) Add the normal times for each element to develop a total time for the job.

(iv) Compute the standard time as given below.

$$\text{Standard Time} = \frac{\text{Total normal Time}}{1 - \text{allowance factor}}$$

3.2.2. *Time Study Equipments*

Whenever the time study is conducted there is a requirement of time study equipments, following are some important time study equipments.

(i) Stop –watch method.

(ii) Study board method.

(iii) Time study form.

(i) Stop - Watch Method: -There are two main types of watch in general use for time study- the mechanical and the electronic. Mechanical watches can again be subdivided into the fly back and non-fly back types, with a third type — the split-hand stop-watch.

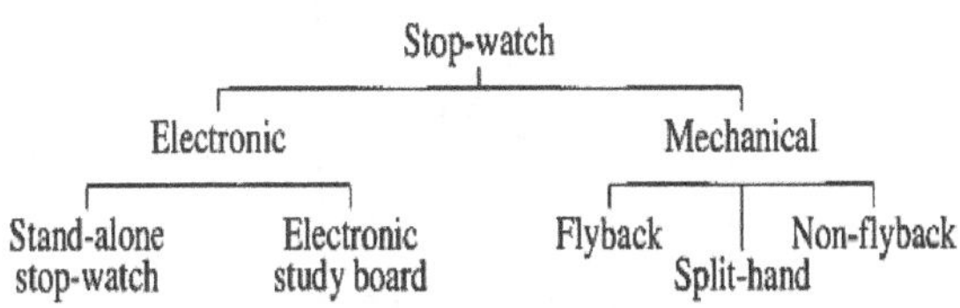

Mechanical-Type Watches may be Obtained with Any One of Three Graduated Scales

- Recording one minute per revolution by intervals of one-fifth of a second, with a small hand recording 30 minutes .
- Recording one minute per revolution calibrated in 1/lOOths of a minute, with a small hand recording 30 minutes (the decimal-minute watch) .
- Recording 1/lOOth of an hour per revolution calibrated in l/10,000ths of an hour; a small hand records up to one hour in 100 divisions (the decimal-hour watch).

In general, with the help of stop watch method we can recording the time up to 0.01 min. It consists of two hands large hands; revolving range is about one **revolution / min.** The complete dial is divided into hundreddivisions. The small watch having small hand that is situated inside the main dial that revolving speed is about one **revolution / 30 min.**The stop watch is kept at right hand top corner of time study board.

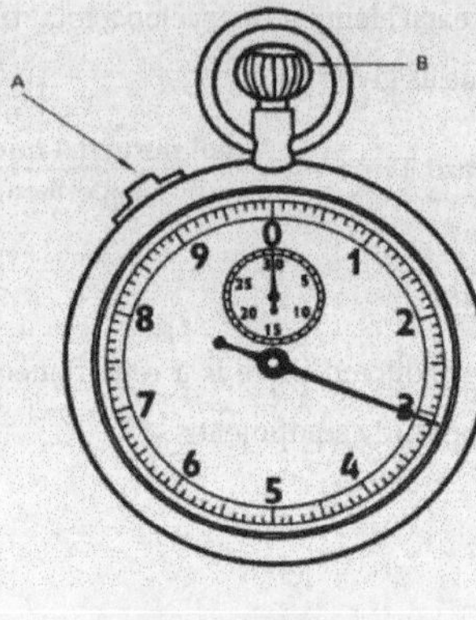

Stop – watch

A = Slide for stopping and starting the movement.

B = Winding knob. Pressure on this knob returns both the hands to zero.

Time – Study Board

3.2.3. Method of Timing by Stop Watch

In stop watch method, time recording for any particular operation is performed by following method.

(i) Continuousmethod:- In this method , the stop watch is starts in the point of operation is being started & continuously run at the end of operation then reading of stop watch is note upon the time record sheet , it is a time of part of a complete operation , if we want to calculate the time of element of operation substract the previous reading from next .

(ii) Snap back method: -In this method at each operation process, stop watch is started and at the end of operation stop watch is stopped, after recoded the reading we have to again start the stop watch for next operation .

(iii)Accumulated method:- In this method we can recorded the time for each operation directly with the help of stop watch , in this method two – stop watch is used to fulfill the required purpose , that are connected by mechanical lever mechanism when operation is started one stop watch is run on the same time other is stop , when operation ended 1st stop watch is stopped automatically & 2nd stop watch is run automatically for next operation , this method is called accumulated method . Thismethod is very easy & help full when recording is complicated process during the operation.

(iv) The time study boardmethod:- The study board is simply a flat board, usually of plywood or of suitable plastic sheet, needed for placing the time study forms. It should be rigid and larger than the largest form likely to be used. It may have a fitting to hold the watch, so that the hands of the work study person are left relatively free and the watch is in a position to be read easily. forefinger or middle finger of the left hand used to press the winding knob .Some work study persons prefer to attach their watches with strong rubber bands or leather thongs around the two middle fingers of their left hands and to hold them at the top of the board in that way.

(v) Time study forms: -Taking a time study requires the recording of substantial amounts of data. These data are in a regular form consisting of element codes or descriptions, ratings and element durations Although the data could be recorded on plain paper, it is more convenient to use pre-printed forms which ensure that each study is of the same consistent format .

Forms Used on the Study Board

- **Time study top sheet:** The top and introductory sheet of a study, on which is recorded all the essential information about the study, the elements into which the operation being studied has been broken down, and the breakpoints used. It may also record the first few cycles of the study itself.
- **Continuation sheet:** This form is used for further cycles of the study. These two forms are the ones most generally used. Together they are adequate for most general time study work. For the recording of short-cycle repetitive operations.

- **Working sheet**: - Foranalyzing the readings obtained during the study and obtaining representative times for each element of the operation. Many time study persons prefer to use simple lined sheets, of the same size as the study sheets, for making their analyses, clipping these to the study sheets when complete.
- **Study summary sheet** to which the selected or derived times for' all three elements are transferred, with the frequencies of the elements' occurrence. This sheet, as its name suggests, summarizes neatly all the information which has been obtained during the course of the study. The heading includes all the details recorded about the operation at the top of the time study top sheet. The completed study summary sheet is clipped on top of all the other study sheets and is thus filed with them.
- **Analysis of studies** sheet on which are recorded, from the study summary sheets, the results obtained in all the studies made on an operation. The analysis of studies sheet records the results of all the studies made of a particular operation, no matter when they were made or by whom.

The use of all these forms, both those employed when actually making the study and those used afterwards to analyses and record it .

Motion picture camera method : In this method , a motion picture camera is used which is moving at constant speed, due to constant speed movement , we can calculate the time for each film and can be done by knowing the no of films covered for any particular operation there by we can calculate exact time , this method is very accurate but very costly. to fulfill the desired purpose following type of motion picture camera is generally used .

(i) Spring driven camera.
(ii) Electric Motor driven camera.
(iii) Slow speed camera.
(iv) Normal speed camera.

3.3. Rating

Rating is that process during which the time study analyst compares the performance (speed) of the operator under observation with the observer own concept of normal performance.

This rating factor will be applied to the selected time value to obtain the normal time for the job.

Systems of rating:-Rating system may be classified in following manners.

3.3.1. *Westing House System of Rating*

A four-factor system, for rating the operator performance , was analyses by Westinghouse system of rating . These four factors are **(1) skill, (2) effort, (3) conditions, and (4) consistency**. A scale of numerical values for each factor was supplied in tabular form, and the selected time obtained from time study by applying the sum of the ratings of the four factors.

Table 3.2: Westinghouse System Factors

Skill			Effort		
+0.15	A1	Super skill	+0.13	A1	Excessive
+0.13	A2		+0.12	A2	
+0.11	B1	Excellent	+0.10	B1	Excellent
+0.08	B2		+0.08	B2	
+0.06	C1	Good	+0.05	C1	Good
+0.03	C2		+0.02	C2	
0.00	D	Average	0.00	D	Average
-0.05	E1	Fair	-0.04	E1	Fair
-0.10	E2		-0.08	E2	
-0.16	F1	Poor	-0.12	F1	Poor
-0.22	F2		-0.17	F2	
Conditions			Consistency		
+0.06	A	Ideal	+0.04	A	Perfect
+0.04	B	Excellent	+0.03	B	Excellent
+0.02	C	Good	+0.01	C	Good
0.00	D	Average	0.00	D	Average
-0.03	E	Fair	-0.02	E	Fair
-0.07	F	Poor	-0.04	F	Poor

Example:

Let;

Selected time = 0.50 minutes,

Therefore,

Normal time=0.50 x 1.13 = **0.565 minutes**.

Excellent skill, B2 +0.08
Good effort, C2 +0.02
Good condition, C +0.02
Good consistency, C +0.01
Total= +0.13

3.3.2. *Synthetic Rating*

In this method, performance of operator is calculated on the basis of pre-determined time system, in respect of operator's speed. In this system actual time is compared with pre-decided standard time. A ratio can be established between the predetermined time value for the element and the actual time value for that element. This ratio is the performance index or rating factor.

$$R= P/A$$

Where,

R-- Performance rating factor.

P-- Predetermined time standard for an elements.

A-- Average or actual time is consumed.

3.3.3. *Objective Rating*

The objective rating method is combination of two rating system. First the operator's speed is rated against a single standard pace which is independent of job difficulty. Only considering the operator's speed of movement, rate of activity, paying no attention to the job itself.

In 2nd stage there will be an **addition of allowances with pace rating** regarding the job difficulties. Further Job difficulty is divided into six classes, these are

(1) Amount of Body Used.
(2) Foot Pedal.
(3) Bi-manualness.
(4) Eye-hand coordination.
(5) Handling Requirements.
(6) Weight of the Job.

3.3.4. *Physiological Evaluation of Performance Rating*

In this rating system , we will have to considered about the **Heart rate in beats per minute** and **oxygen consumption in calories per minute** throughout measuring the physiological work . If **heart rate & oxygen consumptionincreases**that will associated with the physiological cost of the work performance.

3.3.5. *Performance Rating*

This rating system is used to observing the pace or speed of operator activity's with respect to their normal pace. It is ratio between observed speeds to expected speed.

Normal speed = observed time **x** [observed speed / expected speed]

3.4. Allowances

Allowances is very important from subject point of view , In general the normal time taken by the operator for performing any particular operation does not contain any kind of allowance due the short period of operation time with normal pace but whenever the length of operational time is relatively long or the operator engage for the full day of operation , then manufacturing firms will provided some kind of facility for refreshment or in other word The operator may take time out for personal needs, for rest, and for reasons beyond his or her control . Such kind of facility is called allowances. Allowances plays very important role to achieving standard time for the operation,the standard time must include time for all the elements in the operation and in addition it must contain time for all necessary allowances.

The equation form of allowances is given below.

Standard time= Normaltime + Allowances

Following are the important type of allowances from the subject point of view, these are given below

(i) Personal allowance.

(ii) Fatigue allowance.

(iii) Delay allowance.

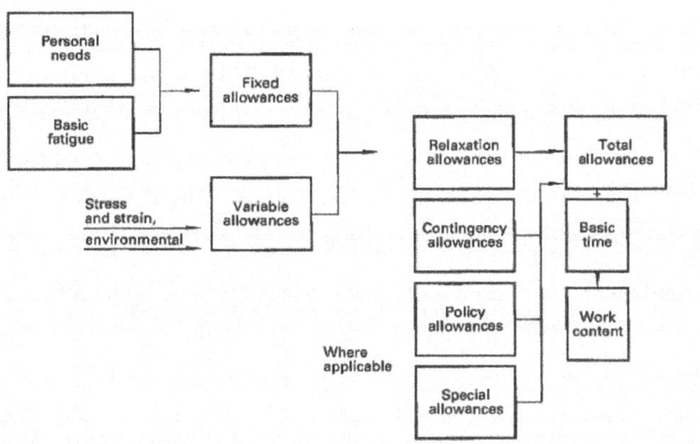

Diagram 3.1: Correlation Diagram of Various Allowances

3.4.1. Personal Allowance

The manufacturing firm provided a facility to operator / worker for fulfilling thepersonal needs such kind of allowance is called personal allowance .this is depends upon the all day time studies. It includes washing, drinking of water etc. Such kind of allowance is depends upon the kind of work , For light work, where the operator works 8 hours per day without organized rest periods, 2 to 5 percent (10 to 24 minutes) per day is about all that the average worker will use for personal time. In case of heavy work and un favourable working condition the amount of excesses time is given to operator / worker it is about 5% more including light work allowance, for personal time. The range of such allowance is depends upon the type of operator for female operator it is about 5%, for male operator it is about 7%.

3.4.2. Fatigue Allowance

Fatigue allowance is provided for eliminating fatigue or recovery of physical fitness aswellas psychological fitness. That is comes on the front line when operator work regularly without interruption. it is depends upon the length of the working day , and the length of the working week but the manufacturing organization & engineers always tries to minimized it . It can be possibly performed by developing the machinery, mechanical handling equipment, tools, and fixtures so that the day's work is more easily done and the employee works in greater physical comfort than formerly. the heavy work , internal environment like heat , humidity , noise is also responsible for physical exertion thus in order to relieve these condition operator / worker taking a rest for some specific period of time if not adverse effect will be imposed on the operator health & their efficiency or productivity definitely has been in decreasing order. Time needed for rest varies with the individual, with the length of the interval in the cycle during which the person is under load, with the conditions under which the work is done, and with many other factors. Organized rest periods, during which time all employees in a department are not permitted to work. The most common plan is to provide one rest period during the middle of the morning and one during the middle of the afternoon. The length of these periods ordinarily varies from 5 to 15 minutes each.

(i) Constant allowance: -This is a basic allowance provided for the worker,who perform the work under ideal condition, the intensity is about 4 to 5 % for both men and women.

(ii) Variable allowance: -whenever the working condition is vary with respect to time variable allowance is provided. such type of situation is arises due to following reasons.

- **Standing position:** whenever operator works in standing position, he / she experienced tiredness to sustain this 2% allowances is provide for operator or worker.
- **Application of force:** whenever operator is handling the weight / load, the application of force is takes place & consumed the muscular energy; the work may be lifting the load, pushing the load, pulling the load, or transporting the load manually. in this case the firms provided the allowances that is depend upon the intensity of load . it is about 10 % for 2 to 5 kg & 20 % for 50 kg weight .
- **Light:** proper lighting is also necessary if not some kind of allowance is also provided.

- **Atmospheric condition:** Atmospheric condition always affects the efficiency & health of the operator for example – air ventilation,temperature,fumes, amount of dust. In this case 10 % to 20% allowance is provided for the operator.
- **Visual strains:** whenever the operation is going on , operator facing the visual strain due inspecting the work-piece directly or by some devices that is depends upon the lighting facility . if light is clear – 4 % allowance is provided on the other hand it is about 10 % for poor lighting facility .
- **Manual or mental strains:** manual strains is due to regularly use of hand , legs or other limb of the body that's create the manual strains , in similar manner when operation is going on before it , it is activity for performing the task mind is completely think about it and hence it create mental strain, **this required a allowance for operator decided by the firm .**

3.4.3. Interference Allowance

When operator operates or handling more then two machines, then there is interference in time period. Thus we have idleness inmachines, due to handling process of machine. In order to compensate the idleness, allowances is provided called interference allowance

3.4.4. Process Allowance

Some kind of allowances is associated with the process , means between the process if any kind of interruption or idleness is obtained then process allowance is provided by the firm , this may be due to power failure , un standardized material , faulty tools / equipments , machine-break down .

3.4.5. Contingency Allowance

This allowance is provided when irregular or in-frequent items of work or delay occurs , which can be measured for that allowance of less than 5% is given .

3.5. Special Type of Allowance

In previous clauses of allowances , management concentrated up on operation cycle & operator , but in this case management fixed a an allowances for worker that does not related to the operation cycle like- start up time , cleaning , change-over tool , setting-up , oiling and some other .

1. **Delay allowance:** -It is expected that machine and equipment will be kept in good repair. When there is a breakdown or when repairs are necessary, the operator is

usually taken off the job and such delays do not enter into the time standard. In such cases the operator is usually paid for waiting time at the hourly base rate .Sometimes, there are minor adjustments, breakage of tools such as drills and taps, or lost time due to occasional variation in material and interruptions by supervisors, and there must be included in the time standard. The kind and amount of delays for a given class of work can best be determined from all-day time studies.

Provision Applying the Allowances:- Personal allowance & fatigue allowance is applied as a percentage of the normal time , but in case of fatigue allowance situation has been changed thus this allowance should apply only to those elements during which the operator works, and not to the machine time during which the machine works .Delays allowance are applied as a percentage of the normal time that is based on machinedelay .

Standard Time = Normal time x [(100)/(100 – Allowances in %)]

3.6. No of Cycle to be Timed

whenever operator / worker engage with the any kind of operation, for performing the elements of operation. it is observed that time for an elements of operation changed from cycle to cycle , this variation is achieved amongst the operator time performance thus this can be reduced by using highlystandardized material , good tools &equipments , good working conditions and qualified & trained worker . This variation cannot be avoided

The following general principles govern the number of cycles to get the representative average cycle time.

a) Greater the accuracy desired in the results, larger should be the number of cycles observed.
b) The study should be continued through sufficient number of cycles so that occasional elements such as setting-up machine, cleaning of machine or sharpening of tool are observed for a good number of times.
c) Where more than one operator is doing the same job, short study (say 10 to 15 cycles) should be conducted on each of the several operators than one long study on a single operator. It is important that enough cycles are timed so that reliable average is obtained.

Following techniques are used to determine the number of cycles to be timed.

- **Use of Tables :-** On the consideration of the cost of obtaining the data and the desired accuracy in results, most companies have prepared their own tables for the use of time study people, which indicate the number of cycles to be timed as a function of the cycle time and the frequency of occurrence of the job in the company.

- **Statistical Methods:** - On the basis of the requirements of the particular situation involved, accuracy and confidence level are decided (An accuracy of a confidence level of 95% is considered reasonable in most cases). A preliminary study is conducted in which some (say N) cycles are timed. Standard deviation o of these (N) observations is calculated as,

$$N = \left\{ \frac{40\sqrt{M\Sigma X^2 - (\Sigma X)^2}}{\Sigma X} \right\}^2$$

Experience has shown that 95% confidence level and ± 5% accuracy is sufficient for time study. From statistics, for this condition, number of cycles to be timed can be found by the following relation. Where N = Required number of observations to predict the little time with in ±5% accuracy and 95% confidence level.

X = each stop watch reading.

M = No. of stop watch readings.

3.7. The Standard Time

The standard time for the job will be the sum of the standard times for all the elements of which it is made up, due regard being paid to the frequencies with which the elements recur, plus the allowance .

Standard time is the total time in which a job should be completed at standard performance.

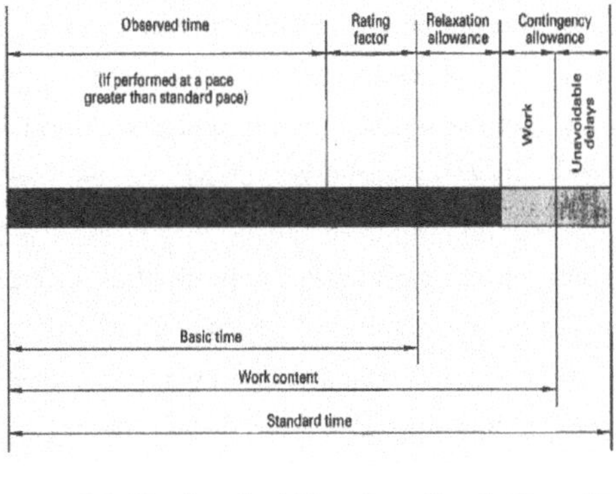

Diagram 3.2: The Standard Time for a Simple Manual Job

3.8. Selection of Jobs

Selection of job is on prime – consideration in work study and always it's try to improving the way of carried out of the work in the work- shop , to reduce production time aswellas improvement of productivity . But the selection of job is depends upon following factor which is very important from subject point of view

1. Economic or cost-effective considerations.
2. Technical considerations.
3. Human considerations.

   a) **Economic considerations :-** It is obviously a waste of time to start or to continue a long investigation if the economic importance of a job is small, orif it is one that is not expected to run for long. Questions that should always beasked are: "Will it pay to begin a method study of this job?" or "Will it pay tocontinue this study?"Obvious choices for study are:

- Key profit-generating or costly operations, or ones with the largest scrap/waste rates.
- Bottlenecks which are holding up other production operations, or lengthy operations that consume a great deal of time.
- Operations involving repetitive work using a great deal of labourand ones that are likely to run for a long time.
- Movements of material over long distances between workstations,those involving the use of a relatively large proportion of labour or whichrequire repeated handling of material.

One of the easiest & simple techniques that can be used to identify key operations is performed by Pareto analysis or ABC analyses or value analyses.

   b) **Technical or technological considerations:-** One of the most important considerations is the desire by management to acquire more advanced technology, whether in equipment or in processes. Thus management may want to computerize its office paperwork or its inventory system, or to introduce automation in the production operations. Before such steps are taken, a method study can point out the most important needs of the enterprise in this respect. If the paper work is being replaced by computerized system then office efficiency will definietly increases . A common term used by information systems specialists in this case is "useless material in yields useless material out". The introduction of new technology should therefore constitute an important factor in the choice of methods of work to be investigated.

c) **Human considerations:-** Certain operations are often a cause of dissatisfaction by workers. They may induce fatigue or monotony or may be unsafe or clumsy to operate. The level of satisfaction should point to a need for method study. Thus an operation which may be perceived as effective by management may, on the other hand, generate a great deal of resentment by the workforce. If suchoperations are addressed by work study specialists as part of an overall work study programme, the benefits of work study can become more apparent to the workforce.

3.9. Breaking a Job Into Elements

Once the study person has recorded all the information about the operation and the operative needed for proper identification in the future, and is satisfied that the method being used is the correct one or the best possible in the prevailing circumstances, it must be broken down into elements.

An element is a distinct part of a specified job selected for convenience of observation, measurement and analysis and A work cycle is the sequence of elements which are required to perform a job or yield a unit of production. The sequence may sometimes include occasional elements.

A work cycle starts at the beginning of the first element of the operation or activity and continues to the same point in a repetition of the operation oractivity.

A Detailed Breakdown Into Elements is Necessary

(i) To ensure that productive work (or effective time) is separated from unproductive activity (or ineffective time).

(ii) To permit the rate of working to be assessed more accurately than would be possible if the assessment were made over a complete cycle. The operative may not work at the same pace throughout the cycle, and may tend to perform some elements more quickly than others.

(iii) To enable the different types of element to be identified and distinguished, so that each may be accorded the treatment appropriate to its type.

(iv) To enable elements involving a high degree of fatigue to be isolated and to make the allocation of fatigue allowances more accurate.

(v) To facilitate checking the method so that the subsequent omission or insertion of elements may be detected quickly. This may become necessary if at a future date the time standard for the job is queried.

(vi) To enable a detailed work specification to be produced.

(vii) To enable time values for frequently recurring elements, such as theoperation of machine controls or loading and unloading work pieces from fixtures, to be extracted and used in the compilation of standard data.

Type of elements: -Eight types of element are generally considered for any particular operation these are repetitive, occasional, constant, variable, manual, machine, governing, and foreign elements.

- A **repetitive element** is an element which occurs in every work cycle of an operation.
- An **occasional element** is an element which does not occur in every work cycle of an operation but which may occur at regular or irregular intervals.
- A **constant element** is an element for which the basic time remains constant whenever it is performed.
- A **variable element** is an element for which the basic time varies in relation to some characteristics of the product, equipment or process, e.g. dimensions, weight, quality, etc.
- A **manual element** is an element performed by a worker.
- A **machine element** is an element performed automatically by any process, physical, and chemical or otherwise that, once started, cannot be influenced by a worker except to terminate it prematurely.
- A **governing element** is an element occupying a longer time within a work cycle than that of any other element which is being performed concurrently.
- A **foreign element** is an element observed which does not form a part of the operations being studied.

3.10. Work Sampling

Work sampling is a method of randomly observing work over a period of time to obtain a distribution of the activities that an individual or a group of employees perform. Work sampling determines the proportion of time spent doing certain activities on a job. It can be used to determine the percentage of idle time and also as a means of assessing nonproductive time to determine performance ratings or to establish allowances. Work sampling is based on statistics, and like all statistical procedures, it can suffer from sampling error. Or in other word we can say that,

Work sampling is a method of finding the percentage occurrence of a certain activity by statistical sampling and random observations.

Work sampling is also sometimes known as **"activity sampling" or "ratio-delay study"**, or **"random observation method"**In order to obtain a complete and accurate picture of the productive time and idle time of the machines in a specific production area, it would be necessary to observe continuously all the machines in that area and to record when and why any of the machines were stopped. But it is very unsuitable aswellas un-favourable situation .

In another condition, 80 per cent of the machines were working and 20 per cent were stopped. If this action was repeated 20 or more times at different times of the day and if each time the proportion of machines working was always 80 per cent, it would be possible to say with some confidence that at any one time there were always 80 per cent of the machines working.

In above both the cases it is impossible to detect that how much amount of time are productive time and idle time .The best method of observation is to making the tours of the factory at random intervals, noting which machines are working and which are stopped, and noting the cause of each stoppage. This is the basis of the **work sampling technique**. Sampling is mainly based on probability. Probability has been defined as "the extent to which an event is likely to occur". Work sampling is based on two basic objectives, these are given below

(i) Calculation of fair day's work , means to measure activities and delays while a man is working and percentage of that he is not working.

(ii) To calculate manual task that is to establish time standards for any specific operation.

Confidence level: Whenever work sampling process is carried out, there will be a requirement of desired confidence level for perfect analyses &calculation. The confidence level is represented on 2-σcurve, this curve called the curve of normal distribution. Curves of normal distribution may be of many shapes.

They may be flatter, or more rounded. To describe these curves we use two attributes: $\bar{x}$, which is the average or measure of central dispersion; and sigma (σ), which is the deviation from the average, referred to as standard deviation. For example the normal distribution curve for any specific operation is given below,

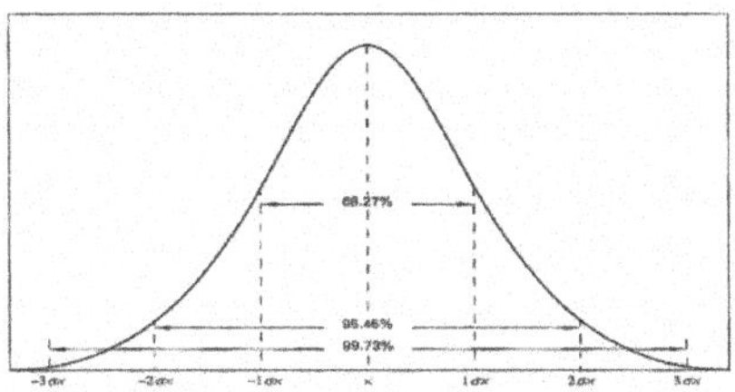

Diagram 3.3: Normal Distribution Curve

In which observation were made randomly for an operation & concluded that,

(i) 95 per cent confidence level or 95 per cent of the area under the curve = 1.96σ_p

(ii) 99 per cent confidence level or 99 per cent of the area under the curve = 2.58σ_p

(iii) 99.9 per cent confidence level or 99.9 per cent of the area under the curve = 3.3σ_p.

In this case we can say that if we take a large sample at random we can be confident that in 95 per cent of the cases our observations will fall within **±1.96σ_p.** In work sampling the most commonly used level is the **95** per cent confidence level.

3.11. Theory of Sampling

It state that the percentage of observation recorded on an operation in any state is reliable estimation of the percentage of time , the operation is in that state provided sufficient no of observation are taken at random ,for example , there total**50** observation is performed amongst **50's** , **46** observation are working observation &**4** observation are idle .

Thus % idle time = (4 / 50) x 100 = 8 %

Working time = (46 / 50)x 100 = 93 %

This study is performed on single operator who is working 8 hours / day , represent that he was idle for 60 min. & working or productive time were 421 min. Of theday.

Determination of Sample Size

The sample size is decided by following method.

Statistical Method

Statistical method

The formula used in this method is:

$$\sigma p = \sqrt{\frac{pq}{n}}$$

where

σp = standard error of proportion
p = percentage of idle time
q = percentage of working time
n = number of observations or sample size we wish to determine.

Before we can use this formula, however, we need to have at least an idea of the values of p and q. The first step is therefore to carry out a number of random observations in the working area. Let us assume that some 100 observations were carried out as a preliminary study and at random, and that these showed the machine to be idle in 25 per cent of the cases (p = 25) and to be working 75 per cent of the time (q = 75). We thus have approximate values for p and q; in order now to determine the value of n, we must find out the value of σp.

Let us choose a confidence level of 95 per cent with a 10 per cent margin of error (that is, we are confident that in 95 per cent of the cases our estimates will be ± 10 per cent of the real value).

At the 95 per cent confidence level:

$$1.96\,\sigma p = 10$$
$$\therefore \quad \sigma p = 5 \text{ (approx.).}$$

We can now go back to our original equation to derive n:

$$\sigma p = \sqrt{\frac{pq}{n}}$$
$$\therefore \quad 5 = \sqrt{\frac{25 \times 75}{n}}$$
$$\therefore \quad n = 75 \text{ observations.}$$

If we reduce the margin of error to ± 5 per cent, we have

$$1.96\,\sigma p = 5$$
$$\therefore \quad \sigma p = 2.5 \text{ (approx.)}$$
$$\therefore \quad 2.5 = \sqrt{\frac{25 \times 75}{n}}$$
$$\therefore \quad n = \frac{25 \times 75}{(2.5)^2}$$
$$= 300 \text{ observations.}$$

Another method of sample size determination for achieving desired accuracy, the no of observation is calculated as

Sx P = k$\sqrt{p(1-p)/N}$

Where k = confidence level,1, 2,3 .... 68 %, 95 %, 99 %

N= sample size or total no of random operation

S = desired relative accuracy

P= % value of activity.

3.12. Requirement of Work Sampling Process

-Before conducting the observation process, we have to analyze the basic objective of the operation, in which we will achieved following conditions.

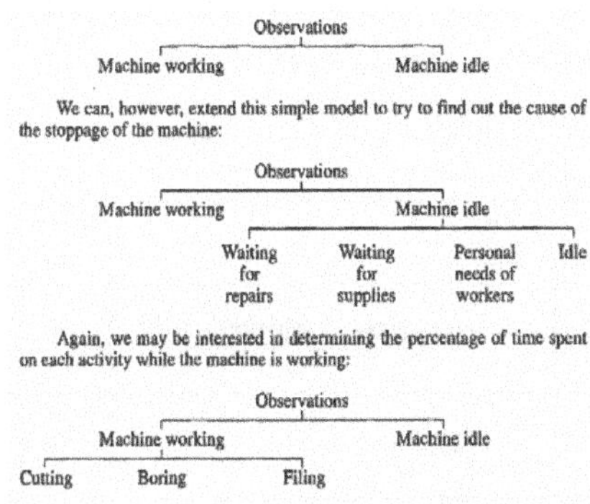

We may also be interested in the percentage time spent by a worker or groups of workers on a given element of work. If a certain job consists of ten different elements, by observing a worker at the defined points in time we can record on which element he or she is working and therefore arrive at an percentage distribution of the time he or she has been spending on each element.

Process of conducting the work sampling: -following are very important steps, from subject point of view that are used in work sampling.

(i) Selecting the job to be studied and determining the objectives of the study.

(ii) Making a preliminary observation to determine the approximate values of p and q.

(iii) In terms of a chosen confidence level and accuracy range, determining n (the number of observations needed).

(iv) State the confidence level.

(v) Estimate the % value of activity or delay measured for one day or two day.

(vi) Design the study.

- Calculate the no of observation to be made
- Calculate the no of observation to be required, select & instruct them.
- Define the no of days or shift is required for observation.
- Taking observation as per plan.
- Design the observation form.

(vii) Determining the frequency of observations, using random tables.

(viii) Designing record sheets to meet the objectives of the study.

(ix) Check the accuracy at the end of the study.

(x) Prepare reports, compile the result.

The analysis of results can be calculated readily on the record sheet. It is possible to find out the percentage of effective time compared with that of

Table 3.3: Performa of Work Sampling Record Sheet

Date:	Observer:	Study No.:	
Number of observations: 75		Total	Percentage
Machine running	卌 卌 卌 卌 卌 卌 卌 卌 卌 卌 卌 卌 ||	62	82.7
Machine idle	卌 卌 |||	13	17.3

Table 3.4: Work Sampling Record Sheet Showing Machine Utilization and Distribution of Idle Time

Date:	Observer:		Study No.:	
Number of observations: 75			Total	Percentage
Machine running		卌 卌 卌 卌 卌 卌 卌 卌 卌 卌 卌 卌 ||	62	82.7
Machine idle	Repairs	||	2	2.7
	Supplies	卌 |	6	8.0
	Personal	|	1	1.3
	Idle	||||	4	5.3

Table 3.5: Work Sampling Record Sheet Showing Distribution of Time on Ten Elements of Work Performed by a Group of Four Workers

Date:	Observer:						Study No.:			
Number of observations: 75										
	Elements of work									
	1	2	3	4	5	6	7	8	9	10
Worker No. 1 Worker No. 2 Worker No. 3 Worker No. 4										

Delays, to analyze the reasons for ineffective time and to ascertain the percentage time spent by a worker, groups of workers or a machine on a given work element.

Advantages and Disadvantages of Work Sampling in Comparison with Time Study.

Advantages

(i) Many operators or activities which are difficult or uneconomical to measure by time study can readily be measured by work sampling.

(ii) Two or more studies can be simultaneously made of several operators or machines by a single study person. Ordinarily a work study engineer can study only one operator at a time when continuous time study is made.

(iii) It usually requires fewer man-hours to make a work sampling study than to make a continuous time study. The cost may also be about a third of the cost of a continuous time study.

(iv) No stopwatch or other time measuring device is needed for work sampling studies.

(v) It usually requires less time to calculate the results of work sampling study. Mark sensing cards may be used which can be fed directly to the computing machines to obtain the results just instantaneously.

(vi) A work sampling study may be interrupted at any time without affecting the results.

(vii) Operators are not closely watched for long period of time. This decreases the chance of getting erroneous results for when a worker is observed continuously for a long period, it is probable that he will not follow his usual routine exactly during that period.

(viii) Observations may be taken over a period of days or weeks. This decreases the chance of day-to-day or week-to-week variations that may affect the results.

(ix) Work sampling studies are preferred to continuous time study by the operators being studied. Some people do not like to be observed continuously for long periods of time.

(x) Work sampling studies are less fatiguing and less tedious to make on the part of time study engineer.

Disadvantages

(i) Work sampling is not economical for the study of a single operator or operation or machine. Also, work-sampling study may be uneconomical for studying operators or machines located over wide areas.

(ii) Work sampling study does not provide elemental time data.

(iii) The operator may change his work pattern when he sees the study person. For instance, he may try to look productive and make the results of study erroneous.

(iv) No record is usually made of the method being used by the operator. Therefore, a new study has to be made when a method change occurs in any element of operation.

(v) Compared to stop watch time study, the statistical approach of work sampling study is difficult to understand by workers.

Application of Work Sampling

(i) For estimating the utilization of available resources like, machine tools,cranks,fork, truck etc.

(ii) For fixation of delay allowance.

(iii) For establishment of time consumption in job activity.

(iv) To obtain a standard time for job.

3.13. Standard Time / Piece Calculation from Work Sampling Technique

work sampling is an another tool for calculating the standard time because with the help of work sampling we will obtained

(i) % of day the worker is working.

(ii) % of day the worker is idle, with the help of work sampling we can determine that , how much work is done during the working portion of the day , it is given as,

$$\frac{\text{(Total time in min.) X (Working time in \%) X (Performance index in \%)}}{\text{Total no items is being manufactured}} \text{ X Allowances}$$

following are some important definition are given below which are related to the standard time,

- **Normal time** is the expected time required to perform some work activity at a normal pace, under normal operating conditions, and using a prescribed method.

- **Allowances** include time for labor fatigue and personal needs, equipment breakdowns, rest periods, information delays, and so on.
- **Standard time** is normal time adjusted for allowances.

Standard time = Normal time (1 + Allowance factor)

Question 3.1 : In a work-sampling study an administrative assistant was found to be working 2,700 times in a total of 3,000 observations made over a time span of 240 working hours.

The employee's output was 1,800 forms. If a performance rating of 1.05 and an allowance of 15 percent are given, what is the standard output for this task?

Solution

Effective number of hours worked = 2,700(240)/3,000 = 216.

Output during this period = 1,800 forms.

Actual time perform = 216(60)/1,800 = 7.2 minutes, or 8.33 forms per hour.

Normal time = Actual observed time X Performance rating = 7.2(1.05) = 7.56 minutes.

Standard time = 7.56(1.15) = 8.694 minutes per form.

Standard output = 60/8.694 = 6.9 forms per hour, or 55 forms per 8-hour day.

Question 3.2

An office worker wants to perform work sampling for task T. It was estimated that Employees are idle 20% of the time. An office worker would like to take a worksample with accuracy of 4% desired confidence level of 95.45%

Solution

Number of Observations, $n = Z^2 P (1-P) / h^2$

Z=2 for 95.45% confidence level

P=Estimate of idle Percentage which is 0.2

h =Desired accuracy of 4 % or 0.04

n = (2)2 (0.2)(0.8) / (0.04)2

= 400 Observations

Question 3.3

An office worker wants to set standard time to complete a task K comprised of three job elements. He clocked work elements and chose to take 6 cycles as shown in table. The allowance for the task is 12% and performance rating for each element is also given in the table. Find the normal time and standard time to complete the task.

Observed Cycles and Performance Rating to Complete task K

Job Element	Cycles observed (minutes)						Performance rating
	1	2	3	4	5	6	
1	15	10	13	17	45	14	100%
2	30	20	26	72	25	23	110%
3	4	3	2	3	5	4	100%

Solution

We can see from the data that some of the observations are unusual like fifth observation for job element 1 and fourth observation for job element 2. These are non-recurring observation which might be result of some disturbance during time study performance. So, delete these unusual observations and compute the average time for all three job elements for rest of the observation

$$\text{Average time for job element 1} = \frac{15+10+13+17+14}{5} = 13.8 \text{ minutes}$$

$$\text{Average time for job element 2} = \frac{30+20+26+25+23}{5} = 24.8 \text{ minutes}$$

$$\text{Average time for job element 3} = \frac{4+3+2+3+5+4}{6} = 3.5 \text{ minutes}$$

Compute the normal time for each job element considering performance rating.

Normal time for job element 1= (13.8) (1.00) = 13.8 minutes

Normal time for job element 2= (24.8) (1.10) = 27.28 minutes

Normal time for job element 3= (3.5) (1.00) = 3.5 minutes

Total normal time to complete the task = 13.8 + 27.28 + 3.5 = 44.58 minutes

The standard time to complete the task considering allowance of 12 %

$$= \frac{44.58}{1-0.12} = 50.66 \text{ minutes}$$

Question for Discussion

1. Define objectives and concepts of work measurement.
2. Explain work sampling,
    (a) What is work factor system.
    (b) Explain the principles of motion economy.
3. What is performance rating ?
4. In a time study for a job done by a worker whose rating is 90, the data are as follows:

    Observed time = 20 minutes,

    Personal needs allowances=4% of Basic time,

    Fatigue allowance = 2.5% of Basic time,

    Contingency workallowance = 2% of Basic time,

    Contingency delay allowance = 1% of Basic time. Calculate

    (a) Basic time.
    (b) Work content.
    (c) Standard time.
5. What is Time Study? How is a job selected for time study ?
6. What do you understand by allowances in the context of time study? What are various allowances and how are they determined? Discuss.
7. Explain about the Work sampling .
8. Bring out the factors affecting work measurement. Also highlight the merits and demerits of work measurement with examples.
9. Under ,what circumstances is work sampling preferred to stop watch time study for establishing work standards?
10. Explain various stop watches used for time study.
11. What are the typical error with time study rating ? How it can be minimized ?
12. Describe the procedure of job design considering behavioral factors by 75 samples of a production cycle gave an average time of 5.28 minute/piece. The performance rating was

estimated as 110% and allowances are 25% of total time available. What is the standard time in minutes/piece?

13. A work sampling study was conducted for 100 hours in the'machine shop in order to estimate the standard time, The total numbers of observations recorded were 2500. No working activity could be noticed for 400 observations. The ratio between manual and machine elements was 2: 1. Average rating factor waS estimated as 1.15 and, the total number of articles produced during the study period were' 6000. Rest and personal allowances may be taken as 120% of the normal time. Compute the standard time in minutes.
14. State the advantages, disadvantages and practical applications of timestandards established by 'Work Sampling Method'.
15. Following table indicates part of the failure data of 1000 components. Find out the question mark items in the table.

Time Interval	No. of Failure	Cumulative Failure	No. of Survivors	Failure Density	Failure Rate	Reliability
0		0	1000			
	130			?	?	
1		130	870			?
	83			?	?	
2		213	787			
	75			?	?	
3		288	712			
	68					

16. Given P0.10 = 0.053 and P0.95 = 0.014, Determine the single sampling plan which exactly meets the consumer's stipulation and comes closer to the producer's stipulation.
17. Determine the acceptance equation for the following double sampling plan. N = 60000 n1 = 80, c1 = 2, r1 = 4 n2 = 160, c2 = 5, r2 = 6. Determine also the probability of acceptance for an incoming process quality of 2% (i.e. 100 p' = 2%) .
18. What is Work measurement? List the various techniques of work measurement and also explain how the Standard time is calculated.
19. State the objectives of work measurement. Name the work measurement techniques.
20. Discuss the procedure establish standard time for completion of a job.
21. Differentiate between method study and work measurement. How is productivity increased by these methods?
22. Explain the following in connection with time study.
    a) Selecting the job.
    b) Breaking the job into elements.

23. Turning gear blanks on centre lathe involves the following elements.

The stop watch data is given. Assuming the rest and the personal allowances as 13% and contingency allowance of 2%, calculate standard time.

Elements	Description	Observation				
1	Pick &place	0.20	1.46	5.22	6.49	14.25
2	Start machine and approach tool	0.30	1.55	5.30	13.10	14.35
3	Turn diameter	1.05	2.31	6.05	13.84	15.10
4	Withdraw tool and stop machine	1.13	2.38	6.14	13.92	15.17
5	Release part and keep it asid	1.28	2.54	6.29	14.06	15.32

Foreign elements: * (1) 2.54 to 5.02 taking to another operator (2) 6.49 to 12.98 away for personal need. Rating factor for element 1 is 90%, element 2 and 4 is 110%, element 3 is 100% (auto cycle), element 5 is 95%

24. What is sampling? How it is useful in statically quality control.
25. Write short note on single and multiple sampling plan.
26. Write short Notes on the following.
    a) Fatique allowance.
    b) Heat stress.
    c) Learning curves.
    d) Industrial noise.
27. What is the difference between method study and work measurement ?List the various objectives of method study.

References

1. Idem, "The Measurement of Work. London", British Institute of Management, 1965.
2. Idem, "Work Methods Training Manual", Los Angeles. Campbell's Book Store. Third Edition, 1950.
3. L. Yerges, "Sound, noise and vibration control", New York: Van Nostrand, 1978.
4. Idem, "Work Sampling. New York", John Wiley and Sons. Second Edition, 1957.
5. Ideni, "Work Measurement Manual", Los Angeles. Campbell's Book Store. Fourth Edition, 1951.
6. J.H. Quick, "Work factor time standards: Measurement of manual and mental work", NewYork and London, McGraw-Hill, 1962.
7. D.Marie, "Adapting working hours to modern needs: The time factor in the new approach to working conditions", Geneva, ILO, 1977.

8. Kanawaty, George (éd.)., "Managing and developing new forms of work organization", (Geneva, ILO, 2nd éd.,), 1981.
9. D. Meyer, "GRASP A Patient Information and Workload Management System", MCS, Morganton, NC, 1978.
10. H.B. Maynard, (ed.), "Industrial Engineering Handbook", 3rd edn, McGraw-Hill, New York. 1971.
11. International Labour Office, Introduction to Work Study, 3rd edn, International Labour Office, Geneva, 1978.
12. ILO 1979, Introduction to Work Study, 3rd edn, International Labour Office, Geneva. International Labour Office (ILO), Job Evaluation, Geneva.
13. Brisley, C.L. 1971, 'Work sampling', in Industrial Engineering Handbook, 3rd edn, H.B. Maynard, McGraw–Hill, New York.
14. C.A. Anderson, "Performance rating", Industrial Engineering Handbook, 3rd edn, W.B, Maynard, McGraw–Hill, New York, 1971.
15. R.M. Crossan and H.W. Nance, "Master standard data: The economic approach to work measurement", (New York and London, McGraw-Hill, 2nd éd.,), 1972.
16. S.M. Genovese, "Work Measurement and Process Improvement Study of Roller Coaster Preventive Maintenance", 2003,
17. R. Rutter, "Work Sampling: As a Win/Win Management Tool", Industrial Engineering, Pp. 30–31, 1994.
18. G.F. Ramsey, "Using Self-Administered Work Sampling in a State Agency", Industrial Engineering, Pp. 44–45, 1993.

CHAPTER 4

INCENTIVE SCHEME

4.0. Incentive Scheme

Incentive scheme is a stimulating process given to the worker for his hard work and working efficiency , in order to encouraging and motivating to do perform the next task very confidently , with very joy & happiness in group team , to achieve the higher productivity or efficiency with high profit to both worker aswellas firms / organization . in this system extra payments is provided with minimum wages or bonus for extra work **OR** The Incentive Schemes are the programs designed to encourage and motivate workmen for higher efficiency and greater output. The Incentives are the monetary rewards given to the workmen in recognition of their achievement of specific results during a specified time period. Employees do not want just money; they want the honor and respect. They will make impossible into possible just you need to trust them and provide them facility with full dignity.

According to International Labor Organization (ILO),

The incentives are "payment by results" but however it is correct to call it as the "incentive system of payment" since it lays more emphasize on motivation, i.e., incentives are imparted to the workers for their outstanding performances. Therefore, the workers are encouraged and motivated to earn more by increasing their productivity.

The incentive scheme are classified as,

(i) **Financial Incentives:** -Money is an important motivator to enhance the efficiency, the financial incentive is given as wages and salaries, bonus, retirement benefits, medical reimbursement, etc Money plays a significant role in satisfying physiological and security/social needs. As money is recognized as a basis of status, respect and power. Financial incentives may further be classified as individual incentives and group incentives. Both are discussed now one by one.

(ii) **Individual incentive scheme:** -In individual incentive scheme, every worker / operator gaining the incentive on the basis of individual output or performance irrespect to aggregate output of his department or firms & payment is done for his / her more output then standard output.

(iii) Group Incentive Schemes:-Group incentive schemes are appropriate where jobs are interdependent and a group of an operator doing the task. In this scheme each employee is paid incentive on the basis of collective performance of his work group following are some elements of Group Incentive Schemes.

- **Profit-sharing:** -In this parts, the organization has been earned a profit not only a single parties efforts but also it's association of so many individual parties. The basic rationale behind profit-sharing is that the organizational profit is an outcome of the co-operative efforts of various parties, therefore, employees should also share in profits as shareholders share by getting dividend on their investment , The very important purpose of introducing profit-sharing is to strengthen the loyalty of employees to the organization. **Profit sharing may be performed in following ways.**
  1. Current.
  2. Deferred.
  3. Combination.

  (i) Current: Under this form, profits are paid to the employees in cash or by cheque.

  (ii) Deferred: Profits are credited to employees' accounts to be paid at the time of retirement.

  (iii) Combination: In this case, a part of employee share of profit is paid in cash or cheque or stock and the remaining part is deferred and credited to his/her account.
- **Co-partnership:** Under co-partnership scheme, employees become shareholders also by having company shares. Now, employ-ees participate in both —profits and management of the company.
- **Scanlon Plan:** The Scanlon plan was developed by Joseph N. Scanlon . in this plan the suggestion & proposals are demanded by workers , employees and managers for cost-cutting methods there by reducing the cost of operation and improving working methods , increase in efficiency on account of cost-cutting and sharing in the gains of increased productivity.

(iv) Non-Financial Incentive :- Such kind of scheme , do not gives any kind of financial facility but provided for a worker / operator to improve his / her efficiency aswellas for motivation .Some of very important elements of Non-Financial Incentive are given below.

- **Job security**:- Nothing can motivate a worker who is appointed as a temporarily in the organization because temporary worker always passing through an insecurity of job , if the worker is very sincere about work and put a great efforts , the organization must be promised for his job security for motivating .

- **Appreciation of Work done:** -Appreciation or praise for work done be it at home, at school/university or at work place, serves as an effective non-financial incentive.
- **Opportunity for Growth:** -operator / worker always need to grow in their career. So, if the employees are provided proper opportunities for further growth and career advancement, they feel much satisfied and become more committed to the organizational goals.
- **Suggestion System :-** Suggestion system is also another non-financial incentive to be used to motivate employees , for suggestion system organization gives supports in term of cash awards , worker's name with his/her photograph in the company's magazine , to motivate other employees of the firms .
- **Competition: -** The management can encourage healthy competition among the employees. This would, certainly, motivate them to prove their capabilities. The management can also rank the employees according to performance. Such of those employees who have performed very well may be given merit certificates.
- **Treat your people as human beings – neither inferior, nor superior:-** Show trust and respect, motivate them for creativity, create a 'safe-to-risk environment', keep them informed of relevant developments inside the organization, mistakes be treated as learning tools instead of blaming them.
- **Training and other improvement programmed**
- **Housing , medical , educational facility for worker & their families**
- **Helpful and cooperative management.**
- **Pension & others benefits.**

4.1. Wages / Incentives Plans

There are so many wage &incentive plans are proposed from time to time by the management gurus , amongs them some are given below having a great importance from subject point of view.

4.1.1. Straight Line Piece-Work System

The Straight line Piece-Work System is very simplest incentive method that is generallyapplied for manual operation for manufacturing an articles of same type , loading , unloading the individual efforts. A system in which the worker is paid according to the number of units produced at a fixed rate per unit during a defined period of time is called a straight line piece-work system or straight line piece-rate system.

In piece rate system ,there are no provision for providing any kind of wages to operators throughout the manufacturing the articles if he / she "s output is below the defined level of output .thus it means that worker's earnings are computed by multiplying the time taken by per unit of time

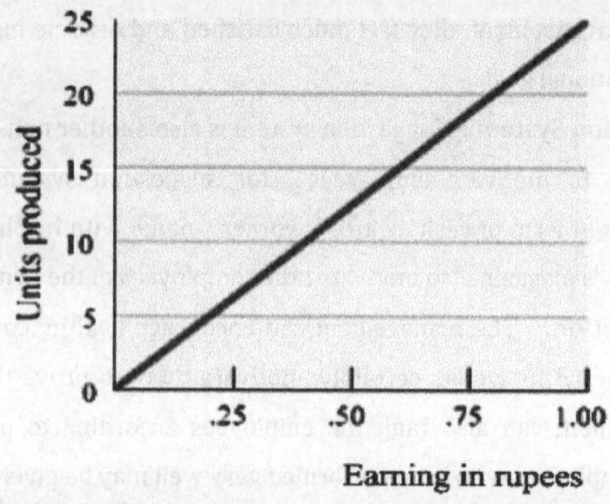

Diagram 4.1: Straight Line Piece-Work System

The straight line piece rate system is graphically represented above that's indicates , operator production per piece rate is fixed & and collecte a payment on the basis of no of articles produces in the defined period time .

Earning of operator = (fixed rate per unit X defined time)

4.1.2. Taylor's Differential Piece-Rate System

Taylor's Differential Piece-Rate System was introduced by F.W. Taylor , he directly point out the standard output level taking as an important parameter for fixation of piece rate & explained that the worker who exceeds the standard output within the stipulated time must be paid a high rate for high production on the other hand if he attains a lower output then standard output within the standard time must be paid the lower rate . Thus, there are two piece-ratessystem combinestogether, one who reach the standard output or exceeds it, is paid 120 percent of the piece rate.

While the one who below the standard level of output, is paid 80 percent of the piece-rate. The graphic arrangement of this system is given below with desired deomostration .

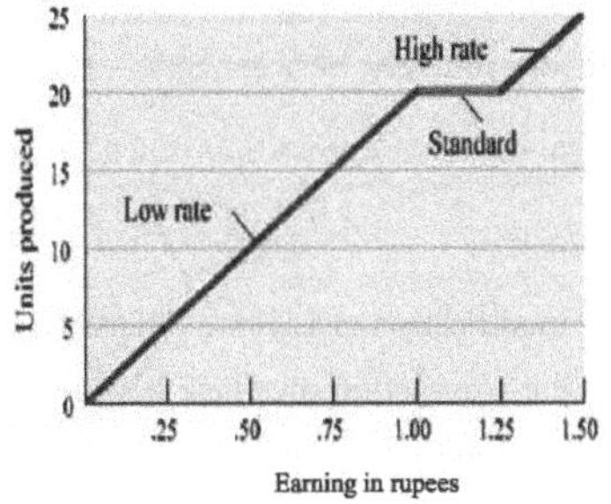

Diagram 4.2: Taylor's Differential Piece-Rate System

The x- axis of the graph repesenting the earning in rupees / piece on the other hand y- axis representing the output of the worker . In first steps straight line that passes throught the center having a blue colour indicate that operator acquired a low rate per piece because he is producing less than standard output produt with respect to standard time , on the other hand the value of standard output is about 20 – piece , if operator produces 20 – piece or more than this value , organization paid high rate of production / piece . means it contains two piece rate system low rate & high rate / piece .

4.1.3. Merrick Differential Piece-Rate System

The Merrick Differential Piece-Rate System is nothing but it is a modification of Taylor's differential piece-rate system which consists of three piece-rates system .

1. **For the beginners.**
2. **For the average / developing workers.**
3. **For the superior/ skilled workers.**

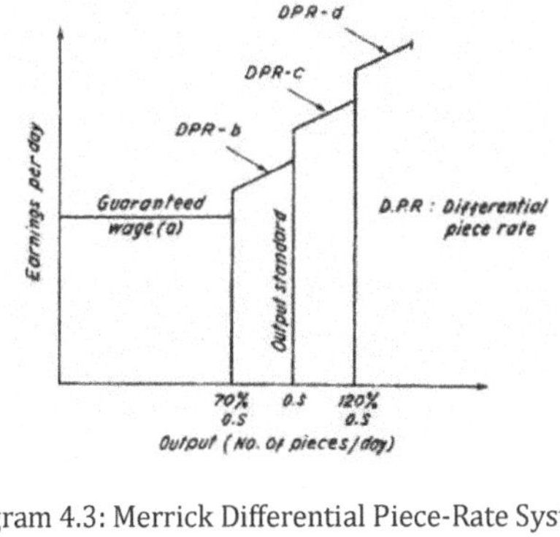

Diagram 4.3: Merrick Differential Piece-Rate System

The worker is paid the straight price rate up to 83% of the standard output.

10 % above the normal rate for producing between 83% – 100%.

20% above the normal rate for producing more than 100% of the standard output.

4.1.4. *Gantt Task System*

Gantt Task System is also a modification of Taylor's differential piece-rate system in which the standard time is fixed for the completion of a task based on a careful time and motion study, workers performance is compared against the standard time to the calculate his efficiency and for making payments of the task . Conclusion of Gantt Task System is given below.

(i) The wages calculation is made on hourly basis instead of Piece-Rate.

(ii) If the worker takes more than the standard time to complete the task, then his efficiency is below 100%, and he is paid only the time wages.

(iii) Bonus are paid to the workers who achieved standard output below standard time , he/ she efficiency is considered as 100 % + paid a bonus of 20% on the wages earned .

(iv) If the workers takes less time than standard time with high level of performance then operator efficiency is considered as more than 100 %, in this case payments is done on the basis higher piece rate.

(v) The standard time is fixed for each operation , that is calculated on basis of scientific work measurement

(vi) The purpose of the Gantt Task System is to improve the performance of operator.

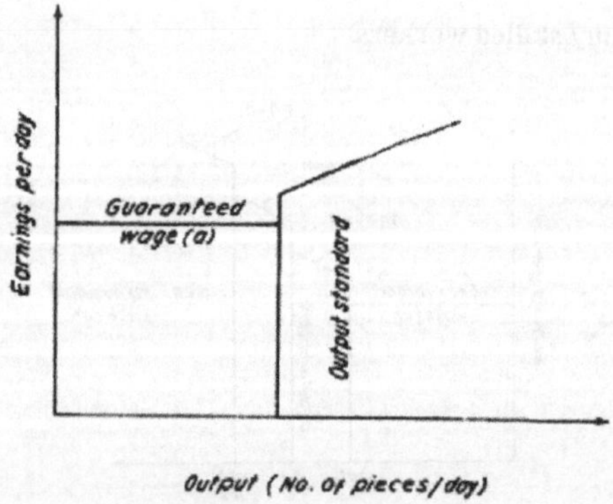

Diagram 4.4: Gantt Task System

In this graph horizontal line represents the output of the operator / day which is a x-axis and earning / day represented by vertical line which is a y- axis.

Vertical line indicating the standard output having the 100 % performance of the operator below this line only time wages will be given , inclined line indicating the "**high level of performance**" of the operatorfor that payment is done on the basis higher piece rate.

Predetermined motion time systems (PMTS) :-PMTS is a work-measurement technique , thereby we can analyze& established a time for completion of job considering the basic human motion for any paticular operation , the motion time is then calculated for each operation and it's accumulation gives the necessaray information regarding the complete time for task finishing at a defined level of performance .

If a complete job is break-up into no of operation and how many repeated or non - repeated motion is being perform aswellas how much time hasbeen taken by the operator for finsishing the operation .

The complete sum up of elemental time for finshing the job is a basic concept of PMTS , PMTS help in to achieve desired level of operator performance .

Large amounts of research, data collection, analysis, synthesis, and validation are required to produce PMTS data and the number of such systems is very low.

The basic advantages of PMTS is that they require a detailed description of the working method - and are thus useful for studying how work is done (and how it can be improved) as well as measuring the time it should take.

Procedure

(1) Study the complete operation cycle few times

(2) Observe and record the activities of one hand at a time.

(3) Record only few symbols at a time.

(4) Recording should be started at a point of time that can be easily distinguished. For examples activity of picking up the such work piece at the beginning of cycle of work is a good point from which recording the movements should be started,

(5) Care should be taken that no activity is left while recording, as it\will affect the method study badly.

(6) Combinations of operations and transport should be avoided, unless they actually occur at the same time.

Predetermined Motion Time Systems (PMTS) is Performed by following Two Method

1) Method time measurement (MTM).
2) Work- factors (WF).

4.2. Method Time Measurement (MTM)

One of the major PMTS systems is MTM (Methods-Times Measurement) .MTM gives values for such basic hand/arm motions as : Reach, Move, Turn, Grasp, Position, Disengage, and Release, together with a small set of full body motions .

MTM is a calculation of time of complete manual operation without using any kind of time measuring devices . for example – In an particular operation motion analyses has been performed means identification of motion, classification and recording & their time value is observed by with the help of MTM data .

Thus the summation of all the time element that is performed the job indicating the complete time taken for the operation . it also help in calculating the standard time for job completion .

There are so many motion are performed by the operator in order to complete the job / task . MTM gives values for such basic hand/arm motions , these are given below .

(i) Reach.
(ii) Move.
(iii) Turn.
(iv) Apply pressure.
(v) Grasp.
(vi) Position.
(vii) Disengage.
(viii)Release.

Except it Some Other Motion is Also Used

1) Side step	**6) kneel on one leg**
2) Turn body	**7) kneel on both leg**
3) Walk	**8) foot motion**
4) Bend	**9) sit**
5) Stop	**10) two-eye movement**

Table 4.1: Motion & Description

Motion	Description
Reach (R)	Reach is the basic element used when the hand or fingers moves to a defined destination or to a general location . Reach depends upon following factor- Nature of destination , Length of motion , Type of reach .
Move (M)	Move is the another basic element employed for transferening the material /equipments /devices to a definite location or place .Nature of destination , distance of motion , Type of weight factors .
Turn (T)	Turn is a motion employed to turn hand while when the hand, either empty or loaded , rotates the hand, wrist, and fore arm about the long axis of the forearm.turn depends upon (i) Degrees turned ,(ii) Weight factor.There are three classification of weight factor i.e. **small load up to 1 kg, medium load from1.1 kg to 5 kg** and **large load from 5.1 kg to 15 kg**
Grasp (G)	Grasp means holding the object / tools / equipments / or others called grasping **G1A** - Indicate easy grasping of object, **G1B** – Grasping an object when surface of base is rough , **G1C** – Grasping is perform when cylindrical object is in contactto each others
Position (p)	Position is employed when engages one object with another object , there are three factors which is affected the position (i) Class of fit. Loose close and Exact(ii) Symmetrical , non symmetrical and semi symmetrical , (iii) Ease of handling. Whether easy to handle or difficult to handle .
Disengage (D)	Disengage is to break contact between one object and another hence the objects are separated away , disengage is depends upon (i) class of fits - loose ,fit , close (ii) Ease of handling – easily hndled or difficult to handling .
Release load (RL)	Release means when the finger or hand does not have a control on object , (i) Normal Release - In this fingers are just opened to let go the object , (ii) Contact Release – no time is allowed for this motion .
Apply pressure (AP)	It is basic elements used to over- come the resistance . (i) **API**- when pressure is applied for regrasping the object or squeezing , (ii) **AP2**- pressure applied against movement by arms and body .

4.2.1. MOST (Maynard Operation Sequence Technique)

Maynard Operation Sequence Technique (MOST) is nothing but it is also a work-measuring technique generally used in work-study for planning & controlling the operation process . the significance of work measurement technique is to improve the productivity , there are so many work measurement technique are availiable thereby we can enhance the productivity but some draw back or diadvantages are imposed by those method , it does't means those metod are not suitable but irrespect of all those the MOST is very accurate and easy method due to M.O.S. T has the data manipulation during operation .

Concept of 'MOST' Work Measurement Technique

M.O.S.T. is a system to measure work, therefore concentrates on the movement of objects. It has been observed that the movements of objects follow certain consistent repeating patterns; such as reach, grasp, move, position object etc. In general the objects can be moved in two ways, either they are picked up and moved freely through space or they are moved while maintaining contact with another surface. Manual work is described by the three movements of objects with manual cranes , these are given below.

(i) The General Move Sequence (for the movement of object freely through air).

(ii) The Controlled Move Sequence (for the movement of object when it remains in contact with the surface).

(iii) The Tool Use Sequence (for the usage of common hand tools).

MOST have three Basic system that are Genereally Applied for Analyses , the Brief Description is Given Below

(i) **Maxi MOST:-** At the highest level, Maxi MOST is used to analyze operations that are likely to be performed fewer that 150 times per week. An operation in this category ranges from less than 2 minutes to more than several hours .

(ii) **Basic MOST:-** At the intermediate level, operations that are likely to be performed more than 150 but less than 1500 times per week should be analyzed with Basic MOST. An operation in this category ranges from a few seconds to 10 minutes .

(iii) **Mini MOST:-**At the lowest level, Mini MOST provides the most detailed analysis. Operations that are likely to be performed more than 1500 times per week should be analyzed with Mini MOST. An operation in this category may have cycle time 10 seconds or less.

Basic Sequence Model for MOST

Table 4.2: Sequence Model for MOST

Activity	Sequence model	Parameter
General Move	ABG ABP A	A – action distance B – body motion G – gain control P – placemen
Controlled Move	ABG MXI A	M – move controlled X – process time I – alignment
Tool Use	ABG ABP * ABP A	F/L – fasten/loosen C – cut S – surface treat M – measure R – record T – think

MOST Procedure

Following are the basic procedure followed by the most technique

1. Watch job/task.
2. Determine sequence(s) to use.
3. Determine index values.

4. Add index values to determine TMU.
5. Multiply TMU by 10.
6. Convert TMU to seconds, minutes, hour.

Importance of MOST Technique

- Latest Benchmarking tool for all type of operations in various industries.
- Streamlines operations and quickly identifies inefficient methods.
- Provides consistent standards and accuracy to within ± 5% with a 95% confidence level.
- Timings can be obtained in advance before actual start of production.
- Universal Application to all types of industries.
- Reduces the time required for data development and standard setting.
- Easy to learn and use.
- Universal Approach, Fast to Apply, Adequate Accuracy.
- Minimum of Paperwork.
- Encourages Method Development and Improvement.

4.3. Work Factor System

work factor is a calculation of time which has been required to perform a work or an particular operation , basically it consists of manual motion time means how much time motion of hand , arm or other limb of body is taken .

The motion may be a repetetive type or may be a simple motion .

Other element is mental process time , means how much time is taken by the operator for thinking about particular operation involving in which manner an operation can be started , its progressive stages & the completion of the task , thus work factor is combined effect of motion both the motion time for an operation .

Work Factor System = (manual motion time + mental process time)

Or in other word we can say that

Work factor is a method of determining of the motion time for a given operation in which we have to consider two things detailed analyses of each motion considering four major variable of the work and standard time taken by each motion that is given in motion time catalogue.

According to the Work Factor system there are four major variables which affect the time to perform manual motions:

1. Body Member used.
2. Distance moved.
3. Manual control required.
4. Weight or resistance involved.

Description of each Variable are Given Below:

(i) Body Member used :- In this Section we will Consider the Motion of the Basic Parts, these are Given Below with Required Notation

- Finger or hand – (F.H)
- Arm – (A)
- Forearm-Swivel – (S)
- Trunk – (T)
- Foot – (FT)
- Leg – (L)

(ii) Distance Moved

In general , shorter the distsnce moved , shorter the time willbe in general farthest the distance moved , with in normal limits . the shorter the time will be per inch movement .If the change of direction is intentionally done during movement it will be taken care of by a work factor, as the distance being measured in a straight line.

(iii) Manual Control Required

The manual control required is specified by work factor in this system. The following classification of the types and degrees of control illustrates the difficulty involved: **Definite Stop Work Factor(D) , directional control work factor (steer)(s) , Care Work Factor (precaution) (p) ,Change of Direction Work Factor (U).**

(iv) Weight or Resistance Involved

The effect of weight on motion time is due to weight of the object handled also . The effect of weight on time is due to the following factors

- The body member used.
- The sex of the operator.
- The two variables, distance and body member are indicated in inches.

In order to understanding the Work Factor principle , describing the motion according to the amount of weight or resistance involved in its performance.

Finally the weight work factors for males and femalesWork Factor Standard Elements are given below Work Factor Standard Elements of Work

Element	Notation
Transport	Move and Reach - TRP
Grasp	GR
Preposition	Pp
Assemble	ASY
Use	Process or Machine Time Manual - US
Disassemble	DSY
MentalProcess	MP
Release	RL

There are so many criteria are availiable on ths basis of that we can classify the work factor system, thses are given below

1. Detailed work Factor (DWF).
2. Simplified work Factor (SWF).
3. Abbreviated work Factor (AWI).
4. Ready work Factor (RWF).
5. Brief work Factor (BWF).

The brief description of those are given below ,

(i) Detailed work Factor (DWF) : Detailed work factor is applicable where manufacturing organization consists of two objectives , these are accurate time standards for any operation and also a savings criteria . DWF finds a wide range of application in many factory and office work , This is a general basic level system.

(ii) Simplified work Factor (SWF) : It has been generally applied for higher level system where medium quantity production is on prime consideration .It is quite effective in preparing estimates before actual production starts and also it is very easy and rapidly apply . time unit is 10 thousands of a minute .

(iii)Abbreviated work Factor (AWI) : It is a higher level system especially suitable for small quantity job shops production . it consists of predetermined time standards in which measurement can be perform very rapidly .The abbreviated time unit is 0.005 (recorded as 5) minute ,The accuracy of the Abbreviated systems is expected to average +12 percent of the detailed system.

(iv) Ready work Factor (RWF) : This higher level modified work factor system is followed by those persons ,who are very familiar with the time study techniques . for example -

People such as product designers ,cost estimators, methods and tool engineers, office supervisors and odthers staff . It is generally useful in determining the time required for a manual work in an industry & also applied for measuring and comparing the time required for manual motions which can be measured medium to long run operations with cycles of 0.15 minutes and greater.

(v) **Brief work Factor (BWF) :** Brief system is relatively new Systems uses a Time Unit of 0.005 minute.

Advantage of work factors :- following are the important advantages of work factor system

- It is very simple & easy process.
- It is performed very quickly & result is achieved with accuracy.
- It is very accurate than METHOD TIME MEASUREMENT (MTM).
- It's eliminate the use of stop watch.
- It consists a manual process as well as mental process.

4.4. Standard Data System

Standard data system is a work-measurement technique having a data base of normal time values , usually associated by work elements that can be used to fixed the standard time for work elements . that has been fixed previously in data base . standard data is a catalog of elemental time standard which had been observed with the help of motion and time study & collected in data base for further application in same type of operation .

Work elements may be classified the standard data system in following ways

(i) Set – up versus production elements.
(ii) Constant versus variable elements.
(iii) Worker – paced versus machine elements.
(iv) Regular versus irregular elements.
(v) Internal versus external elements.

Advantages of Standard Data System

(i) Productivity increases in term of cost saving due to setting the standard.
(ii) It can set standard for any particular operation before prodution.
(iii) There is no requirement of performance rating.
(iv) Consistancy in standard data.
(v) Minimum times is required to develop time standard as compared to time study.

(vi) This method is cost saving method for work elements.

(vii) It Permit the establishment of accurate time standards before the job is performed.

(viii)The Standard data time standards are more accurate.

Question for Discussion

1. "Incentives are necessary for smooth and efficient running of a factory". In this context discuss the significance and relevance of Incentive schemes.
2. Write short notes on the following:
   a) Measurement of physical workload.
   b) Wage incentive plans.
3. Describe some of the common methods of wage incentive plans with suitable examples
4. What is the role of PMTS in work measurement ?
5. Give the list of various advantages of using PMTS systems over the ordinary time study
6. What do you understand' by standard data ? Give various steps for developing standard data and give its uses.
7. Discuss various types of wage incentives. What type of incentive would you recommend for white colored people ?
8. Write short notes on the following:
   a) Methods Time Measurement (MTM).
   b) Physiological effects of noise.
   c) Physical work load.
   d) Mental work load.
9. Define wages and explain in brief various types of incentives.
10. Explain essentials for good wage incentive plan
11. State wage incentive plans. Explain any one in details.
12. Describe the characteristics of good wage plan. Discuss any one wage plan
13. Discuss the incentives which are being given to the entrepreneurs by the government
14. Distinguish between job evaluation and merit rating. Discuss its importance and objectives.

References

1. C.W. Lyme, “Wage Incentive Methods, Their Selection, Installation and Operation”, New York. Ronald Press, 1954.
2. R. Marriott, “Incentive Payment Systems: a Review of Research and Opinion”, London. Staples Press. Second Edition, 1961.

3. E.R.M. Culuu, "Financial Incentives Based on Work Measurement", London. British Institute of Management, 1963.
4. GRACRE, "A Fair Day's Pay", London. Management Publications, 1949.
5. International Labour Office: Payment by Results. Geneva. 1951. Studies and Reports, New Series, Ninth Impression, 1967
6. Idem: Job Evaluation. Geneva. 1960. Studies and Reports, New Series, No.56. Third Impression, 1966.
7. R.M. Currie, "Financial incentives based on work measurement", revised J.E. Faraday (London,British Institut of Management, 2nd éd.,), 1971.
8. Work study, revised J.E. Faraday (London, Pitman, 4th éd.,), 1977.
9. S.M. Lowry, H.B. Maynard and G.J. Stegemerten, "Time Motion Study Formulas for Wage Incentives", (New York London MCGRAW-HILL BOOK COMPANY, INC.), 1940.

CHAPTER 5

HUMAN FACTORS ENGINEERING

5.0. History of Human Factor Engineering

The foundations of the science of ergonomics appear to have been laid within the context of the culture of Ancient Greece. A good deal of evidence indicates that Hellenic civilization in the 5th century BC used ergonomic principles in the design of their tools, jobs, and workplaces. One outstanding example of this can be found in the description Hippocrates gave of how a surgeon's workplace should be designed and how the tools he uses should be arranged . It is also true that archaeological records of the early Egyptians Dynasties made tools, household equipment, among others that illustrated ergonomic principles. The term ergonomics is derived from the Greek words ergon [work] and nomos [natural laws] (The Outline of Ergonomics, i.e. Science of Work, Based on the Truths Taken from the Natural Science). Later, in the 19th century, Frederick Winslow Taylor pioneered the "Scientific Management" method, which proposed a way to find the optimum method for carrying out a given task. Taylor found that he could, for example, triple the amount of coal that workers were shoveling by incrementally reducing the size and weight of coal shovels until the fastest shoveling rate was reached. Frank and Lillian Gilbreth expanded Taylor's methods in the early 1900s to develop "Time and Motion Studies". They aimed to improve efficiency by eliminating unnecessary steps and actions. By applying this approach, the Gilbreths reduced the number of motions in bricklaying from 18 to 4.5, allowing bricklayers to increase their productivity from 120 to 350 bricks per hour. The terms human-factors engineering and human engineering are used interchangeably on the North American continent. In Europe, Japan, and most of the rest of the world the prevalent term is ergonomics, a word made up of the Greek words, ergon, meaning "work," and nomos, meaning "law." Despite minor differences in emphasis, the terms human-factors engineering and ergonomics may be considered synonymous. Human factors and human engineering were used in the 1920s and '30s to refer to problems of human relations in industry, an older connotation that has gradually dropped out of use. Some small specialized groups prefer such labels as bioastronautics, biodynamics, bioengineering, and manned-systems technology; these represent special emphases whose differences are much smaller than the similarities in their aims and goals.The data and principles of human-factors engineering are concerned with human performance, behaviour, and training in man-machine systems; the design and development of man-machine systems; and systems-related biological or medical research. Because of its broad scope, human-factors engineering draws upon parts of such social or physiological sciences as anatomy, anthropometry, applied physiology, environmental medicine, psychology, sociology, and toxicology, as well as parts of engineering, industrial design, and operations research. World War II marked the development of new and complex machines and weaponry, and these made new demands on operators' cognition. The decision-making, attention, situational awareness and hand-eye coordination of the machine's operator became key in the success or failure of a task. It was observed that fully functional aircraft, flown by the best-trained pilots, still crashed. In the decades since the war, ergonomics has continued to flourish and diversify. The Space Age created new human factors issues such as weightlessness and extreme g-forces. How far could environments in space be tolerated, and what effects would they have on the mind and body? The dawn of the Information Age has resulted in the new ergonomics field of human-computer interaction (HCI). Likewise, the growing demand for and competition among consumer goods and electronics has resulted in more companies including human factors in product design. The coining of the term Ergonomics, however, is now widely attributed to British psychologist Hywel Murrell, at the 1949 meeting at the UK's Admiralty, which led to the foundation of The Ergonomics Society. He used it to encompass the studies in which he had been engaged during and after the Second World War.More than twenty technical subgroups within the Human Factors and Ergonomics Society (HFES) indicate the range of applications for ergonomics. Human factors engineering continues to be successfully applied in the fields of aerospace, aging, health care, IT, product design, transportation, training, nuclear and virtual environments, among others. Kim Vicente, a University of Toronto Professor of Ergonomics, argues that the nuclear disaster in Chernobyl is attributable to plant designers not paying enough attention to human factors. "The operators were trained but the complexity of the reactor and the control panels nevertheless outstripped their ability to grasp what they were seeing [during the prelude to the disaster]." Physical ergonomics is important in the medical field, particularly to those diagnosed with physiological ailments or disorders such as arthritis (both chronic and temporary) or carpal tunnel syndrome. Pressure that is insignificant or imperceptible to those unaffected by these disorders may be very painful, or render a device unusable, for those who are. Many ergonomically designed products are also used or recommended to treat or prevent such disorders, and to treat pressure-related chronic pain. Human factors issues arise in simple systems and consumer products as well. Some examples include cellular telephones and other hand held devices

Definition of human factor engineering :- There were so many definition had been proposed in previously by the various scientist , engineers and management gurus , regarding the human factor engineering and ergonomics , some of the important definition are given below those are very important from subject point of view .

(i) This discipline is concerned with designing both products and processes and equipment used in manufacturing so as to maximize their ability to be used comfortably, safely, and effectively by human beings .

(ii) There are two basic objective of human factors engineering these are

1. Consider any man/machine combination as a total system to insure that the equipment operational requirements do not exceed human abilities .
2. Consider the human performance tolerance, thereby insuring optimal speed, accuracy, and quality of performance, eliminating hazards to operating personnel, and maximizing the comfort of the operator .

(iii) Human factors engineering, or human engineering, is concerned with ways of designing machines, operations, and work environments so that they match human capacities and limitations .

(iv) It is concerned with designing for human use and apply information about human characteristics, capacities, and limitations to the design of human tasks, machines, machine systems, and environments so that people can work safely, comfortably, and effectively.

(v) It is that branch of science and technology which includes what is known and theorized about human behavioral and biological characteristics. It serves as a repository and source of data and principles that can be validly applied to the specification, design, evaluation, operation, and maintenance of products and systems that are intended for safe, effective, satisfying use by individuals, groups, and organizations. The term 'human factors' is considered synonymous with the term 'ergonomics.

(vi) It is a systematic approach to studying problems of human-machine interaction and to arriving at practical solutions on a scientific basis.

(vii) In a broad focus, human engineering (or human factors engineering) is only one of several endeavors, the one concerned with the design and layout of equipment, facilities, and environment. A very simple definition of human engineering is "a discipline concerned with designing man-made objects so that people can use them effectively and safely and creating environments suitable for human living and work."

(viii) Human-factors engineering, also called ergonomics or human engineering, science dealing with the application of information on physical and psychological characteristics to the design of devices and systems for human use.

5.1. Effect of Ergonomics in the Workplace

The terms'ergonomics' related to work-place where the operator does the work , for example type writer & operator sitting , should be such that operator can perform the job comfortably, and effectively.Ergonomics in the workplace has to do largely with the safety of employees, both long and short-term. Ergonomics can help reduce costs by improving safety. Further work place may be classified as reactive or proactive on the basis of effective implementation of ergonomics . Reactive ergonomics is when something needs to be fixed, and corrective action is takes place ,Proactive ergonomics associated the searching the area in which development is required . for example equipment design, task design, or environmental design.

(i) Equipment design changes the actual, physical devices used by operators / workers.

(ii) Task design changes in which manner operator handling the equipment.

(iii) Environmental design releated with the modification of the environment in which operators work.

Another important partof ergonomics is an engineering psychology behave as an interdisciplinary part of ergonomics and study have been peformed between the relationships of operator to machines, with the intent to improving such relationships. This consists of re-designing the equipment, how the operator utilizing the machines, or changing the location in which the work takes place. Ergonomics branch also have an important parts called a Macroergonomics which deals with the design, examining organizational environments, culture, history, and work goals ,physical design of tools and the environment .It also deals with the optimization of the designs of organizational and work systems through the consideration of personnel, technological, and environmental variables and their interactions.The goal of macroergonomics is to improved productivity, and employee satisfaction, health, safety etc .It analyzes the whole system, finds how each element should be placed in the system, and considers all aspects for a fully efficient system.

5.2. Sitting Posture for Operator

The best way to reduce pressure and fatigue in the back is to be in a standing position. However, there are times when you need to sit. When sitting, the main part of the body weight is transferred to the seat. Some weight is also transferred to the floor, back rest, and armrests. When the proper sitting & supports could not provided to the operator he suffers with unwanted pressure on the back causing pain .The combination which minimizes pressure on the lower back is having a backrest inclination of 120 degrees and a lumbar support of 5 cm.

The 120 degrees inclination means the angle between the seat and the backrest should be 120 degrees. The lumbar support of 5 cm means the chair backrest supports the lumbar by sticking out 5 cm in the lower back area . A proper kneeling chair creates the open body angle by lowering the angle of the lower body, keeping the spine in alignment and the sitter properly positioned to task. The benefit of this position is that if one leans inward, the body angle remains 90 degrees or wider.In a proper kneeling chair, some of the weight bears on the shins, not the knees, but the primary function of the shin rests (knee rests) are to keep one from falling forward out of the chair. Most of the weight remains on the buttocks. For example - The interaction between operator and machine deserves particular attention, as this is an important aspect of work study. It is so-called "interface" between the worker and the machine represented by by a dotted line . the characteristic of any machine is the way that it conveys information to the operator which is provided through the displays.The basic purpose of the displays is to establish a communication link between the machine and the worker and the process which is being going on . the pictorial view of the such system is given ,

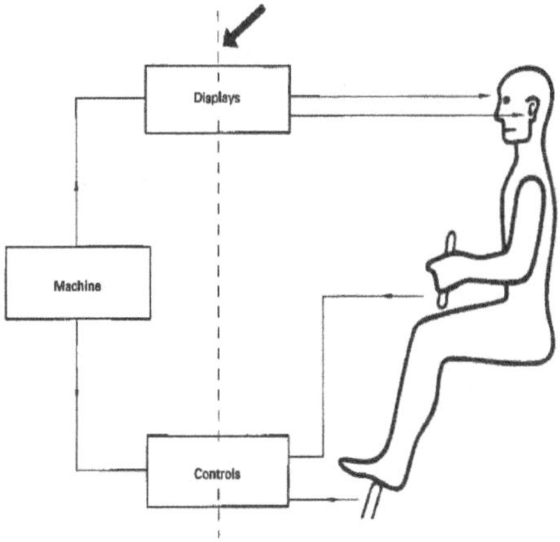

Diagram 5.1: The Operator/Machine Interface

The Suitable Demostration of the Display is Given Below

(i) The display design must be clearly define the task and the way in which the particular kind of operator performs it.

(ii) There are three kinds of display: pictorial, qualitative and quantitative .

(iii) In most tasks the operator receives information both from a real work process and from an artificial display representing it. The artificial display must be so designed as to be compatible with the real display in terms of patterns and relative movements.

(iv) In designing displays for maximum speed and minimum attention, use the kinaesthetic channel.

(v) For maximum attention, use the auditory channel.

(vi) For maximum precision and agreement between operators, use the visual channel.

Various type of Ergonomie display design are very important from subject point of view that had been fullfill the required purpose , these are given below.

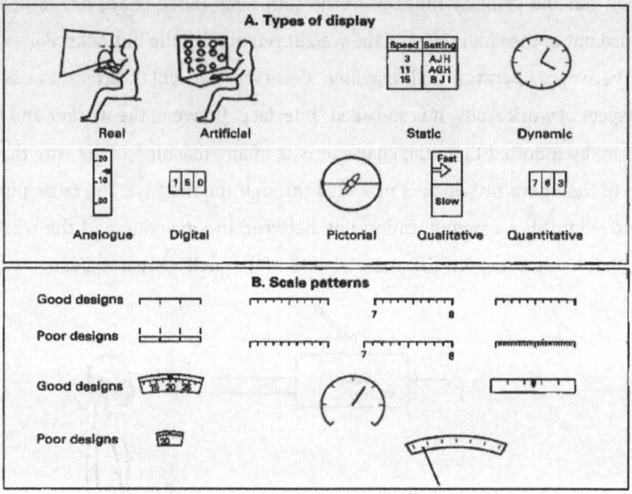

Diagram 5.2: Design of Display System

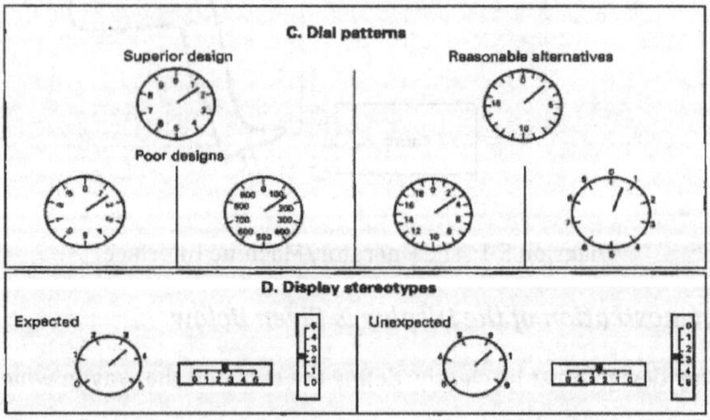

Diagram 5.3: Design of Display System

Ergonomie display design :- Ergonomics is also concerned with the way a worker is able to control the running of operations .(see figure). Basically there are two essential factors involved, time and space: the time it takes for a worker to exert control or react to a system out of control which could be matched with the motor skills of the worker; and the space available to permit freedom of movement when exerting control.

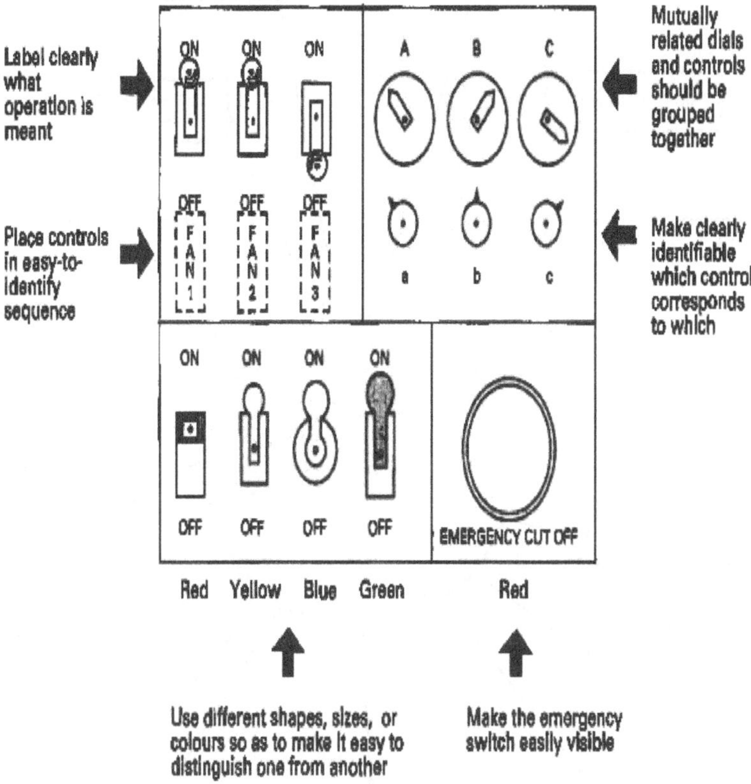

Diagram 5.4: Ergonomie Design of Controls

5.3. Ergonomics

Ergonomics is nothing but it is composition of two branch of science these are Multi-Disciplinary Science and Inter-Disciplinary Science .

The association of various department / disciplines regarding to ergonomics indicates the releationship amongs them that's given in diagramatic arrangement .

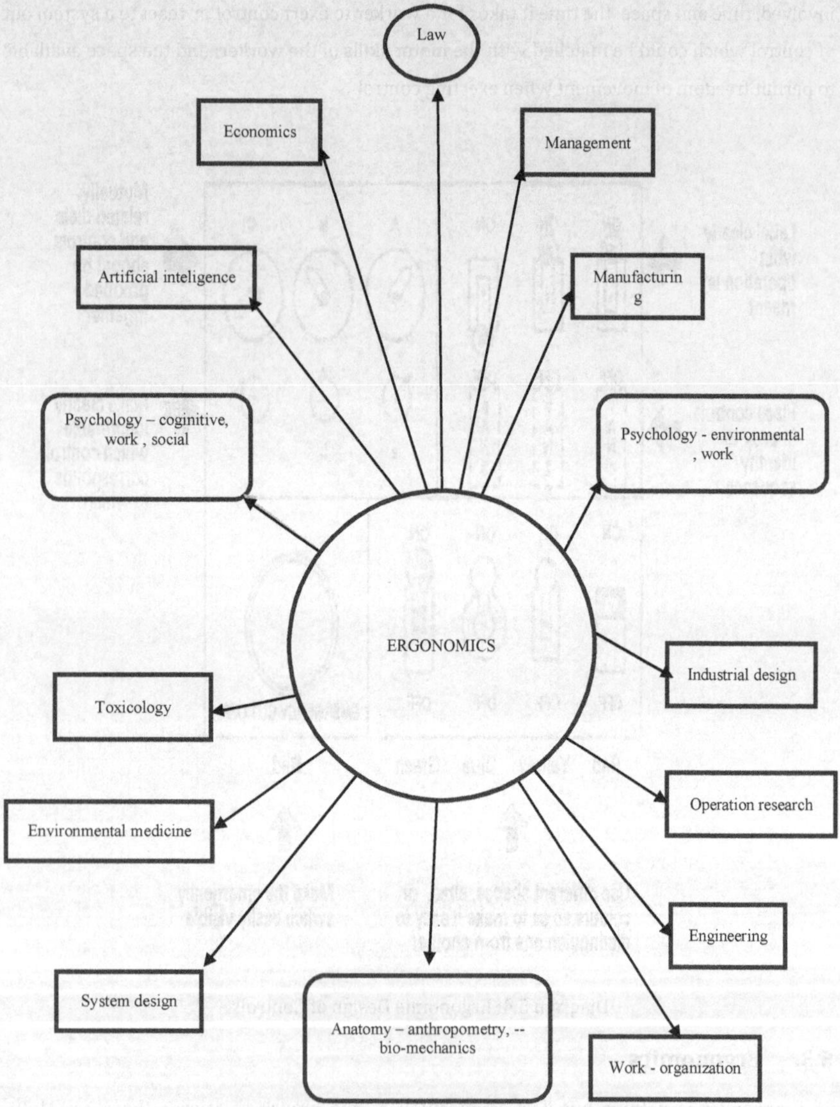

Diagram 5.5: Ergonomic with Various Department

The above system represents the key elements of an ergonomics / human factors engineering except these elements we will also put here the diagramatical arrangement of various field where ergonomics plays an important role.

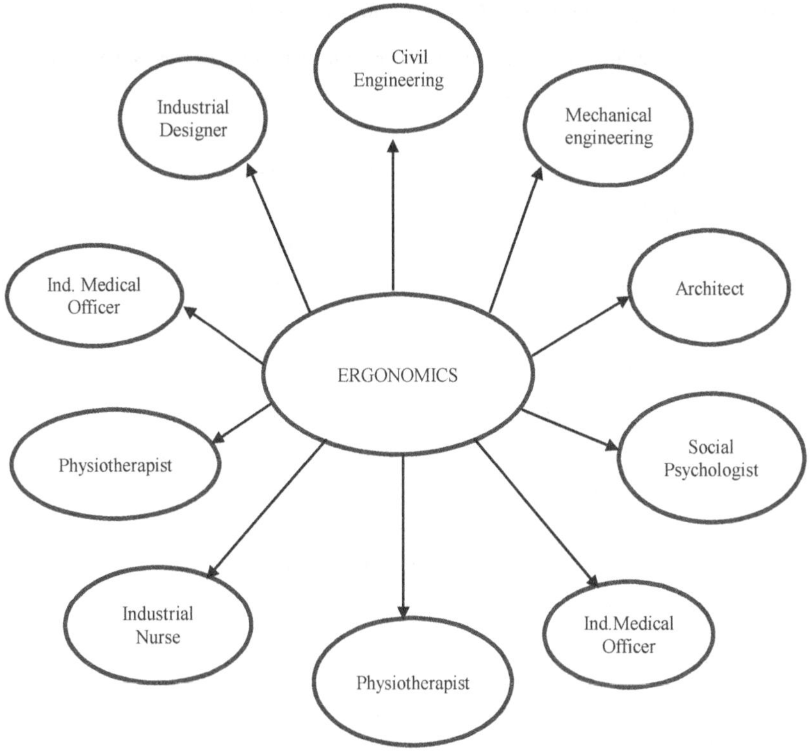

Diagram 5.6: Ergonomic with Engineering & Others Field

Objectives of Ergonomics : Following are Some Important Objectives of the Ergonomics

(i) Ergonmocs improve human well being and overall system performance by create optimization with the integeration of man and machine system .

(ii) It reduces the physical and mental strains so that the operator feel good there by improving productivity .

(iii) It also control the other factors like due to that risk of injury , accidents . illness and error thus improve productivity .

(iv) It improve design of machine , modification in product design , also improve the proposed way of doing the job .

(v) It also helpfull in improving the men – machine coordination .

(vi) Egonomics is a basic tool , there by we can enhance the productivity of the system .

Facility Provided by the Ergonomics to Operator and Staff

(i) To provide facility for operator in order to gain profit for the organization in term of productivity enhancement for example – lighting , climate , ventilation , control noise level .

(ii) To improve working posture and reduces motion of the operator thereby fatigue is in decreasing order .

(iii) To control & handling of machine levers and other control system .

(iv) to create the joy full and happier atmosphere for operator due to that his efficiency will be in increasing order .

Advantages of Ergonomics

(i) higher productivity could be achieved .

(ii) less fatigue experienced by the operator , due to more human comforts .

(iii) it reduces the chance of accident or any kind of risk there by operation may perform safely .

(iv) man-machine coordination is established in better ways .

(v) productivity & efficiency of the operator has been increased.

5.4. Man-Machine-System

A man-machine system is one in which a human operator is in some way reliant on, or integrated with, a tool or machine.

The core concept is that the machine is in some way enhancing the abilities of the human user.One example of a commonly used human-machine system is a forklift that might be used in a warehouse or loading dock. This is a machine that requires a human to control it. In turn, the forklift is a machine that enhances the abilities of the human.Human–machine interactions and processes always occur in realistic contexts , The plant interacts with the human operator through its interfaces and controls.

1. Interfaces are display panels, indicators, decision support tools. They transform the behaviour of the machine in visual, auditory and tactile information. These support the operator in perceiving and understanding the state and dynamic evolution of the system and in developing the strategies for its management and control
2. Controls are means by which it is possible to operate on the system and automation in order to implement the operator's intention and strategy. Interventions of controls are transformed in machine information by actuators.

Another Definition of Man-Machine System may be Given as,

A man–machine system (MMS) can be defined as a composite, at any level of complexity, of personnel, procedures, materials, tools, equipment, facilities, and software. The elements of this composite are used together in the intended operational or support environment to perform a given task or achieve a specific production, support, or mission requirement.

Man-machine interaction or man-machine systems have multidisciplinary or interdisciplinary views and approaches. It consists of three basic domains of man-machine interaction or systems

1. Cognitive science and ergonomics (as the human sciences).
2. Automation and systems engineering (as the systems sciences).
3. Information and communication engineering (as the computer sciences).
4. Organizational and cultural aspects.

There are Four Basic Socio-Technical Elements of Man Machine System , these Four Elements Are:

1. Organisational processes.
2. Personal and external factors.
3. Local working conditions.
4. Defences, barriers and safeguards.

1) Organisational Processes

It is defined as, the culture of an organisation may be defined as the set of rarely articulated, largely unconscious beliefs, values, norms, and fundamental assumptions that the organisation makes about itself,the nature of people in general and its environment.In effect, the culture is the set of "unwritten rules" that govern "acceptable behaviour" within and outside the organisation.

2) Personal and External Factors :- The Definitions of External and Personal Factors are Given Below

- External factors can be considered as all random physical or system contingencies that may alter or impinge on local working conditions and safety measures, so as to foster inadequate system performance and erroneous human behaviour.

- Personal factors are individual, physical, or mental conditions that affect human behaviour, and are specific to each person. They can only be accounted for by a random variable affecting the generic behaviour of large classes or categories of people.

Personal and external factors should be considered as random variables in an overall framework of accident causal path, and their role can only be marginal in a structured analysis of an organisation.

3) *Local Working Conditions*

It is defined as , Local working conditions are the specific factors that influence the efficiency and reliability of human performance in a particular work context Local working conditions affect the performance of tasks by influencing either the interface between operators and control systems and/or cognitive activities . **For example** -- workplace design,interfaces with automation, tools and instruments, protective equipment, job planning, procedures, supervision processes, workload, training, and policies.

4) *Defences, Barriers and Safeguards*

In this section we study about defences, barriers, and safeguards are the measures which have been developed by the organisation aimed at creating awareness, warning, protection, recovery, containment, and escape from hazards and accidents.

Following are Some Important Function of this Section

(i) To create awareness and understanding of the hazards.

(ii) To support the detection process of off-normal conditions.

(iii) To support restoring normal operating conditions.

(iv) To protect from injury.

(v) To contain the consequences of an event.

(vi) To support escape in the case of loss of control or accident.

5.5. Classification of Man – Machine System :- Man Machine System May be Classified as Given Below

(i) Close loop system.

(ii) Open loop system.

(iii) Manual loop system.

(iv) Mechanical system.

(v) Automatic system.

(i) **Close loop system** :- Sometimes, we may use the output of the control system to adjust the input signal. This is called feedback. Feedback is a special feature of a closed loop control system. A closed loop control system compares the output with the expected result or command status, then it takes appropriate control actions to adjust the input signal , a closed loop system is always equipped with a sensor . The output signal is fed back to the input to produce a new output. A well-designed feedback system can often increase the accuracy of the output .

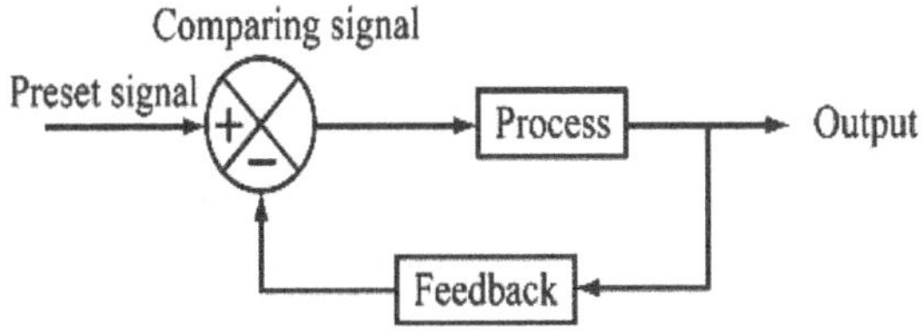

Block Diagram of a Closed Loop Control System

Feedback can be divided into positive feedback and negative feedback. **Positive feedback** causes the new output to deviate from the present command status, **Negative feedback** directs the new output towards the present command status, so as to allow more sophisticated control. For example : - **A driver has to steer continuously to keep his car on the right track.**Most modern appliances and machinery are equipped with closed loop control systems , **For example** -- Air conditioners, Refrigerators, Automatic rice cookers, Automatic ticketing machines, etc.

brief description of closed loop control system :-An air conditioner consists of thermostat to detect the temperature and control the operation of its electrical parts to keep the room temperature at a preset constant.

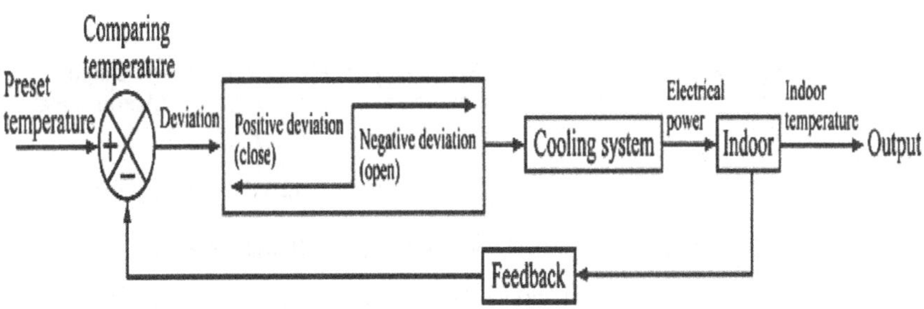

Block Diagram of the Control System of an Air Conditioner

(ii) Open loop system:-- open loop control system, Its operation is quite simple, when an input signal directs the control element to respond, an output will be produced and further no control is being performed aswellas there is no feedback is adopted . Examples of the open loop control systems include washing machines, light switches, gas ovens, etc.

Block Diagram of an Open Loop Control System

Description of washing machine -- An example is that the operation of a washing machine does not depend on the cleanness of the clothes. Both the structure and the control process of an open loop control system are very simple, but the result of the output depends on whether the input signal is appropriate or not.

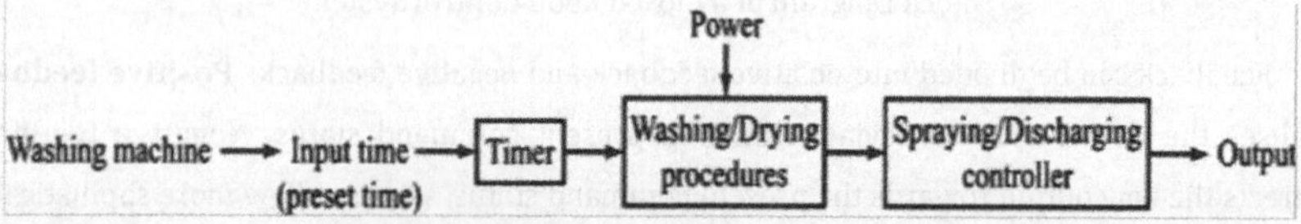

Block Diagram of an Open Loop Control System (Washing Machine)

Another system of open loop control system is the burglar alarm system: -The function of the sensor is to collect data regarding the concerned house. When unauthorized person entered into house electronic sensor is triggered off.

It will send a signal to the receiver. The receiver will then activate the alarm, which will in turn generate an alarm signal. The alarm signal will not cease until the alarm is stopped manually. The block diagram is given below,

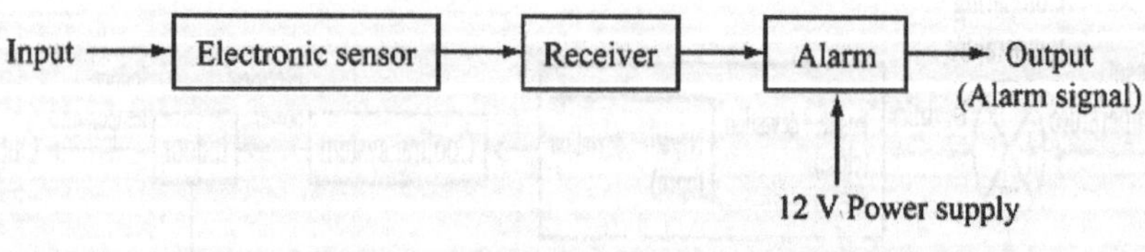

Block Diagram of an Open Loop Control System (Burglar Alarm)

- **Manual loop system:** -To improve the precision of the control systems, operators were engaged to control the machines. They played an important role as sensors and decision-makers. They compared the inputs with the status needed, then provided feedback and made decision (their brains). Afterwards, they adopted some procedures to stabilize the systems and minimized the errors. Lastly, the outputs were close to the requirements. Therefore, manual operation in the system is a kind of closed loop control system.

- **Automatic control system:-**Under the conditions of clear and repeated procedures, and those procedures which are operated by automatic adjustment system instead, automatic control machinery is more suitable for use. Therefore, those automatic controlled machines are suitable for boring and repeated works. For example, it is better for a temperature sensor involved in the control of the switching on or off the compressor of the air conditioner.
- **Mechanical system :-** It consists of man with power driven machine , the machine is operated by electric power and perform its function with little variation , operator controls the operation with the help of levers , handles , push button , start button , stop button , gear shifting , etc.

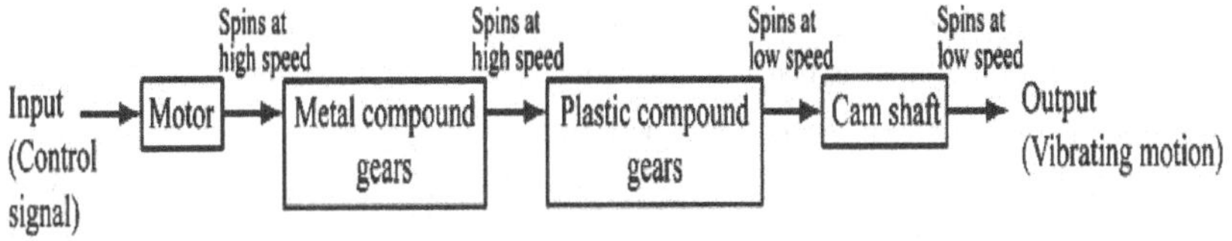

Block Diagram of a Mechanical System

5.6. Goals of Man Machine System

Following are some important goal of man machine system.Man-machine interaction is goal-oriented. The overall goals of man-machine systems are mainly

(i) Productivity goals.

(ii) Safety goals.

(iii) Humanization goals.

(iv) Environmental compatibility goals.

The productivity goals include economic as well as product and production quality goals. The importance of the safety goals is strongly influenced by the application domain.

This goal class dominates all others in many large-scale systems and, particularly, in risky systems. The humanization goals comprise team and work organization, job satisfaction, ergonomic compatibility, and cognitive compatibility.

The latter includes the sub-goals of transparency and human understanding. The environmental compatibility goals refer to the consumption of energy and material resources as well as to impacts on soil, water, and air.

(v) The improvement of safety in any technological system demands that appropriate measures are developed aiming at creating awareness, warning, protection, recovery, containment, and escape from hazards and accidents. In case of designing the MMS , three fundamentally principle are used these are:

- **Supervisory Control.**
- **User-Centered Design (UCD).**
- **Systems' Usability.**
- Adequate models of human–machine systems and interactions must be applied for simulating dynamic interplays of humans and machines, as events embedded and dependent on the socio-technical contexts in which they are generated and evolve.
- In particular, four areas of application must be considered, namely: design, training, safety assessment, and accident investigation.

(i) **The design** of human–machine interactions implies implementing basic principles of human centered automation in the design process.

(ii) **Training** intends to increase the ability of operators to manage safety critical situations, and to capture and notice those factors and indicators of the context that favour the occurrence of errors or mismatches between human situational awareness and system performance.

(iii) **Safety assessment** of plants and organizations represents a basic requirement and the most complete method by which prevention and control of possible accidents can be performed.

(iv) **Accident/incident** investigation aims at identifying systemic and socio-technical root causes that generate accidents.

- The implementation of effective MMS, variety of tools and approaches are applied for the continuous verification that adequate safety conditions exist and are maintained before and during the lifetime of a system, **at the stages of design, training, safety assessment, and accident investigation.** Thus we can say that this study point is representing the co-relation between above factors four.
- The last point is related to the definition of appropriate safety levels of a plant and ways to regularly assess them. Check all the indicators, markers, and parameters time to time, in order to measurement of the safety level of a system.

The continuous measurement and assessment of safety levels of MMSs is essential for ensuring minimum risk and effective performance of a plant. This process requires the identification of safety critical indicators, as guiding elements of safety assessment, and the performance of recurrent safety audits throughout the whole socio-technical system and organization.

5.7. Important Aspect of Man Machine System

work station is place where a worker occupies when performing a job . some example of work station is - work stand or work table for machine operation , assembly or inspection , a well designed work station is important for preventing disease related to poor working condition as well as for ensuring that the work is productive . every work station should be design with both the worker and task in mind so that the work can be performed , smoothly , efficiently and comfortably . if work station is properly design , the worker should be able to maintain a correct and comfortable body posture . Uncomfortable work posture create a variety of problem,

(i) Back injury.
(ii) Development and aggravation of RSIs.
(iii) Circulatory problem in legs.

Following are some important main cause of such a problems,

(i) Poorly designed seating.
(ii) Standing for long period.
(iii) Reaching.
(iv) Inadequate lighting arrangement.

Characteristics of Man-machine System: - followings are some important Characteristics of Man-machine System.

(i) The man-machine system consists of the man, the machine and system environment.
(ii) It is essentially artificial by nature and is specifically developed to fulfill some purpose or specific aim.
(iii) It has specific inputs and outputs which are appropriately balanced.
(iv) It is variable in size and complexity and is dynamic in performance.
(v) Sub-systems of man machine system interact with and affect the other parts.
(vi) The man-machine system becomes more efficient when inputs and out puts are adequately balanced.
(vii) Environmental factors or system environment effects system performance.

5.8. Human Tasks with Automation and Control

It is important to consider cognitive capabilities and information processing behavior of human users interacting with the technical systems, as well as the tasks to be performed within any human-machine system, before and beyond all technological possibilities. Task orientation and human-centered designs are necessary for improved operational use of technical processes and their human-machine interaction. Any user-oriented design of interactive products, such as human-machine systems, should start with definitions of user classes, overall goals and means for their achievement, as well as with task analyses, in order to have a solid basis for user requirements and systems specifications, particularly also for the functional specifications of the human-machine systems components and functionalities.

User requirements should consider the goals-means-tasks relationships, and should be based on a task-oriented perspective, two principal task categories can comprise all human activities in human-machine systems, namely controlling and problem solving. All other tasks such as monitoring and communicating can be classified as subtasks or supporting tasks. Figure shows a schematic block diagram of a conceptual human performance model together with the automated technical system with their interactions via control devices and displays for controlling and problem solving with respect to prescribed goals which have to be matched by the system's outputs.

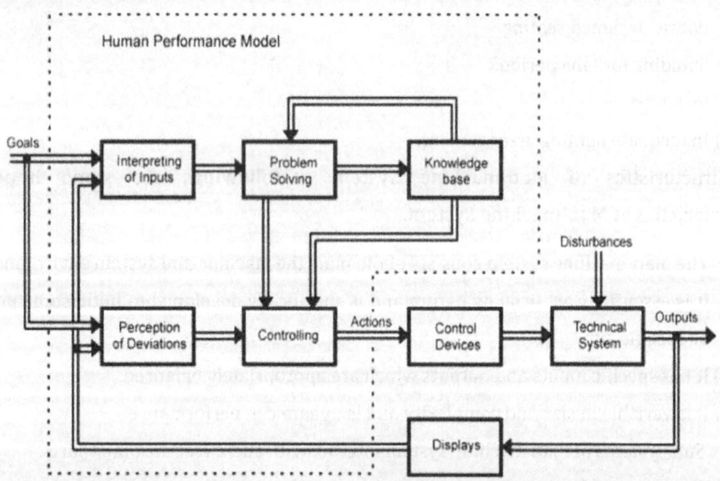

Diagram 5.7: Human Performance Model with Controlling and Problem Solving, Interacting with the Automated Technical System

Controlling comprises continuous and discrete tasks of open-loop and closed-loop activities. The higher cognitive tasks of problem solving include fault management (with fault detection, diagnosis, compensation, and correction) as well as planning. The two task categories require a different selection of information about goals and systems outputs. While the perception of deviations between goals and outputs is normally sufficient for controlling, both types of inputs have to be interpreted more consciously for problem solving tasks. Further, the problem solving tasks are much more knowledge-based activities than the controlling tasks.

Roughly speaking, the two human task categories of controlling and problem solving correspond to the two categories of computer use in advanced automated technical systems, namely computer control (by supervision and control systems) and computer support (by decision support systems) . This includes the more traditional levels of automation systems as well as all levels of knowledge-based systems. Only since the latter became a more mature technology over the last years, the possibilities of function and task allocations between human operators and computers allow us to relate both human task categories to corresponding categories of computer use. That means also that truly human-centered designs of human-machine systems can now better be pursued and implemented with the latest technology of knowledge support, The interactions between supervision and control systems, decision support systems, and human operators have to be carried out in such a way that goals and their sub goals are achieved with respect to the technical process and the overall system. The goals and appropriate goal structures are either prescribed or derived by the human operators. Thus, some of the major responsibilities of human operators are the selection of goals and sub goals, the formation of goal structures, and the checking of goal accomplishment and sub goal compatibility. A closer look at the two task categories of controlling and problem solving shows similarities and differences between their human form and its automated counterpart. The behavior of human operators and, in principle, also of automation systems can be classified into the three information processing phases of categorization, planning, and action. These three phases have to be accomplished in all controlling and problem solving tasks, either on a state-oriented, context-oriented or structure-oriented level. Here, context refers to the fields of information involved. The three levels correspond with the cognitive levels of skill-based, rule-based, and knowledge-based human behavior.

Human operators of technical systems can much more flexibly maneuver through all these different possibilities than the automation systems. However, they may be under time pressure or otherwise impaired or overloaded in certain, particularly critical situations. Therefore, the state of computerization of the technical systems has often been developed quite far. This again has sometimes led to the so-called "ironies of automation". For example, boredom may occur in normal situations with highly automated systems, and boredom may lead to loss of vigilance, poor job satisfaction, and possibly human errors. Human-centered designs of automated technical systems have to avoid these effects by choosing a flexible function and task allocation between human operators, automation systems for supervision and control, and knowledge-based systems for decision support under all normal, abnormal, and emergency conditions . The issues of human-machine interaction are not only concerned with the interaction between a single human and the dynamic technical system, the machine. Real work situations in many application domains require that several humans of different human user classes frequently interact with the same technical system and communicate with each other. Some examples of such technical systems are chemical plants, cement plants, power plants, transportation systems, and discrete manufacturing systems.

In plant-wide control and communication, several partially overlapping groups exist. Their official hierarchical structure is often complemented by an informal hierarchical or even flexible opportunistic organization, dependent on company cultures, social and personal preferences, and specific work requirements. As an example, the application domain of the cement industry shows that several persons from different human user classes need to co-operate, in the plant control room or, also, in office rooms during group meetings. These people from different human user classes share their overlapping skills and knowledge, and interact with each other, at least partially, via the human-machine interfaces or directly face-to-face, in addition to communicating with the technical system also via the human-machine interface .

5.9. Control Devices

A very broad spectrum of control functionalities as possibilities of control input in human-machine interaction exists. Different forms of control can be exerted, either direct (e.g., gestural as with the conductor of an orchestra), via bio-signals or via control devices. Control via bio-signals is of particular value for handicapped people, e.g., in the operation of an arm prosthesis. Speech and motion (displacements, forces, and gestures) can be used, directly or via control devices. Speech input and gestural control will eventually become new means of control in the future. A few operational systems are already in use. Speech recognition allows the automatic interpretation of the human voice, based on models for phonemes which are the small units of sound in any word and sentence. This technology has great potential for human control of computers and machines, once the recognition rate and the speaker independence can be further increased. Gestural control has been developed with music technologies and in engineering. Currently, it is an emerging subject of research including gesture sensing/capture technologies, intention/emotion detection, mapping strategies, etcThe predominant form of control in human-machine interaction is the one using control devices. Important types of control devices are keyboards (with alphanumeric or with functional keys), mouse, joystick, trackball, steering wheel, pedals, knobs, and switches. Force-displacement characteristics of these control devices need to be appropriately designed. Also, some virtual control devices exist which can be operated on display screens, e.g., buttons, sliders, and menu options.

5.10. Position of Work Station

There are some important aspects are given below related to work station.

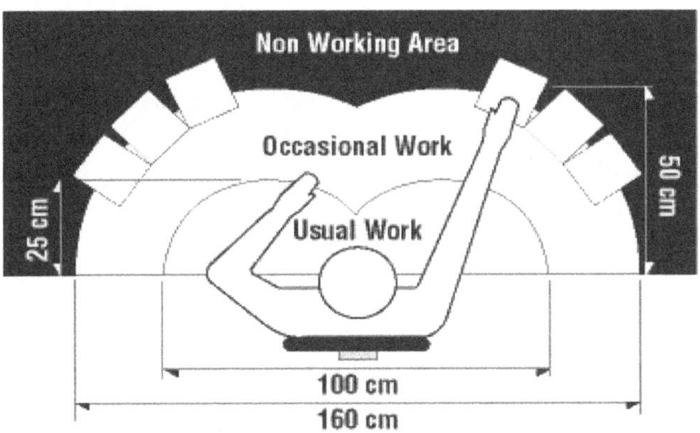

Diagram 5.8: Position of Work – Station

Head Height

(i) Allow adequate space for the tallest possible work.

(ii) Positiondisplay at or below eye level because peoples always look slightly downward.

Shoulder Height

(i) Control panel should be placed between shoulder and wristheight.

(ii) Avoid placing objects above shoulders height objects or control that are used often.

Arm reach : -equipments / devices or others necessary things should be placed nearer to reach , there by arm movement is to be minimized so that over-stretching of arm may be avoided when arm movements takes place .

(i) All the devices must be settle according to the height of operator, height may be classified as – shortest height operator, medium height operator and tallest.

(ii) In any particular , particular operation how much & which kind of material is being used aswellas which kind of tools / equipments / devices will be used should be nearer or in front of the operator or may be easy reach .

Elbow height: - job table or work table should be at the height of elbow or less the permissible height of elbow; therefore the operator efficiency / productivity remain constant.

Hand height: -smallest equipments / actuating devices / general purpose others usable / related material devices are kept between hand & shoulders height.Chair height should according to leg length and height of the work table / surface, foot rest should be provided in order to avoid dangling aswellas enough space must be provided in front of chair, for leg outstretched.Hand grips should be according to the job work & hand size with enough space for their movements.

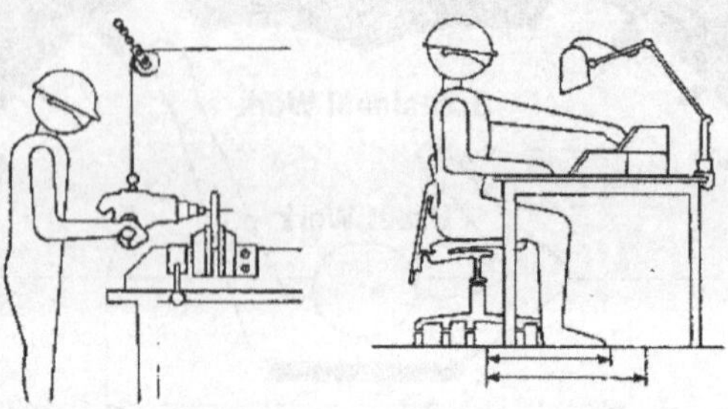

Diagram 5.9: Position of Work-Station

(iii) Work layout and tool must be suitable their needs according to the operators hands working, for example – left hand workers or right hand workers.

(iv) Eliminates glare and shadow, good lighting is essential.

Point Must be Remembers for Good Work – Station Design

(i) The work station is a place a worker occupies when performing a job.

(ii) A well designed work station, save us from diseases & improve the productivity of operators.

(iii) Work station designed as per the operator task.

(iv) work station must be designed according to the correct & comfortable body posture.

(v) All the necessary ergonomics factors must be considered while designing the work-station.

5.11. Sitting and Chair Design: Followings are Some Basic Highlights Regarding the Sitting Work

(i) The worker sitting should be such that, there will not be any kind of stretching & twisting.

(ii) Sitting should be straight in front of and close to the workers.

(iii) Sitting and work- table height must be equivalent to elbow height.

(iv) Back of body should be straight and shoulders is to be in relaxed position & if there are requirement of any kind of supports for elbow or for arm then provision should be made for that by manufacturing departments .

Position of work chair: -The following ergonomics point must be kept in mind while the work chair is designed.

(i) Work chair should be appropriated to the job task & work bench or work table.

(ii) Height of seat should be according to the operators, back – rest must be adjustable.

(iii) Chair should be flexible towards forwards &backwards.

(iv) Inside the table sufficient space must be availablefor relaxing the legs

(v) Footrest must be provided with the sitting. Forrelaxing the feet, back & knees.

(vi) Seat should be curve slightly down-ward.

(vii) For better stability of chair there should be a provision of five legs in a table.

(viii)Chair must be covered in such amanners to preventslipping of the chairs.

This Example Gives Information Regarding the Sitting

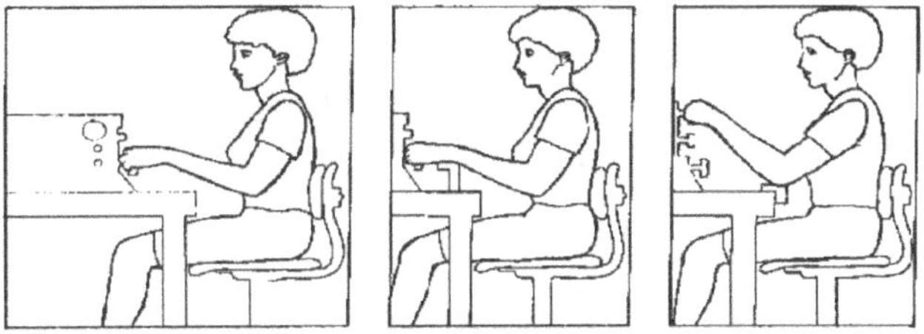

Bad design　　Good design　　Good design

Diagram 5.10: Position of Sitting Design Comparative Design

5.12. Standing Work Station

There are two types of working conditions generally adopted by operators, such as sitting condition and standing condition. Sitting condition may be favourable but in case standing for long extent of time creates back pain, leg swelling, blood circulation problem, back fatigue, tiredness in others limbs of body. Following are some important features of the standing works.

(i) During the job position, operator should have sit on stool or chairs between the certain intervals of time throughout the standing work.

(ii) Workers should be use upper arms & their sides with permissible bending &twisting of the back.

(iii) Height of worktable must be according to the height of operators.

(iv) The work table or surface must be adjustable; it may be adjusted according to the height of operator.

(v) Provision of foot rest is also necessary.

(vi) Operator should be wear low height heel shoes during the standing work.

(vii) Enough space should be there for relaxing the body limbs throughout the standing works

(viii) Work should be performed 8 to 12 inch in front of the body.

Diagram 5.11: Suitable Standing Position of Operator

5.13. Information & Communication Ergonomics

The technical development has led to the interlinking of the computers, assigned in the companies internal as "Intranet" and worldwide as "Internet". In these networks in a unconceivable manner information is transported, processed and made available from the point of view of the user's as good as detached from a material supporter (of course, if this is not correct actually!). The hypermedia organization of the information, the investigation of human search strategies, whose result is a requirement for the creation of search engines and again the technical representation of the so gained information, is a main object of the already mentioned information ergonomics. The concept information ergonomics is often shortened related to the creation of Internet-based information systems, here in special the navigation design. In the foreground here stands the support of search strategies (searching) which should lead fast to the required information.

Communication Ergonomics

Whereas the information ergonomics refers partly substantially to the use of the Internet, the use of the Intranet belongs to the so-called communication ergonomics. Concerning this it must be highlighted that user's interfaces differ in the production area generally considerably from such in information systems. While Internet based information systems are based as a rule on information tied up to links (hypermedia) which opens for their part a huge number of further information, by contrast as a rule user interfaces in the production area allow access on a high detailed, but limited thematic area.

5.14. Human Information Processing

A central topic in human-machine interaction is the way in which humans process information. Following are some important steps of the human information system.

Information system	Description
Sensory processing	The input (called stimuli) comes from all the human senses (**also known as sensory systems**).
Short-term sensory store	Each sensory system has a mechanism which prolongs any stimulus for a short time after the stimulus has occurred.
Perceptual encoding	The stimuli are assigned to a single perceptual category. This is also known as detection, recognition, identification, categorization, pattern recognition, etc .
Decision making	After categorizing the stimulus, you must decide what to do with it.
Response execution	The response (action) is executed.
Feedback and information flow	You monitor the consequences of your action.
Attention	You select information sources to be processed. (For example, when you hear a suspicious sound, you decide to pay more attention to your hearing than to your eyesight.)

Human behaviour :- In general , there are three categories of human behaviour regarding the ergonomics , which has been very important from subject point of view.

Skill-based behavior (SBB):-it concerns the basic motor performance of your body. You generally don't think about this kind of behavior and it's just something which you learn. Skill-based behavior is caused by (continuous) signals; SBB is a very fast type of behavior.

Rule-based behavior (RBB) :- When applying **rule-based behavior (RBB)**, you are using stored rules. These rules are often learned by experience or have been communicated by others.

For example -- rule-based behavior occurs when a child suddenly crosses the road in front of your car.

Knowledge-based behavior (KBB) :- Sometimes simple rules won't get you to your goal. In this case, **knowledge-based behavior (KBB)** is **used**,an example of knowledge-based behavior is planning a route to a place you haven't been before.KBB is presented by symbols.

$$H \quad \text{Log}_2 N$$

5.14.1. Information Input and Processing

Information is nothing but it is defined as the reduction of uncertainty,Sometimes it is also known as cognitive psychology, cognitive engineering, and engineering psychology. For example - the occurrence of highly certain events do not convey much information (such as fasten seat belt sign). But the **occurrence** of highly unlikely events convey more information (such as high temperature warning in a car). Because the fasten seat belt sign conveys less information because it is expected. Although it is an important message, the importance is not directly considered in the information definition. Because this message comes on every time the car is started, it contains little (or no) information in the context of information theory.

Measurement of Information: In general information is measured in**bits**which is indicated by letter '**H**'. A bit gives information about the amount of information required to decide between two equally likely alternatives. Following three types of information may be possible according to the information, these are given below

(i) If the probabilities of N alternatives are equal , then amount of information is calculated as,

$$H = \log_2 N$$

(ii) When the alternatives are not equally likely then amount of information is calculated as

$$hi = \log_2 \frac{1}{p_i}$$

Where " I " Information conveyed by event

And " p_i " is probability of those events.

(iii) The average information of events having different probabilities is given as –

$$Hav = \sum_{i=1}^{N} p_i\left(\log 2\frac{1}{pi}\right)$$

The maximum possible information occurs when all the alternatives have equal probability, but when the information decreases from maximum as the difference between the probabilities of the events is increased , then reduction in information from the maximum is known as redundancy which is represented in % value the mathematically , it is calculated as

$$\%\,\text{Redundancy} = \left(1 - \frac{Hav}{H\max}\right) \times 100$$

Human information input and processing depends on the sensory reception of relevant external stimuli which contain the information of some object, event, or environmental condition. These information may comes out directly or indirectly.

Indirect sensing: -Actually indirectsensing has-been associated with human factors. In this case, the new distal stimuli may be coded stimuli or reproduced stimuli. In both cases the coded or reproduced stimuli become the actual distal stimuli to the human sensory receptors. In this system display is a term that applies to any indirect method of presenting information

Types of information presented by displays: -Display system may be following type that dependents on information gathered & its type,

(i) **Dynamic information:** -Dynamic information changes continuously or is subject to change through - - time. Examples are: traffic lights, radar displays, temperature gauges

(ii) **Static information:**-Static information remains fixed over time. Examples are: alphanumeric data, traffic signs, charts, graphs, labels. Note that static information presented through VDT is being considered as static information

(iii) **Quantitative:** - such as temperature or speed.

(iv) **Qualitative:** - represents approximate value, trend or rate of change.

(v) **Status:** - reflects the condition of a system (such as on or off, and traffic lights).

(vi) **Warning and signal:**-indicating danger or emergency.

(vii) **Representational:** - pictorial or graphical representation of objects, areas, or other configurations. Such as photographs, maps, heartbeat oscilloscope.

(viii) **Identification:** - used to identify a condition, situation or object. Such as traffic lanes and colored pipes.

(ix) **Alphanumeric and symbolic:** - such as signs, labels, and printed material and computer print outs.

(x) **Time-phased: -** display of pulsed or time-phased signals. The duration and inter-signal intervals are controlled

Coding of information : -Coding of information is a very important, part of processing to achieve a desired output. it is a symbolic representation on displayed , where coding takes place when the original stimulus information is converted to a new form , these Information may be coded along many dimensions , such as varying the size, brightness, colour and Variation in the frequency, intensity, or on-off pattern of an audio signal.

Characteristics of a good coding system : - following are special characteristics of coding system.

(i) **Delectability of codes:-** The used stimulus must be detectable by human sensory mechanisms under the present environmental conditions.

(ii) **Dis-criminability of codes:-** Every code symbol must be discriminable from other symbols. In this case, the number of coding levels is important

(iii) **Meaningfulness of codes: - A** coding system should use codes meaningful to the user. Meaning could be inherent in the code or learned. The meaningfulness is related to the concept of conceptual compatibility

(iv) **Standardization of codes:-**When a coding system is to be used by different people in different situations, it is important that the codes be standardized and kept the same. The meaning of the colour red to different populations.

(v) **Use of multidimensional codes: -** this can increase the number and discrimin-abilty of coding stimuli used.

5.15. Sensory Input Devices

sensor can be defined as " A device which provides a usable output in response to a specified measured and. " Here, the output is usually an 'electrical quantity' and measured and is a 'physical quantity, property or condition which is to be measured'. Thus in the case of, say, a variable inductance displacement element, the quantity being measured is displacement and the sensor transforms an input of displacement into a change in inductance. Sensors in manufacturing industries are basically employed to automatically carry out the production operations as well as process monitoring activities. Sensor technology has the following important advantages in transforming a conventional manufacturing unit into a modern one.

(i) Sensors alarm the system operators about the failure of any of the sub units of manufacturing system. It helps operators to reduce the downtime of complete manufacturing system by carrying out the preventative measures.

(ii) Reduces requirement of skilled and experienced labors.

(iii) Ultra-precision in product quality can be achieved.

Transducer:- It can also be defined as a device that converts a signal from one form of energy to another form. A wire of Constantan alloy (copper - nickel 55-45% alloy) can be called as a sensor because variation in mechanical displacement (tension or compression) can be sensed as change in electric resistance. This wire becomes a transducer with appropriate electrodes and input-output mechanism attached to it. Thus we can say that 'sensors are transducers'.

Potentiometer Sensors: -The potentiometer can be of linear or angular type. It works on the principle of conversion of mechanical displacement into an electrical signal. The sensor has a resistive element and a sliding contact. The slider moves along this conductive body, acting as a movable electric contact. The resistive element is a wire wound track or conductive plastic. The track comprises of large number of closely packed turns of a resistive wire. Conductive plastic is made up of plastic resin embedded with the carbon powder,during the sensing operation; a voltage VS is applied across the resistive element the output voltage is proportional to the displacement of the slider over the wire.

Tachogenerator :- Tachogenerator works on the principle of variable reluctance , It consists of an assembly of a toothed wheel and a magnetic circuit . Toothed wheel is mounted on the shaft or the element of which angular motion is to be measured. Magnetic circuit comprising of a coil wound on a ferromagnetic material core.

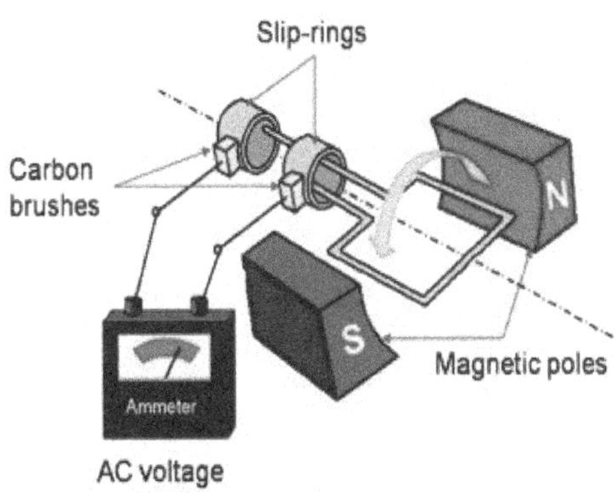

Diagram 5.12: Tachogenerator

As the wheel rotates, the air gap between wheel tooth and magnetic core changes which results in cyclic change in flux linked with the coil. The alternating EMF generated is the measure of angular motion. A pulse shaping signal conditioner is used to transform the output into a number of pulses which can be counted by a counter. In general, these sensors exhibit nonlinearity error of about ± 0.15% and are employed for the rotationsup to about 10000 rev/min.

Pyro-electric sensors: -These sensors work on the principle of pyro-electricity , which states that a crystal material such as Lithium tantalite generates charge in response to heat flow.

In presence of an electric field, when such a crystal material heats up, its electrical dipoles line up.

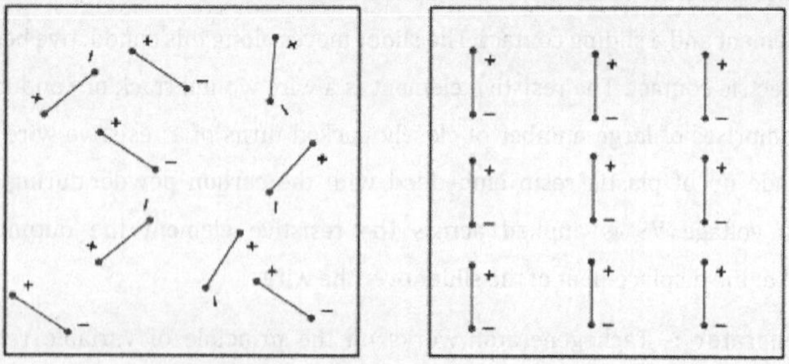

Diagram 5.13: Pyro-Electric Sensors

This is called as polarization. On cooling, the material retains its polarization. In absence of electric field, when this polarized material is subjected to infra red irradiation, its polarization reduces. This phenomenon is the measure of detection of movement of an object.

Applications of Pyroelectric Sensors

(i) Intrusion detector.

(ii) Opto-thermal detector.

(iii) Pollution detector.

(iv) Position sensor.

(v) Solar cell studies.

(vi) Engine analysis.

Strain Gauge as force Sensor :-Strain gauge based sensors work on the principle of change in electrical resistance.

When, a mechanical element subjects to a tension or a compression the electric resistance of the material changes. This is used to measure the force acted upon the element.

2. Strain Gauge as force Sensor

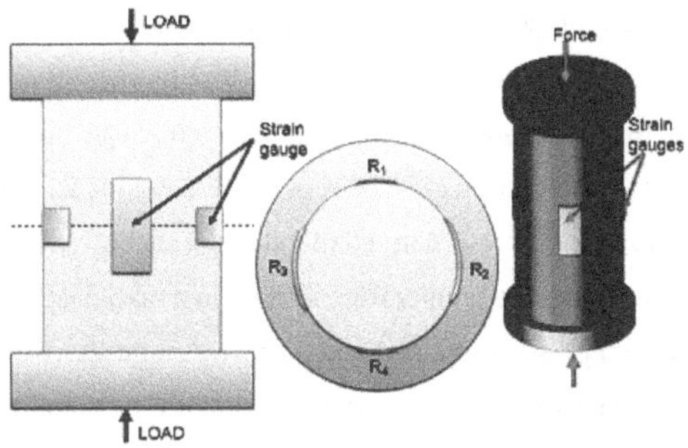

Diagram 5.14: Strain Gauge as Force Sensor

Piezo-electric sensor : -Piezoelectric sensor is used for the measurement of pressure, acceleration and dynamic-forces such as oscillation, impact, or high speed compression or tension. It contains piezoelectric ionic crystal materials such as Quartz.

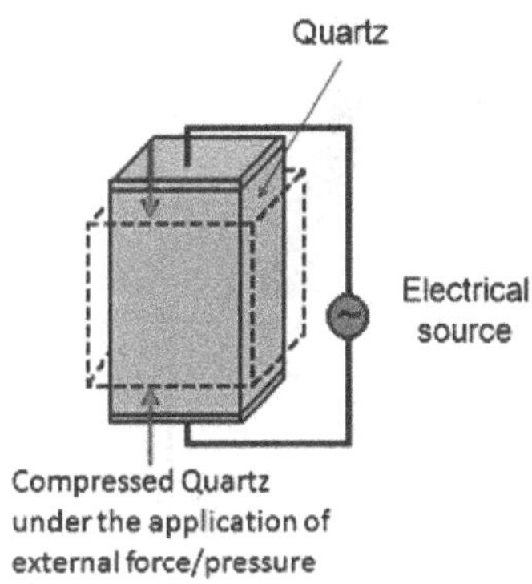

Diagram 5.15: Working of Piezo-Electric Sensor

On application of force or pressure these materials get stretched or compressed. During this process, the charge over the material changes and redistributes. One face of the material becomes positively charged and the other negatively charged. The net charge q on the surface is proportional to the amount x by which the charges have been displaced. The displacement is proportion to force. Therefore we can write,

$$q = kx = SF$$

Where k is constant and S is a constant termed the charge sensitivity.

Bi-metallic strip: -Temperature conveys the state of a mechanical system in terms of expansion or contraction of solids, liquids or gases, change in electrical resistance of conductors, semiconductors and thermoelectric EMFs. Temperature sensors such as bimetallic strips, thermocouples, thermistors are widely used in monitoring of manufacturing processes such as casting, molding, metal cutting etc. The construction details and principle of working of some of the temperature sensors are discussed in following sections. Bimetallic strips are used as thermal switch in controlling the temperature or heat in a manufacturing process or system. It contains two different metal strips bonded together. The metals have different coefficients of expansion. On heating the strips bend into curved strips with the metal with higher coefficient of expansion on the outside of the curve

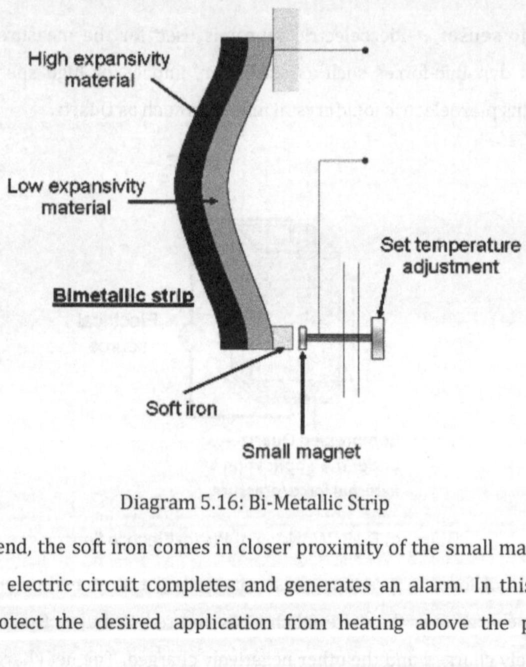

Diagram 5.16: Bi-Metallic Strip

As the strips bend, the soft iron comes in closer proximity of the small magnet and further touches. Then the electric circuit completes and generates an alarm. In this way bimetallic strips help to protect the desired application from heating above the pre-set value of temperature.

Resistance temperature detectors (RTDs):- RTDs work on the principle that the electric resistance of a metal changes due to change in its temperature. On heating up metals, their resistance increases and follows a linear relationship.

$$R_t = R_0(1 + \alpha T)$$

Where R_t is the resistance at temperature T (°C) and R0 is the temperature at 0°C and α is the constant for the metal termed as temperature coefficient of resistance. The sensor is usually made to have a resistance of 100 Ω at 0°C .RTDs are used in the form of thin films, wire wound or coil. They are generally made of metals such as platinum, nickel or nickel-copper alloys. Platinum wire held by a high-temperature glass adhesive in a ceramic tube is used to measure the temperature in a metal furnace. Other applications are:

(i) Air conditioning and refrigeration servicing.

(ii) Food Processing.

(iii) Stoves and grills.

(iv) Textile production.

(v) Plastics processing.

(vi) Petrochemical processing.

(vii) Micro electronics.

(viii)Air, gas and liquid temperature measurement in pipes and tanks.

(ix) Exhaust gas temperature measurement.

Thermistors :- Thermistors follow the principle of decrease in resistance with increasing temperature. The material used in thermistor is generally a semiconductor material such as a sintered metal oxide (mixtures of metal oxides, chromium, cobalt, iron, manganese and nickel) or doped polycrystalline ceramic containing barium titanate (BaTiO3) and other compounds. As the temperature of semiconductor material increases the number of electrons able to move about increases which results in more current in the material and reduced resistance.

Applications of Thermistors

(i) To monitor the coolant temperature and/or oil temperature inside the engine.

(ii) To monitor the temperature of an incubator.

(iii) Thermistors are used in modern digital thermostats.

(iv) To monitor the temperature of battery packs while charging.

(v) To monitor temperature of hot ends of 3D printers.

(vi) To maintain correct temperature in the food Handling and processing industry equipments. To control the operations of consumer appliances such as toasters, coffee makers, refrigerators, freezers, hair dryers, etc.

Thermocouple: - Thermocouple works on the fact that when a junction of dissimilar metals heated, it produces an electric potential related to temperature. As per Thomas See beck (1821), when two wires composed of dissimilar metals are joined at both ends and one of the ends is heated, then there is a continuous current which flows in the thermoelectric circuit. The net open circuit voltage (the See beck voltage) is a function of junction temperature and composition of two metals.

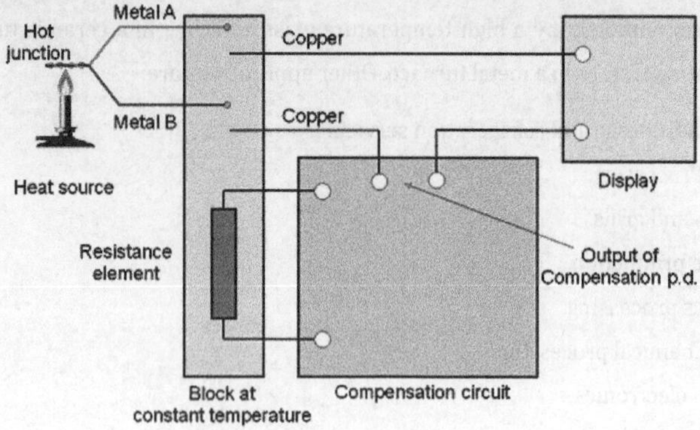

Diagram 5.17: Thermocouple

Light sensors: -A light sensor is a device that is used to detect light. There are different types of light sensors such as photocell/photo resistor and photo diodes being used in manufacturing and other industrial applicationsPhotoresistor is also called as light dependent resistor (LDR). It has a resistor whose resistance decreases with increasing incident light intensity. It is made of a high resistance semiconductor material, cadmium sulfide (CdS).The resistance of a CdS photo resistor varies inversely to the amount of light incident upon it. Photo resistor follows the principle of photoconductivity which results from the generation of mobile carriers when photons are absorbed by the semiconductor material.

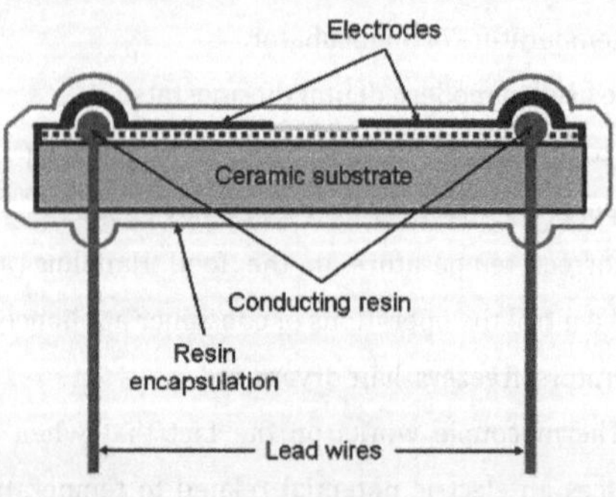

Diagram 5.18: Light Sensors

The CdS resistor coil is mounted on a ceramic substrate. This assembly is encapsulated by a resin material. The sensitive coils electrodes are connected to the control system though lead wires. On incidence of high intensity light on the electrodes, the resistance of resistor coil decreases which will be used further to generate the appropriate signal by the microprocessor via lead wires.

Applications of Photo Resistor

Computers, wireless phones, and televisions, use ambient light sensors to automatically control the brightness of a screen, Barcode scanners used in retailer locations work using light sensor technology. In space and robotics: for controlled and guided motions of vehicles and robots. The light sensor enables a robot to detect light. Robots can be programmed to have a specific reaction if a certain amount of light is detected.

(i) Auto Flash for camera.

(ii) Industrial process control.

Photo diodes:-Photodiode is a solid-state device which converts incident light into an electric current. It is made of Silicon. It consists of a shallow diffused p-n junction, normally a p-on-n configuration. When photons of energy greater than 1.1eV (the band gap of silicon) fall on the device, they are absorbed and electron-hole pairs are created. The depth at which the photons are absorbed depends upon their energy. The lower the energy of the photons, the deeper they are absorbed. Then the electron-hole pairs drift apart. When the minority carriers reach the junction, they are swept across by the electric field and an electric current establishes. Photodiodes are one of the types of photo detector, which convert light into either current or voltage. These are regular semiconductor diodes except that they may be either exposed to detect vacuum UV or X-rays or packaged with a opening or optical fiber connection to allow light to reach the sensitive part of the device.

Applications of Photo Diodes

- **Camera:** Light Meters, Automatic Shutter Control, Auto-focus, Photographic Flash Control
- **Medical:** CAT Scanners - x ray Detection, Pulse Oxi-meters, Blood Particle Analyzers
- **Industry** - Bar Code Scanners, Light Pens, Brightness Controls
- **Encoders--** Position Sensors, Surveying Instruments,Copiers - Density of Ton**er**
- **Safety Equipment-- Smoke**Detectors, Flame Monitors, **Security** Inspection Equipment - Airport X ray Intruder Alert - Security System.

- **Automotive--** Headlight Dimmer, Twilight Detectors, Climate Control - Sunlight Detector
- **Communications --** Fiber Optic Links,Optical Communications,Optical Remote Control.

5.16. Signal Conditioning Operations

Transducers sense physical phenomenon such as rise in temperature and convert the measured and into an electrical signal viz. voltage or current, The signals given by a transducer may be nonlinear in nature or may contain noise. Thus before sending these signals to the mechatronics control unit it is essential to remove the noise, nonlinearity associated with the raw output from a sensor or a transducer. It is also needed to modify the amplitude (low/high) and form (analogue/digital) of the output signals into respective acceptable limits and form which will be suitable to the control system. These activities are carried out by using signal conditioning devices and the process is termed as 'signal conditioning'.

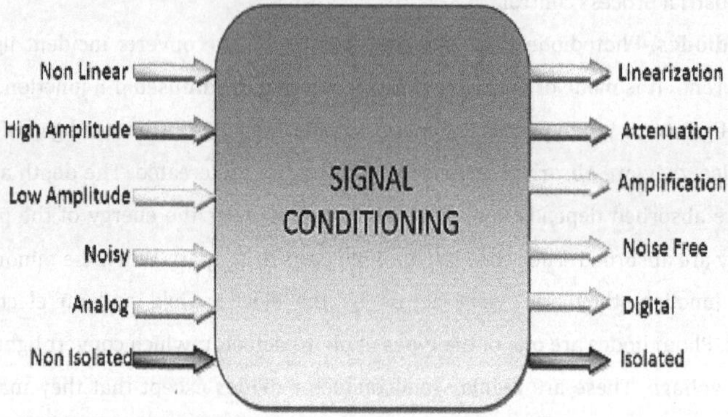

Diagram 5.19: Signal Conditioning System

Signal conditioning system enhances the quality of signal coming from a sensor in terms of,

Protection --To protect the damage to the next element of mechatronics system such microprocessors from the high current or voltage signals.

Right type of signal --To convert the output signal from a transducer into the desired form i.e. voltage / current.

Right level of the signal --To amplify or attenuate the signals to a right /acceptable level for the next element.

Noise -- To eliminate noise from a signal.

Manipulation --To manipulate the signal from its nonlinear form to the linear form.Sensory input devices and sensory output devices are usually applied for improving automatic systems, methods, and apparatuses are provided , the sensory input devices and sensory output devices are used to measure and evaluate at least one of cognition, knowledge, operation skills, physical properties, sensation, vision and on the other hand these sensory input devices and **sensory** output devices are integrated to enhance communications aswellas to optimize performance of devices / measuring instruments .Sensory input devices, sensory output devices, systems, and methods of their operation enhance the user experience through computing systems. Sensory input devices receive and process auditory, gustatory, kinesthetic, olfactory, tactile, vestibular, visual input from a user, sensory output devices provide auditory, gustatory, kinesthetic, olfactory, tactile, vestibular, visual stimulation to a user. Collectively, these devices, methods, and systems improve the use of automatic systems, methods, and apparatuses to measure and evaluate at least one of vision, cognition, knowledge, operation skills, physical properties, and sensation of the masses. Such enhanced measurements and evaluations engage those with lingual differences as well as incapacitated individuals.

(i) An automatic system is one wherein the means for identifying a subject includes at least one of auditory recognition, biological recognition, biometric recognition, bodily circulation part recognition, bodily sound recognition, bodily temperature recognition, bodily touch recognition, brain pattern recognition, chemical recognition, dental recognition, eye recognition etc. And such a automatic system generally used for testing, evaluating, screening, training, or any combination thereof for at least one of vision,

(ii) cognition, knowledge, operation skills, physical properties, or sensation of a subject comprising , a computer for processing data wherein a computing environment is conducted over a network, thin client, or interconnected means for data exchange and storage; an input device for recording response information from a subject in response to test stimuli; a mechanism to offer an option for feedback a linking device for linking data to at least one institution whereby said data can be automatically stored, shared, reviewed, referred, and/or studied; a display device configured for presenting test stimuli to a subject and for displaying the test data received by the input device .

(iii) An automatic system automatically performs at least one of a display, response, data storage, data transmission, interpretation of data, or any combination thereof in the form of at least one of a hologram, computer-generated hologram, X-ray imaging,

Magnetic resonance imaging (MRI), and nuclear magnetic resonance imaging (NMRI), electrical impulse imaging, visual imaging, or any combination thereof.

(iv) An automatic system is that wherein the input device is selected from at least one of an accelerator pedal, brake pedal, camera, computer mouse, hologram, horn, joystick, keyboard, physical measure device, microphone, physical property device, sensor, sensory device, sensory input device, steering wheel, telepathic input device, touch screen, video device, virtual reality device, wireless mouse and pad, X-ray, control device fitted with motion sensors, interactive screen, any other device capable of detecting an individual's response to testing stimuli, or any combination thereof.

5.17. Information Reception and Processing

Information reception and processing , both plays an important role in any manufacturing system or production for improving the productivity / efficiency , the data reception and processing is defined as:

(i) **Data receiving: -** The basic goal of most of this methods is to get the information / data that will be machine-readable data. Machine readable data is created for processing by a computer, instead of the presentation to a human user. The structure of such data relates to contained information, and not the way it is displayed eventually.

(ii) **Data processing: -** Data processing is, generally, the collection and manipulation of items of data to produce meaningful information. it can be considered a subset of information processing, the change (processing) of information in any manner detectable by an observer **or** it may be also defined as A data processing system is a combination of machines, people, and processes that for a set of inputs produces a defined set of outputs. The inputs and outputs are interpreted as data, facts, information, depending on the interpreter's relation to the system.

Following are the Some Basic Steps Applied for Data Processing

(i) Data conversion is the conversion of computer data from one format to another. Throughout a computer environment, data is encoded in a variety of ways , the conversion may require processing by the use of a special conversion program, or it may involve a complex process of going through intermediary stages .

(ii) Data validation is the process of ensuring that a program operates on clean, correct and useful data. It uses routines, often called "validation rules" "validation constraints" or "check routines", that check for correctness, meaningfulness, and security of data that are input to the system. Data validation is intended to provide certain well-defined guarantees for fitness, accuracy, and consistency for any of various kinds of user input into an application or automated system.

(iii) Sorting is any process of arranging items systematically, and has two steps – ordering is arranging items in a sequence ordered by some criterion. Categorizing is grouping items with similar properties.

(iv) In descriptive statistics, summary statistics are used to summarize a set of observations, in order to communicate the largest amount of information as simply as possible.

(v) In statistics, aggregate data are data combined from several measurements. When data are aggregated, groups of observations are replaced with summary statistics based on those observations.

Automatic data processing: - The term automatic data processing was applied to operations performed by means of unit record equipment.

Electronic data processing: - Computerized data processing or Electronic data processing represents a later development, with a computer used instead of several independent pieces of equipment.

Commercial data processing: - Commercial data processing involves a large volume of input data, relatively few computational operations, and a large volume of output.

Data analysis: - For science or engineering, the terms data processing and information systems are considered too broad, and the more specialized term data analysis is typically used. Data analysis uses specialized and precise algorithms and statistical calculations that are less often observed in a typical general business environment.

Information system: -An information system is defined as "any organized combination of people, hardware, software, communications networks, and data resources that stores and retrieves, transforms, and disseminates information in an organization.

5.18. Factors Affecting the Information Reception and Processing

Following are important factors which affect the information reception and processing.

(i) Software obsolescence.

(ii) Hardware obsolescence.

(iii) Liveware (people) obsolescence.

(i) Software Obsolescence: -Software obsolescence occurs when software is rendered obsolete due to a number of reasons as stated below. The obsolescence of many commercial applications and operating systems depends on the "end of support" date, since the software becomes a security risk when security patches are no longer provided by the software vendors.The reasons for software obsolescence can be listed according to the categories of obsolescence:

- **Technological obsolescence :-** The reasons for termination of sales and support for software are due to the software no longer being sold by the original vendor, being unable to expand or renew licensing agreements, and the software no longer being supported by the original vendor and third parties under software maintenance agreements , Software applications also become obsolete when they are retired from use and taken off the market due to technology advancements and decrease in product
- **Functional obsolescence: -** The functionality of the software will become obsolete due to hardware requirements or other software changes to the system. Software vendors produce new software to run on improved hardware and make older software versions obsolete, Business strategies of organizations can change and the software changes in line with the new business strategies, causing functional obsolescence of the software
- **Logistical Obsolescence:-** Access to software is limited or terminated due to digital media obsolescence, formatting, or degradation.

(ii) Hardware Obsolescence :-Hardware obsolescence can be categorized according to the types of obsolescence. The availability of more technologically advanced hardware will create technological obsolescence of existing hardware. Software upgrades that do not execute correctly on existing hardware will be a factor that creates functional obsolescence of hardware. When replacement parts cannot be obtained, this causes logistical obsolescence of hardware.

(iii)Liveware Obsolescence :- Liveware obsolescence occurs in the form of skills and knowledge of computer hardware and software becoming obsolescent. The extent of liveware obsolescence can be observed by the quantity of training and certifications available for people to learn new hardware and software The productivity of end users and information systems specialists is affected by skills obsolescence. IT professionals must always upgrade their professional knowledge and skills because their existing intellect will become obsolete in a few years . As new technology changes, the roles of IT professionals will be constantly changing and they must be able to adapt and be willing to learn new skills.

Factors Affecting and Affected by Information Systems Obsolescence

(i) **Technology advancements: -** Vendors create new software due to technology advancements. Organizations upgrade or replace their existing software with the newer software, leading to the obsolescence of the existing software.

(ii) **Low product popularity: -** Software applications are retired from use and taken off the market by vendors due to decrease in product popularity .

(iii) **No vendor support: -** Vendors of software applications do not provide support for older versions. Lack of vendor support would cause organizations to change their software because they would not be able to obtain necessary updates, technical assistance, and other support

(iv) **Competition in business environment: -** Competition in the business environment will cause organizations to replace or upgrade their existing software in order to meet the new challenges brought on by competition .

(v) **Changes in Business Strategies: -** The business strategies change and the software changes with the new business

(vi) **Mergers and acquisitions: -** Mergers and acquisitions cause the software used by a business to be changed (Expert survey). The acquired organization may have to change its existing software if it is not compatible with the software that is in use in the acquiring organization.

(vii) **New operating systems: -** Newer operating systems may not support older application software. This would require the application software to be changed or upgraded.

(viii) **New application software: -** Newer applications may not run efficiently or at all on older operating systems. Therefore the operating system would have to be upgraded.

(ix) **Different software versions: -** Different versions of the same application software may be incompatible. This would require the organization to discontinue the use of one incompatible version, usually the older version

(x) **Hardware Obsolescence Factors: -**Hardware obsolescence is affected by the following factors : -

(xi) **Technology advancements: -** Technology advancements give rise to new hardware. This hardware is marketed as being more efficient than existing hardware. Therefore, organizations would replace their existing hardware with the new hardware.

(xii) **No vendor support : -** Spares to support in-service hardware become unavailable because some component is no longer in production or is no longer in demand by the marketplace. Therefore the entire hardware unit may have to be replaced by the organization, in case a component malfunctions.

(xiii) **Incompatible hardware: -** New hardware is incompatible with old hardware components. This would require the older hardware components also to be replaced .

(xiv) **New software: -** Software upgrades require an increased specification computer for the software to run efficiently. Older hardware systems may not be able to handle the requirements of the new software.

(xv) **Lower cost new hardware: -** Hardware replacement costs are sometimes lower than costs to upgrade specific components. This would mean that entire hardware systems would be replaced, rather than upgrading the required components.

(xvi) **Mergers and acquisitions : -** Organizations may have to change their hardware systems due to major changes in the business environment, such as mergers and acquisitions . Affected by the component of information systems. e.g. Skills and knowledge is affected by liveware obsolescence .

Live-ware Obsolescence Factors :- The following are the liveware obsolescence factors.

(i) Skills and knowledge: - Employees require training when new software applications or operating systems are introduced to the organization, because they have to acquire the new skills and knowledge to operate the systems. If they do not update their skills and knowledge, they will become obsolescent.

(ii) Need for new employees: - When new information systems are introduced organizations may hire new employees to operate these systems. This may cause the existing employees to become obsolescent if their skills and knowledge are not up to date.

Question for Discussion

1. What is mean by applied physiology in work place ? Explain the features design standing workers, seated workers.
2. Describe the principles for design of visual and auditory displays.
3. What do you understand by 'display' and hence describe various types of visual display and visual indicators.
4. Discuss the ergonomic considerations for the design of work place layout.
5. Discuss the various factors which are considered for ergonomic design controls.

6. How are Industrial Engineering and productivity related ? 'Define the various types of productivity ?
7. From the ergonomic design point of view ,give recommendations for posture both for standing and slated at work.
8. Discuss the general steps in the measurement of mental work load.
9. Detail the principles of seat design with relevant sketches.
10. What do you understand by man machine considerations in designing? What should be designing considerations for case of maintenance?
11. How the human factor engineering plays its role in designing the controls? Give suitable examples.
12. Why is identification of control is so important? Discuss different type of control devices and the function which they perform.
13. What do you meant by posture? What are different posture evaluation tools explain in detail?
14. Give the list of various advantages of using PMTS systems over the ordinary time study.
15. Explain human aspects of work – study.
16. Define ergonomics.? What role does it play in job design? Illustrate with examples.
17. Discuss the criterion of selecting the seat height of operators with respect to work surface height.
18. Explain in brief the normal work area and work place design.
19. Discuss the ergonomic considerations for the design of work place layout.
20. Illustrate and explain the model of human information processing.

References

1. G. Salvendy, "Handbook of human factors", New York, John Wiley, 1987.
2. T.P. Peck, "Occupational safety and health: A guide to information sources", Management information guides. No. 28, Detroit, Michigan, Gale Research Co., 1974.
3. International Occupational Safety and Health Centre (CIS): CIS abstracts (Geneva, ILO-CIS; published periodically).
4. E. Hovenga, "Performance measurement", Management in the Australian Health Care Industry, C.Clinton & D. Scheiwe (eds), Harper Educational, 1995.
5. W.K. Hodson, "Fundamentals of indirect measurement", Industrial Engineering Handbook, 3rd edn,H.B. Maynard, McGraw-Hill, New York, 1971.
6. R.W. Hammond, "The history and development of industrial engineering", Industrial Engineering Handbook, 3rd edn, H.B. Maynard (ed.), McGraw-Hill, New York, 1971.

7. P.K. Robinson, "Aging and Technological Advances", Presented At NATO Conference Series III, Human Factors, Vol. 24, 1971.
8. C.G. Drury, B. Paramore, H.P. Van Cott, S.M. Grey and E.N. Corlett, "Task analysis", In Handbook of Human Factors, edited by G Salvendy. New York: Wiley, 1987.
9. A. Freivalds, "The ergonomics of tools", Int Rev Ergon, Vol. 1, Pp. 43-75, 1987.
10. B.T. Davies and H. Shahnawaz, "CIS 77-1774 An anthropo meter for use in developing countries", Ergonomics, Vol. 20, No. 3, Pp. 317-320, 1977.
11. D. Meister, "Guide to human engineering design for visual displays", Canoga Park, CA: Bunker-Ramo, 1969.
12. D. Kryter Karl, "The effects of noise on man", New York: Academic, 1970.
13. "American National Standards for Human Factors Engineering of Visual Display Terminal Workstations", ANSI/HFS Standard No. 100, 1988.
14. C.M. Harris, "Handbook of noise control", New York: McGraw-Hill, 1957.
15. M.S. Sanders and E.J. McCormick, "Human factors in engineering and design", 6th ed., New York: McGraw-Hill, 1987.
16. C. Wickens, "Engineering Psychology and Human Performance", New York: Harper Collins, 1992.
17. J.R. Wilson, "Participation: A framework and foundation for ergonomics", J Occup Psychol., Vol. 64, Pp. 67-80, 1991.
18. J.R. Wilson and E.N. Corlett, "Evaluation of Human Work: A Practical Ergonomics Methodology", London: Taylor & Francis, 1990.
19. J. Womack, T. Jones and D. Roos, "The Machine That Changed the World", New York: Macmillan, 1990.
20. W.E. Woodson, B. Tillman and P. Tillman, "Human Factors Design Handbook", New York: McGraw-Hill, 1991.
21. V. Zinchenko and V. Munipov, "Fundamentals of Ergonomics", Moscow: Progress, 1989.
22. S.H Hsu and Y. Peng, "Control /display relationship of the four-burner stove: A re-examination", Human Factors, Vol. 35, Pp. 745-749, 1993.
23. W.H. Cushman and D.J. Rosenberg, "Human Factors in Product Design", Amsterdam: Elsevier, 1991.
24. D. Gopher and E. Donchin, "Workload An examination of the concept", In Handbook of Perception and Human Performance, edited by K Boff, L Kaufman, and JP Thomas. New York: Wiley, 1986.

25. A. Chapanis, "Ethnic Variables in Human Factors Engineering", Baltimore: Johns Hopkins University, 1975.
26. G. Grote, "A participatory approach to the complementary design of highly automated work systems", In Human Factors in Organizational Design and Management, edited by G Bradley and HW Hendrick. Amsterdam: Elsevier, 1994.
27. J.H. Staramler, "The Dictionary of Human Factors Ergonomics. Boca Raton: CRC Press, 1993.

23. A. Chapanis, [illegible] in Human Factors Engineering, Baltimore: Johns Hopkins University, [illegible].

24. [illegible] participatory approach to the [illegible] design of highly automated work systems", in Human Factors in Organizational Design and Management, edited by G. Brown and H.W. Hendrick, Amsterdam: Elsevier, [illegible].

25. [illegible], The Dictionary of Human Factors Ergonomics, Boca Raton: CRC Press, [illegible].

CHAPTER 6

ANTHROPOMETRY

6.0. Anthropometry

Anthropometry is nothing but it is a fundamental branch of physical anthropology. It represents the quantitative aspect. A wide system of theories and practice is devoted to defining methods and variables to relate the aims in the different fields of application. In the fields of occupational health, safety and ergonomics anthropometric systems are mainly concerned with body build, composition and constitution, and with the dimensions of the human body's interrelation to workplace dimensions, machines, the industrial environment and clothing.

Basically Anthropometry consists of two syntax these are Anthrop & Metrics which is a Greek word delivering the meaning 'Anthrop means men/human and metrics means measurements' is related to the study of human physical dimensions, thus it is the study of measurements or proportions of the human body according to sex, age, height, weight, width, length of reach , dimensions of bone, muscle, and adipose (fat) tissue etc. for identification purposes & understanding human physical variation.

Anthropometric variables: -An anthropometric variable is a measurable characteristic of the body that can be defined, standardized and referred to a unit of measurement. Linear variables are generally defined by landmarks that can be precisely traced on the body. Landmarks are generally of two types: skeletal-anatomical, and virtual landmarks that are simply found as maximum or minimum distances .Anthropometric variables have both genetic and environmental components and may be used to define individual and population variability. The choice of variables must be related to the specific research purpose and standardized with other research in the same field. Anthropometric variables are mainly linear measures, such as heights, distances from landmarks with subject standing or seated in standardized posture; diameters, such as distances between bilateral landmarks; lengths, such as distances between two different landmarks; curved measures, namely arcs, such as distances on the body surface between two landmarks; and girths, such as closed all-around measures on body surfaces, generally positioned at at least one landmark or at a defined height. Other variables may require special methods and instruments. For instance skin-fold thickness is measured by means of special constant pressure calipers. Volumes are measured by calculation or by immersion in water. To obtain full information on body surface characteristics, a computer matrix of surface points may be plotted using bio-stereometric techniques.

Instrumentsused: Although sophisticated anthropometric instruments have been described and used with a view to automated data collection, basic anthropometric instruments are quite simple and easy to use.

Much care must be taken to avoid common errors resulting from misinterpretation of landmarks and incorrect postures of subjects.The standard anthropometric instrument is the anthropometer a rigid rod 2 meters long, with two counter-reading scales, with which vertical body dimensions, such as heights of landmarks from floor or seat, and transverse dimensions, such as diameters, can be taken. A stadiometer is a fixed anthropometer, generally used only for stature and frequently associated with a weight beam scale.

For transverse diameters a series of calipers may be used: the pelvimeter for measures up to 600 mm and the cephalometer up to 300 mm. The latter is particularly suitable for head measurements when used together with a sliding compass ,Skin-fold thickness is measured with a constant-pressure skin-fold caliper generally with a pressure of 9.81 x 104 Pa (the pressure imposed by a weight of 10 g on an area of 1 mm2).For arcs and girths a narrow, flexible steel tape with flat section is used. Self-straightening steel tapes must be avoided.

6.1. Systems of Variables

A system of anthropometric variables is a coherent set of body measurements to solve some specific problem.

In the field of ergonomics and safety the main problem is fitting equipment and work space to humans and tailoring clothes to the right size. Equipment and work space require mainly linear measures of limbs and body segments that can easily be calculated from landmark heights and diameters, whereas tailoring sizes are based mainly on arcs, girths and flexible tape lengths. Both systems may be combined according to need.

Basic set of anthropometric variables:- Ergonomics always considered the anthropometrics variable, regarding to the subject point of view following are some important variable that had been proposed by the various scientists / engineers, these are listed below.

Below listed elements are the important data or variables that is used in ergonomics, except it some others is always used in the anthro-pometry. Here under we are also putting pictorial view of anthropometry data for sitting and standing position of operators.

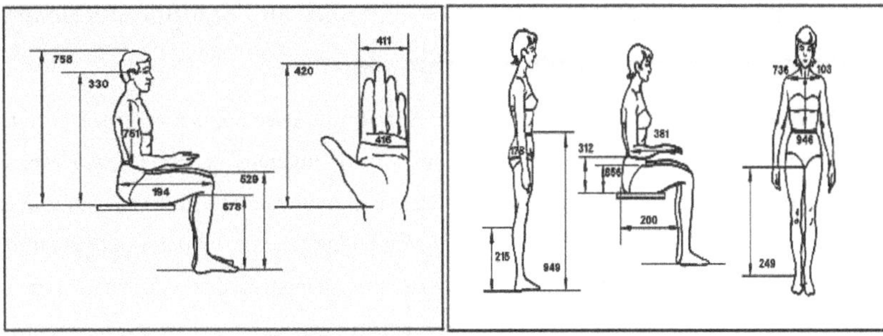

Diagram 6.1: Sitting and Standing Position of an Operators According to the Anthropometric Variable . All Dimension are Given in CM

Table 6.1: Anthropometric Variable

No.	Variable	No.	Variable
01	Stature (vertical distance from floor to head vertex)	19	Shoulder height (from floor to acromion)
02	Eye height (from floor to inner eye corner)	20	Elbow height (from floor to radial depression of elbow)
03	Crotch height (from floor to pubic bone)	21	Finger tip height (from floor to grip axis of fist)
04	Hip breadth, standing (the maximum distance across hips)	22	Finger tip height (from floor to grip axis of fist)
05	Eye height, sitting (from seat to inner corner of the eye)	23	Sitting height (from seat to head vertex)
06	Elbow height, sitting (from seat to lowest point of bent elbow)	24	Shoulder height, sitting (from seat to acromion)
07	Lower leg length (height of sitting surface)	25	Knee height (from foot-rest to the upper surface of thigh)
08	Body depth, sitting (seat depth)	26	Forearm-hand length (from back of bent elbow to grip axis)
09	Elbow to elbow breadth (distance between lateral surface of the elbows)	27	Buttock-knee length (from knee-cap to rearmost point of buttock)
10	Index finger breadth, proximal (at the joint between medial and proximal phalanges)	28	Hip breadth, sitting (seat breadth)
11	Index finger breadth, distal (at the joint between distal and medial phalanges)	29	Index finger length
12	Hand length (from tip of middle finger to styloid)	30	Hand breadth (at metacarpals)
13	Wrist circumference	31	Foot breadth
14	Foot length	32	Heat circumference (at glabella)
15	Sagittal arc (from glabella to inion)	33	Head length (from glabella to opisthocranion)
16	Head breadth (maximum above the ear)	34	Bitragion arc (over the head between the ears)
17	Waist circumference (at the umbilicus)	35	Tibial height (from the floor to the highest point on the antero-medial margin of the glenoid of the tibia)
36	Cervical height sitting (to the tip of the spinous process of the 7th cervical vertebra).		

6.2. Anthropometric Principles in Work-Space and Equipments Design: - Following Principle Leads to Anthropometry

(i) **Designing for a population of users:** -Specify the user population (to a group of people sharing common ancestors,common occupations, common geographical locations or age groups.) and then todesign to accommodate as wide a range of users as possible and acknowledge andallow for the inherent variability of the user population.

(ii) **Sources of human variability**: - Biological anthropologists distinguish four types of human adaptation. Over many lifetimes, genetic changes may occur as a result of natural selection. Over the course of a lifetime, organisms exhibitplasticity (literally the capability of beingmoulded). Over the short term, organisms can exhibit acclimatization, and over there by short termbehavioural adaptation. Only the last two of these forms of adaptation are reversible.

(iii) **Implications for ergonomics:** -These findings have far-reaching implications for ergonomists attempting to design to accommodate a wide range of people. The structure of populations and their livingconditions are changing in many parts of the world. This means that anthropometric data captured in the past may no longer be representative. When designing for international markets, then, each target country has to be considered separately.

Applications of Anthropometric Data

(i) **Equipment and Facilities :-** Criteria of judgment of design features of physical equipment and facilities: comfort, work performance, safety, physical effects. Human body measurements are clearly relevant to the design of equipment or facilities for human use. Such equipment is of three kinds: spaces into which men fit, whether for work, rest or protection; tools or controls which men use; and protective clothing and devices (e.g. helmets, masks, goggles, parachutes) which men wear. Although many basic body measurements like stature, sitting height and weight are generally applicable, each of the three different kinds of end product mentioned above requires some distinct measurements. Clothing design, for example, utilizes several body circumferences, special limb lengths (spine-shoulder-fingertip; crotch height), and details of the head, face, hands and feet.

(ii) **Spaces: -** One of the applications of anthropometric data is the design of work spaces, including the work-space envelopes. The limits of the work-space envelope for seated personnel are determined by functional arm reach, which is influenced by such factors as the presence of any restraints, apparel worn, angle of back rest and especially by the direction of arm reach and the nature of the manual activity to be performed. Improper postures caused by poorly designed facilities and methods bend the spine and increase the pressure between the vertebrae. The discs serving as cushions between the vertebrae are then subject to wear and tear (pathological degeneration) which constitutes one of the main reasons of backaches of workers.

(iii) The objective in applying anthropometric data will be to improve the design of things and spaces for people to use so that they are more comfortable, efficient, easy to use and safer than previous designs. In order to do this it is necessary to have some idea of the dimensions that are really important for the particular design under consideration, the problems that have been encountered by others and the recommendations, design rules or regulations that are available.

6.3. Display Systems

Displays provide information about the status of equipment. Displays may apply to the operator's visual sense (lights, scales, counters, cathode-ray tubes, flat panel electronics, etc.), to the auditory sense (bells, horns, recorded voice messages, electronically generated sounds, etc.) or to the sense of touch (shaped controls, Braille, etc.). Labels, written instructions, warnings or symbols ("icons") may be considered special kinds of displays.

The Four basic Properties of Displays Are

(i) Display only that information which is essential for adequate job performance.

(ii) Display information only as accurately as is required for the operator's decisions and actions.

(iii) Present information in the most direct, simple, understandable and usable form.

(iv) Present information in such a way that failure or malfunction of the display itself will be immediately obvious.

6.3.1. Visual Displays

A visual display is most appropriate if the environment is noisy, the operator stays in place, the message is long and complex, A wide range of technology is used with visual displays. If we leave aside the hand printed and handwritten signs that we find in everyday life, we see that the printed media are very much to the forefront. As well as printed documentation, there is also colour printing and associated processes that are used to produce signs used in buildings, on roads and in other contexts. Related to this, there are various labels and packaging that are associated with everyday and work objects.Then there are devices like lights and illuminated signs that we use to form part of a machine interface or to enhance a sign that has to be used in all visibility conditions. There are displays that are specifically designed to carry changing parameter states to us. These are the dials and the screens that we are familiar with from everyday objects such as cars or we have seen depicted in films and television programmes about complex systems such as aircraft control cabins. Very few people will be familiar with the instrument arrays used in an aircraft cockpit, but most will have seen these illustrated. A common impression is that they are highly complex environments with a wide range of dials, lights and, nowadays, computer screens. Increasingly, the classic ergonomics of lights and dials has given way to concerns with computer screen technology. This latter technology either mimics these devices in a more flexible format, or it may present information in a new way. The ever-present computer screen has found its way, not only into offices, but also into control rooms and vehicle control stations such as cockpits and ship's bridges. Early computer displays were rudimentary with letters and numbers presented in very simple forms. The current display technology enables high quality displays of a whole range of variables. These displays may also represent the real world either from recorded sources or direct information in terms of closed circuit television pictures.Given the variety of media that have emerged to convey information, it will be foolhardy to predict what might be possible in the future. The important thing for ergonomists is to establish good principles of display design which will transcend particular forms of technology. General principles of display design can then be translated into specific principles within a particular medium. To take a simple example from the early days of printing, the maximum contrast that could be obtained from the printed word involved using black symbols/letters on white paper. There would always have been some variation in this depending on ink and paper quality; only in recent times has very white paper been available for this process. The darkness of the print similarly would be limited by contemporary knowledge of dyes and inks. The almost universal use of this principle was not applicable in the early days of computer screen technology. Initially, it was much easier to present the lettering as the bright source against the screen as the dark source. This was the reversal of the normal principle (i.e., light on dark). Added to this were difficulties that the words, the light source, were usually in light green and the background was usually in dark green. Current computer screen technology has re-established the principle of dark on light. The lesson that still has to be learned is that computer screens, for a number of reasons, are not an ideal reading media. Using a computer screen to mimic a book has not been particularly successful. In addition, the principle has yet to be learned in relation to the enormous variety of colours that can now be produced on computer screens. Many people in designing their web pages choose colours that they think go well together, but very often these combinations produce problems for reading text. This is not to say that web pages and computer screens cannot do many things differently and better than standard printed work. The question is what things can they do better and under what conditions. It is striking that the printed word not only remains very much with us, but also remains very little altered in terms of display characteristics from the earliest handwritten book.

6.3.2. Visual Display Technology

The range of technology available for visual display of information is very broad. The use of a form of technology in a particular context will depend upon such factors as cost, compatibility with existing hardware, preferences, and most importantly of all, the appropriateness of the technology for the particular visual display problem. Many of the choices in the design process will become self-evident as a project develops. Many of the choices will be technology-based in terms of such factors as hardware/software compatibility, preferences and suitability for the task in question. In certain applications, where two technology solutions are equal in their technical implications, then a choice of one over the other may be determined by human factors issues.

Similarly, duplication of function will also be a possibility, depending upon the nature of the task demands. For example, a digital clock will enable a very accurate reading of time. An analogue clock, with hands and numbers, will enable quick estimates of the amount of time before or until a particular timed event. For one or other of these time-related tasks, one or other type of timepiece will be optimal. If both of these tasks are commonly performed, then it would be better to display both functions on the same timepiece.

There are Three Basic Types of Visual Displays

(i) The check display indicates whether or not a given condition exists (for example a green light indicates normal function).

(ii) The qualitative display indicates the status of a changing variable or its approximate value, or its trend of change (for example, a pointer moves within a "normal" range).

(iii) The quantitative display shows exact information that must be ascertained (for example, to find a location on a map, to read text or to draw on a computer monitor), or it may indicate an exact numerical value that must be read by the operator (for example, a time or a temperature).on of an object.

6.3.3. *Auditory Display*

Where information is presented to you via your ears. These displays are good at attracting your attention when vision can't be used (at night or with someone with limited sight, for example). Because of the ability of audio alarms to attract attention, they are often used to draw your attention to something (a car indicator clicking), and as emergency alarm systems (a fire alarm).

6.3.4. *Tactile Display*

Where information is presented to you via your sense of touch. The human tactile system is not as sensitive to differences or changes in stimulus as either the visual or auditory systems. An example of a tactile display is the Braille alphabet for the blind. The tactile sense can also tell you subtle information like temperature or surface condition (wet, sticky, slimy etc.) Have you ever felt a radiator to see if it is warming up

6.3.5. *Quantitative Display*

Quantitative displays show exact information. Digital quantitative displays present information directly as numbers, for example, the clock on your computer. Analogue quantitative displays can also be used where a length or angle represents the information, for example, a thermometer where the length of mercury or alcohol represents the temperature. The use of a particular quantitative display depends on the kind of information that is required. If you need a precise reading, then digital indicators are most easily read.

6.3.6. *Qualitative Display*

Qualitative displays give information about particular states, for example, hot or cold, alarm or no alarm. These displays can provide information about rate of change or direction of deviation from a desired value. These displays may include indicators and warning devices. They can be used in circumstances where you only need to know that a certain condition exists, for example, when the temperature is too hot or too cold, as in the case of an indicator light on an iron which goes out when the iron is up to temperature. The specific value is not needed, although that may be conveyed to you by other, quantitative visual displays. Check-reading displays are a specific type of qualitative display in which you determine whether the value of a continuously changing variable is normal, or within an acceptably normal range, for example, car fuel gauges and tyre pressure gauges. Check-reading displays should have clearly distinguishable characteristics to identify the neutral or normal satisfactory condition, or the undesirable condition; perhaps green marking for an 'OK' level and red for 'out-of-limits'.

6.3.7. Represented Display

These displays can portray either working models or simplified diagrams of a complex process, system, or machine. They enable you to perceive the functioning of each part of the system or machine in correct relation to the whole system. London Underground and Ordnance Survey maps, and railway signal box mimic diagrams are examples of representational displays. Representational displays should aim at schematic representation of information (keep it simple) rather than a complete representation or the actual representation.

Design Guidelines for Visual Displays Are

(i) Arrange displays so that the operator can locate and identify them easily without unnecessary searching. (This usually means that the displays should be in or near the medial plane of the operator, and below or at eye height.)

(ii) Group displays functionally or sequentially so that the operator can use them easily.

(iii) Make sure that all displays are properly illuminated or illuminant coded and labelled according to their function.

(iv) Use lights, often coloured, to indicate the status of a system (such as ON or OFF) or to alert the operator that the system, or a subsystem, is inoperative and that special action must be taken. Flashing red indicates an emergency condition that requires immediate action. An emergency signal is most effective when it combines sounds with a flashing red light.

For more complex and detailed information, especially quantitative information, one of four different kinds of displays are traditionally used:

(i) A moving pointer (with fixed scale).

(ii) A moving scale (with fixed pointer).

(iii) A counters or "pictorial" displays.

Especially computer-generated on a display monitor.It is usually preferable to use a moving pointer rather than a moving scale, with the scale either straight (horizontally or vertically arranged), curved or circular. Scales should be simple and uncluttered, with graduation and numbering so designed that correct readings can be taken quickly. Numerals should be located outside the scale markings so that they are not obscured by the pointer. The pointer should end with its tip directly at the marking. Numbers should increase left to right, bottom to top or clockwise. Mechanical displays with pointers and printed scales were increasingly replaced by "electronic" displays with computer-generated images, or solid-state devices using light-emitting diodes.Monochrome displays have only one colour, usually green, yellow, amber, orange or white (achromatic). If several colours appear on the same chromatic display, they should be easily discriminated. It is best to display not more than three or four colours simultaneously (with preference being given to red, green, yellow or orange, and cyan or purple). All should strongly contrast with the background. In fact, a suitable rule is to design first by contrast, that is, in terms of black and white, and then to add colours sparingly.

Following Guidelines Must be Followed by the Manufacturer Regarding the Display

(i) Blue is a good colour for backgrounds and large shapes. However, avoiding for text, thin lines or small shapes.

(ii) The colour of alphanumeric characters should contrast with that of the background.

(iii) When using colour, use shape as a redundant because redundant coding makes the display much more acceptable for users who have colour-vision deficiencies.

(iv) As the number of colours is increased, the sizes of the colour-coded objects should also be increased.

(v) Red and green should not be used for small symbols and small shapes in peripheral areas of large displays.

(vi) Using opponent colours (red and green, yellow and blue) adjacent to one another or in an object/background relationship is sometimes beneficial and sometimes detrimental.

(vii) Avoid displaying several highly saturated, spectrally extreme colours at the same time.

Warnings:-Ideally, all devices should be safe to use. In reality, often this cannot be achieved through design. In this case, one must warn users of the dangers associated with product use and provide instructions for safe use to prevent injury or damage. It is preferable to have an "active" warning, usually consisting of a sensor that notices inappropriate use, combined with an alerting device that warns the human of an impending danger. Yet, in most cases, "passive" warnings are used, usually consisting of a label attached to the product and of instructions for safe use in the user manual. Such passive warnings rely completely on the human user to recognize an existing or potential dangerous situation, to remember the warning, and to behave prudently.

Labels and signs for passive warnings must be carefully designed by following the most recent government laws and regulations, national and international standards, and the best applicable human engineering information. Warning labels and placards may contain text, graphics, and pictures—often graphics with redundant text. Graphics, particularly pictures and pictograms, can be used by persons with different cultural and language backgrounds, if these depictions are selected carefully. However, users with different ages, experiences, and ethnic and educational backgrounds, may have rather different perceptions of dangers and warnings. Therefore, design of a safe product is much preferable to applying warnings to an inferior product.

6.4. Safety Signs and Colour Code at Work-Place

Guidance is for employers and duty holders, and others who have responsibility for the control of work sites and premises, or operating equipment requiring verbal and/or non-verbal communications , Safety signs and colour code are very useful & important tools to help protect the health and safety of employees at workplace

Significance of safety sign :- safety sign is also gives a clear view about a particular insecurity , whenever the operator is working at work-place or operating the equipments / machine . Some point will clarify about the safety sign.

(i) Draw attention to health and safety hazards.

(ii) Point out hazards that may not be obvious.

(iii) Provide general information and directions.

(iv) Remind employees where personal protective equipment must be worn.

(v) Show where emergency equipment is located.

(vi) Indicate where certain actions are prohibited.

Following three type of sign categories are generally applied at work-place,

(i) Regulator.

(ii) Warning.

(iii) Information.

Colour attracts attention and can be used extensively for safety purposes. For example, colour can be used as an additional safety measure to identify the contents of pipes and the nature of the hazard.

The choice of colour also draws attention to the probability of a hazard causing harm. For example, the colour red is used to indicate a definite hazard. A potential hazard is communicated by the colour yellow.

When employees are aware of the hazards around them and take the necessary precautions, the possibility of an injury, illness or other loss can beminimized. Following tables gives complete details of colour indication & their purpose of in industries.

Table 6.2: Safety Sign

COLOR	MEANING	APPLICATION
Red	Danger Stop	Safety cans and signs Emergency stop bar or button on machinery. Identification of fire equipment
Fluorescent Orange, Orange-Red	Bio-safety	Labels and containers for blood and infectious waste. (Warning labels must be fluorescent orange or orange-red with the bio-safety symbol in a contrasting color.)
Orange	Warning	Parts of machinery or energized equipment that may cut, crush or otherwise injure. Inside of transmission guards for pulleys, gears, etc
Green	Safety	Location of first aid equipment. Location of safety equipment, respirators, safety showers, etc.
Blue	Information	Signs and bulletin boards. Specific railroad warnings against starting, using or moving equipment being repaired.
Black, White, Yellow or Combination of Black with White or Yellow	Boundaries	Traffic or housekeeping markings. Stairways, directions and borders.
Magenta or Purple on Yellow	Radiatation caution	X-ray, alpha, beta, gamma, neutron and proton radiation.

6.5. Factorial and Graphic Display

Factorial display is nothing but it is representation of positive integer (value more than zero) value of two data , which is being compare with the help of plot on X- axis and Y- axis .Graphical display is also a representation of data which being compared & conclusion is made ,Most charts are a variation of one of four basic types: pie charts, bar charts, time series charts and scatter plots , a Pareto chart, a normal plot, and a half-normal plot. Choosing the right type of chart depends on the characteristics of the data and the relationships you want displayed . But In present scenario it is a combination of monitor and video adapter that makes a computer capable of displaying graphics.

Pie Charts

Pie charts are used to represent the distribution of the categorical components of a single variable. Note that as a general rule, multivariate comparisons provide for more meaningful analyses than do single variable distributions and for this and other reasons pie charts should be rarely used, if at all.

Rules for Pie charts

(i) Avoid using pie charts.

(ii) Use pie charts only for data that add up to some meaningful total.

(iii) Never ever use 3D pie charts; they are even worse than 2D pies.

(iv) Avoid forcing comparisons across more than one pie chart.

(v) Pie charts should rarely be used. Pie charts usually contain more ink than is necessary to display the data and the slices provide for a poor representation of the magnitude of data points.

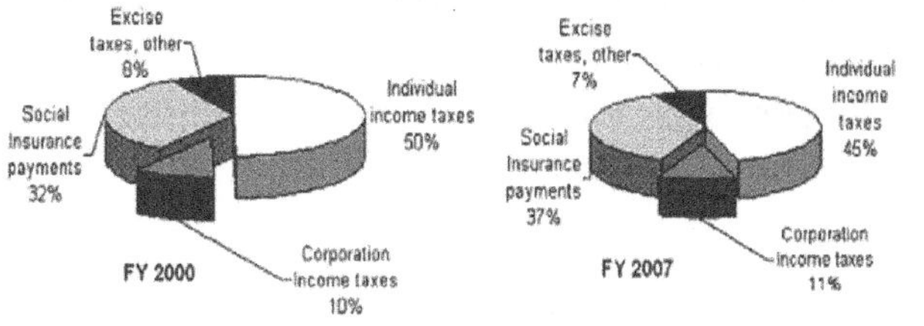

Diagram 6.2: Pie Charts

Bar Charts

Bar charts typically display the relationship between one or more categorical variables with one or more quantitative variables represented by the length of the bars. The categorical variablesare usually defined by the categories displayed on the x-axis and, if there is more than one data series, by the legend.

Rules for Bar Charts

(i) Minimize the ink, do not use 3D effects.

(ii) Sort the data on the most significant variable.

(iii) Use rotated bar charts if there are more than 8 to 10 categories.

(iv) Place legends inside or below the plot area.

(v) With more than one data series, beware of scaling distortions.

Bar charts often contain little data, a lot of ink, and rarely reveal ideas that cannot be presented much more simply in a table. Minimizing the ink-to-data ratio is especially important in the case of bar charts. Never use a 3D bar chart. Keep the gridlines faint. Display no more than seven numbers on the y-axis scale. If there are fewer than five bars, consider using data labels rather than a y-axis scale; it doesn't make sense to use a five-numbered scale when the exact values can be shown with four numbers.

Time Series Line Charts

The time series chart is one of the most efficient means of displaying large amounts of data in ways that provide for meaningful analyses. The typical time series line chart is a scatter plot chart with time represented on the x-axis and the lines connecting the data points.

Rules for Time Series (Line) Charts

(i) Time is almost always displayed on the x-axis from left to right.

(ii) Display as much data with as little ink as possible.

(iii) Make sure the reader can clearly distinguish the lines for separate data series.

(iv) Beware of scaling effects.

(v) When displaying fiscal or monetary data over time, it is often best to use deflated data (e.g. inflation-adjusted or % of GDP .

Scatter Plots

The 2D scatter plot is the most efficient medium for the graphical display of data. A simple scatter plot will tell you more about the relationship between two interval-level variables than any other method of presenting or summarizing such data.

Rules for Scatter Plots

(i) Use two interval-level variables.

(ii) Fully define the variables with the axis titles.

(iii) The chart title should identify the two variables and the cases (e.g. cities or states).If there is an implied causal relationship between the variables, place the independent variable (the one that causes the other) on the x-axis and the dependent variable (the one that may be caused by the other) on the y-axis.

(iv) Scale the axes to maximize the use of the plot area for displaying the data points.

(v) It's a good idea to add data labels to identify the cases.

With good labeling of the variables and cases and common-sense scaling of the x and y-axes, there's not a lot that can go wrong with a scatter plot, although extreme outliers on one or more of the variables can obscure patterns in the data.

Box Plots

John W. Tukey invented the box plot as a convenient method of displaying the distribution of interval-based variables.

Rules for Box Plots

(i) A simple box plot plots the median and four quartiles of data for an interval level variable.

(ii) Box plots are best used for comparing the distribution of the same variable for two or more groups or two or more points.

(iii) Box plots are an excellent means of displaying how a single case compares to a large number of other .

Pareto Chart

Select to determine the magnitude and the importance of an effect. The chart displays the absolute value of the effects and draws a reference line on the chart. Any effect that extends beyond this reference line is statistically significant.

Normal

Select to compare the magnitude and statistical significance of main and interaction effects from a 2-level factorial design. The fitted line indicates where you would expect the points to fall if the effects were zero. Significant effects have a label and fall toward the left or right side of the graph. The normal probability plot displays negative effects on the left side of the graph and positive effects on the right side of the graph.

Half Normal

Select or to compare the magnitude and statistical significance of main and interaction effects from a 2-level factorial design. The fitted line indicates where you would expect the points to fall if the effects were zero. Significant effects have a label and fall toward the right side of the graph. The half normal plot displays the absolute value of all effects, positive and negative. Instead of putting negative effects to the left and positive effects to the right, all the significant effects are on the right side, which emphasizes their relative magnitudes.

Cathode ray tube (CRT): It is modern device used for displaying the result on screen actually it was associated by the branch of science (physics) that is applied in modern industrial age , there are different display systems like **cathode ray tubes (CRTs), liquid crystal displays (LCDs), electroluminescent displays (ELDs), plasma displays and light emitting diodes (LEDs)** are available in the present technology. Here we are going to discuss about the Cathode Ray Tube (CRT).

Operation of CRT :- Cathode Ray Tube (CRT) is a computer display screen, used to display the output in a standard composite video signal. The working of CRT depends on movement of an electron beam which moves back and forth across the back of the screen. The source of the electron beam is the electron gun; the gun is located in the narrow, cylindrical neck at the extreme rear of a CRT which produces a stream of electrons through thermionic emission. Usually, A CRT has a fluorescent screen to display the output signal. A simple CRT is shown in below.

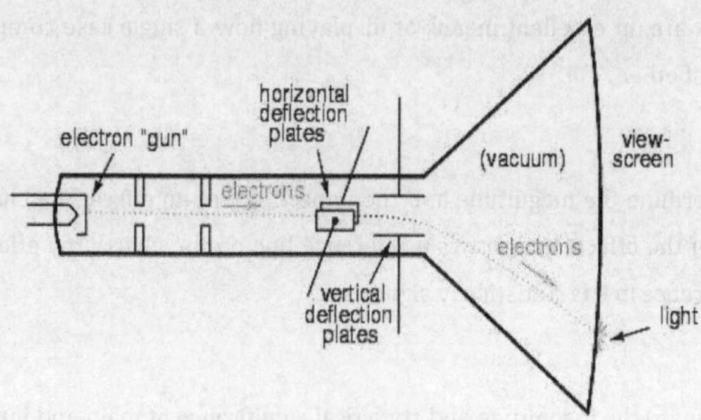

Diagram 6.3: Cathode Ray Tubes

The operation of a CRT monitor is basically very simple. A cathode ray tube consists of one or more electron guns, possibly internal electrostatic deflection plates and a phosphor target. CRT has three electron beams – one for each (Red, Green, and Blue) is clearly shown in figure. The electron beam produces a tiny, bright visible spot when it strikes the phosphor-coated screen. In every monitor device the entire front area of the tube is scanned repetitively and systematically in a fixed pattern called a raster. An image (raster) is displayed by scanning the electron beam across the screen. The phosphor's targets are begins to fade after a short time, the image needs to be refreshed continuously. Thus CRT produces the three color images which are primary colors. Here Main parts of the cathode ray tube are cathode, control grid, deflecting plates and screen.

Cathode

The heater keeps the cathode at a higher temperature and electrons flow from the heated cathode towards the surface of the cathode. The accelerating anode has a small hole at its center and is maintained at a high potential, which is of positive polarity. The order of this voltage is 1 to 20 kV, relative to the cathode. This potential difference creates an electric field directed from right to left in the region between the accelerating anode and the cathode. Electrons pass through the hole in the anode travel with constant horizontal velocity from the anode to the fluorescent screen. The electrons strike the screen area and it glows brightly.

The Control Grid

The control grid regulates the brightness of the spot on the screen. By controlling the number of electrons by the anode and hence the focusing anode ensures that electrons leaving the cathode in slightly different directions are focused down to a narrow beam and all arrive at the same spot on the screen. The whole assembly of cathode, control grid, focusing anode, and accelerating electrode is called the electron gun.

Deflecting Plates

Two pairs of deflecting plates allow the beam of electrons. An electric field between the first pair of plates deflects the electrons horizontally, and an electric field between the second pair deflects them vertically, the electrons travel in a straight line from the hole in the accelerating anode to the center of the screen when no deflecting fields are present, where they produce a bright spot.

Screen

This may be circular or rectangular. Screen is coated with special type of fluorescent material. Fluorescent material absorbs its energy and re-emits light in the form of photons when electron beam hits the screen. When it happens some of them bounces back just like bouncing of cricket ball from a wall. These are called as secondary electrons. They must be absorbed and returned back to cathode, if it is not so they accumulate near screen and produce space charge or electrons cloud. To avoid this, aquadag coating is applied on funnel part of CRT from inside.

Advantages of CRT

(i) CRT's are less expensive than other display technologies.
(ii) They operate at any resolution, geometry and aspect ratio without decreasing the image quality.
(iii) CRTs produce the very best color and gray-scale for all professional calibrations.
(iv) Excellent viewing angle.
(v) It maintains good brightness and gives long life service.

Auditory Display: -The displays that characterize human-machine interfaces are predominantly visual in nature. However, other sensory modalities provide additional resources for the transmission of information. Auditory signals provide a number of advantages particularly in complex visual environments. While visual displays require direct observation, humans are sensitive to sounds from any direction. Auditory displays can be used to orient attention to events outside the user's field of view. Auditory displays can also utilize speech to communicate Information. Speech can be perceived directly or through communication systems, e.g., radios or telephones, from recordings, e.g., telephone message tapes or digital recordings, or from synthesized speech from computers. In addition to these sources, a wide variety of signals can be used to give us information. The invention of the bell is perhaps a good example of early technology that fulfilled a particular purpose. The ringing of bells in religious buildings would be one way of drawing attention of an individual or group to a key event. Messages can in turn be codified in terms of the way in which a bell or bells are used. Newer technology gives rise to various buzzers, hooters, sirens, etc. These can transmit information by manipulating a variety of attributes including the amplitude (volume), frequency (pitch) and timing between component parts of the signal. An intermittent signal that increases in amplitude, pitch and the number of pulses per minute provides a powerful means of communicating the time course of a particular process. Like the whistle on steam kettle, such alarms can be considered the auditory analogue of an egg timer, providing information about the occurrence of a forthcoming event as well as signalling the requirement for a timely response. Although there is a tendency to think of auditory displays in terms of alarms, it is also important to be familiar with the way auditory signals are utilised by everyday office equipment and domestic appliances. Sound provides a particularly useful mechanism for feedback, indicating that an accepted input has been made to a device. Auditory icons are now used to signal particular events such as a key stroke, the selection of a file or the opening of a new application. Auditory events such as a whoosh to signal the sending of a message provide a medium that is intuitively simple as well as computationally less demanding than a visual equivalent. Like the visual icons traditionally used in computer interfaces, auditory icons are designed to map virtual events to the sounds associated with similar objects in the real-world (Brewster, 2002). The tendency of the auditory system to integrate separate sounds into a coherent percept (a series of tones is often perceived as a rhythmic pattern) also provides a basis for effective pattern recognition. Adaptation or 'habituation' means an auditory sequence that signifies the routine functioning of a particular system will quickly become background noise as long as it's presented at a comfortable volume. Sounds that are at odds with the established pattern provide a highly salient cue to changes in routine function and provide the user an easily detectable cue to potential problems .auditory displays having the following steps -- detection, discrimination, identification, recognition and comprehension.One issue with auditory signals is that they should be presented at a level that exceeds the ambient levels of sound (approximately 10 dB louder than background noise). Whilst an office is a fairly quiet environment, in many workplaces, machinery or noise from other people may make the detection of a particular signal more difficult. Effective auditory displays require signal-to-noise ratios that are appropriate to the context and environment in which they are used. Producing a startle reflex may be desirable for fire alarms but the same reaction to an in-car alarm might have disastrous consequences.

The temporal structure of sound also produces certain limits on its use in displays. Speech for example, is sequentially organised so the listener has to attend the whole message for accurate comprehension. Maintaining information that is no longer present in the display entails a working memory load that may be absent in visual displays. The spatial resolution of the auditory system is also less than that of the visual system, with objects tending to merge when they are not highly discriminable on non-spatial dimensions. Masking and the reflection of sound off different surfaces can also render the localization of auditory signals difficult. In open country, it is easy to discern the direction of an approaching emergency vehicle from its siren. In built up areas, however, where the sound will echo from buildings and other vehicles, discerning the direction from which an emergency vehicle is coming can be difficult. To counter this, modern sirens are often interspersed with a loud broad-band noise. Sounding like an old fashioned fog horn, these sounds include frequencies that are more easily localized than the high pitched wails or alternating tones traditionally used. Emergency sirens are also prone to masking, with individuals often listening to loud music within sound-proofed cars. Emergency vehicle drivers often complain about being ignored by people who clearly are not aware of their presence, even though they are sounding a very loud alarm.

Factors Influencing Visual Display Effectiveness

(i) **Physical location:** -The positioning of a visual display is of key importance in determining its effectiveness, the assumption can be made that the physical location of the display is optimized for the purpose for which it is required. If a user has to make frequent references to the current time, then the positioning of the clock within their normal field of view would be self-evident. For a user who only makes infrequent use of the current time a more peripheral location can be used.

(ii) **Display arrangements:** -In many advanced applications, it is likely that a number of displays will need to be grouped in close proximity to each other. This will be the case with motor car instruments, aircraft instruments etc. Similarly, items appearing on a computer screen for a set of tasks will need to all fall into the general field of view of the user. At the same time, they will need to be clearly distinguishable from each other.

(iii) **Lighting conditions:** -The extent to which the display is usable will depend upon the extent to which there is sufficient light falling on or emitted by the display. In the case of warning lights and computer displays, they have light sources within themselves and although adjustments may be made for 'brightness', there should be no problem over viewing them, even in a darkened room. Supplementary lighting may be required for other displays. There is usually background lighting for car instruments, which in turn can be adjusted for brightness, but for many displays, the natural or artificial light within the workspace is what will determine whether they can be viewed or not , Lighting conditions are a good example of the interactive nature of ergonomics. Optimizing a display for different conditions, for example day and night time usage, requires an understanding of the way the visual system responds to variability in light and the ability to utilize technology to maximize the user's ability to perceive and interpret the information portrayed.

(iv) **Static versus dynamic displays :-** Another obvious way in which displays can differ from each other is the extent to which they function purely as a static source, typically in the case of notices, signs, labels and instructions and the extent to which they represent dynamically changing features, typically called parameters. Analogue clocks display dynamic changes though it may be difficult to see the hands moving. Other displays associated with processes and changing states are much more dynamic in nature. Sometimes the changes are discrete as in the changing indication of how many miles/kilometers a car has travelled. Sometimes they are continuous in the way that the speedometer needle moves up and down in relation to changing speed. The complexity of a modern aircraft cockpit reflects the number of parameters that are being simultaneously measured and displayed for the pilot's attention. Dynamic displays can provide real-time information to the user about multiple parameters. Poorly designed dynamic stimuli in contrast, run the risk of confusing the user by overloading their perceptual and attentionresources.

6.6. Guiding Principles

There are some importantguiding principles which is generally used for display design , the brief discussion of followings are given below :-

(i) **Make display legible/audible :-** The contrast of the display, the visual angle (location), illumination, noise, masking, etc the salience of the display should be clearly visible by drivers.

(ii) **Avoid absolute judgment limits :-** Instead of showing gradually changing colors, it is better to show discrete changing of colors, e.g. divided into 5 distinct levels, so it's easily distinguishable.

(iii) **Top-down processing :-** People perceive & interpret signals according to what they expect on the basis of their past experience, e.g. if a series of buttons have "pressed=on", then the last button can't be "pressed=off".

(iv) **Redundancy gain :-** A very good example is the traffic light, as the 3 states of information are coded into both positions (top middle bottom) and colors (red amber green).

(v) **Discriminability. Similarity causes confusion: use discriminable elements.** Using texts in displaying information while driving may run into this problem, e.g. "there's a traffic 3125 meters ahead" and "there's a traffic 2125 meters ahead" are too similar.

(vi) **Principle of pictorial realism:-** A good example is showing high and low speed on a vertical scale instead of horizontal scale, because top to bottom is more easily understood as high to low, compared to right to left.

(vii) **Principle of moving part :-** A good example is showing more lines in order to indicate more distance to the car ahead and less lines to indicate less distance, instead of the other way around.

(viii) **Minimizing information access cost :-** A driver should not spend too much time for finding the correct information source. For example, if the color changes of a display are the most important source of information, then it should be the most salient. Another example of making the lowest access cost is to have the display area quite small so that user doesn't have to find display elements in a too wide area.

(ix) **Proximity compatibility principle :-** This is something to do with mental model. For example, similarly functioning buttons are supposed to be closer together than other buttons, or maybe more similar colors or patterns. This principle can complete the principle 8 above of having close proximity or smaller area of searching for a display element.

(x) **Principle of multiple resources :-** This is something to do with multimodality. Two information sources visually and auditorily can more easily be perceived than using both visual channels. For example, a navigation map showing a city name and speech saying "you are 500 meters from your destination" are easily perceived compared to putting the "500 meters from your destination" as a text over the map.

(xi) **Replace memory with visual information: knowledge in the world** .This can be called the principle of ecological display. Showing something that directly resembles what's happening on the real world. The more similar to the real world, the more informative it is and user can more easily decode the information.

(xii) **Principle of predictive aiding :-** Predictive display would pro-actively showing something to users about possible future happening, instead of reactively respond to user's actions. This is especially important during driving, as it's time critical. For example, a navigation system can show a segmented or dotted line showing the direction of the car towards a place where a road disturbance is happening. This way, the driver can quickly understands and try to change the path to other roads.

(xiii) **Principle of consistency :-** This is somehow related to long-term memory information. Previous knowledge of a certain way of information display would be used again when users meet another information display of the same signal. Using a commonly known display element/symbol in order to convey different information, simply would not work.

6.7. Basic Characteristics of Display System

From the subject point of view, any kind of measuring display system must have following basic characteristics, which are given below.

(i) Resolution: -Resolution is defined as the smallest discernible or measurable detail in a visual presentation. For example, the resolution of a CRT display can be expressed by the maximum number of lines that can be displayed in a given space, as usually done with the resolution of photographic films. One can also describe the minimum spot size that a device can display at a given luminance (brightness). The smaller the minimum spot, the better the device. Thus, the number of dots of minimum size (picture elements—also known as pixels) per inch (dpi) represents the quality of the device, e.g., a 72 dpi device is inferior to a 200 dpi display.

(ii) Addressability: -Addressability describes the number of individual points in the field that the device is capable of specifying.

(iii) Flicker: -Flicker is the perception of brightness varying over time. The severity of flicker depends on various factors such as the characteristics of the phosphor, size and brightness of the flickering image, etc. Recent research shows that refresh rates up to 90 Hz may be needed to satisfy 99 per cent of users, while in earlier research, refresh rates well below 50 Hz were thought to be satisfactory. Some modern devices offer an adjustable refresh rate; unfortunately, higher refresh rates are coupled with lower resolution or addressability. The ability of a device to display high "resolution" images with high refresh rates can be assessed by its video bandwidth. For displays with high quality, the maximum video bandwidth lies above 150 MHz, while some displays offer less than 40 MHz.

(iv) Jitter: -Jitter is the result of spatial instability of the image; a given picture element is not displayed at the same location on the screen after each refresh process. The perception of jitter cannot be separated from the perception of flicker. Jitter may have its cause in the VDU itself, but it can also be induced by interaction with other equipment at the workplace, such as a printer or other VDUs or devices that generate magnetic fields.

(v) **Contrast:** -Brightness contrast, the ratio of the luminance of a given object to its surroundings, represents the most important photometric feature for readability and legibility. While most standards require a minimum ratio of 3:1 (bright characters on dark background) or 1:3 (dark characters on bright background), optimum contrast is actually about 10:1, and devices of good quality achieve higher values even in bright environments. The contrast of "active" displays is impaired when the ambient light is increased, while "passive" displays (e.g., LCDs) lose contrast in dark environments. Passive displays with background lighting may offer good visibility in all environments under which people may work.

(vi) **Sharpness:** - Sharpness of an image is a well-known, but still ill-defined feature. Hence, there is no agreed-upon method to measure sharpness as a relevant feature for legibility and readability.

(vii) **Legibility and readability:**-Readability refers to whether a text is understandable as a series of connected images, while legibility refers to the perception of single or grouped characters. Thus, good legibility is, in general, a precondition for readability. Legibility of text depends on several factors: some have been investigated thoroughly, while other relevant factors like character shapes are yet to be classified. One of the reasons for this is that the human eye represents a very powerful and robust instrument, and the measures used for performance and error rates often do not help to distinguish between different fonts

(viii) **Fonts and readability:** -A font is a family of characters, designed to yield either optimum readability on a given medium, e.g., paper, electronic display or projection display, or some desired aesthetic quality, or both. While the number of available fonts exceeds ten thousand, only a few fonts, numbered in tens, are believed to be "readable". Since legibility and readability of a font are also affected by the experience of the reader—some "legible" fonts are believed to have become so because of decades or even centuries of use without changing their shape—the same font may be less legible on a screen than on paper, merely because its characters look "new". This, however, is not the main reason for the poor legibility of screens.

(ix) Radiation: -Radiation is the emission or transfer of radiant energy. The emission of radiant energy in the form of light as the intended purpose for the use of VDUs may be accompanied by various unwanted by-products such as heat, sound, infrared and ultraviolet radiation, radio waves or x rays, to name a few. While some forms of radiation, like visible light, may affect humans in a positive way, some emissions of energy can have negative or even destructive biological effects, especially when the intensity is high and the duration of exposure is long. Thus it is very necessary that the intensity of radiation should have with in permissible limit.

(x) Radio-frequencyradiation :- Electromagnetic radiation with a frequency range from a few kHz to 109 Hertz (the so-called radiofrequency, or RF, band, with wavelengths ranging from some km to 30 cm) can be emitted by Visual Display; however, the total energy emitted depends on the characteristics of the circuitry however, the field strength of this type of radiation is likely to be small and confined to the immediate vicinity of the source.

(xi) Optical radiation :- Optical radiation covers visible radiation (i.e., light) with wavelengths from 380 nm (blue) to 780 nm (red), Infrared from 3x1011 Hz to 4x1014 Hz, wavelengths from 780 nm to 1 mm; ultraviolet from 8x1014 Hz to 3x1017 Hz).Visible radiation is emitted at 100 cd/m2 and ultraviolet radiation that must be below the occupational exposure standards.

(xii) Physical characteristics of the display device (e.g., resolution in terms of the total number of pixels, and the physical size of the display).

(xiii) Acuity of the human visual system (e.g., the limits of distinguishability of the human eye for different image features like color, orientation and size, and the visual angle subtended by elements on the viewer's eye).

(xiv) Visualization technique (e.g., the methods used to map a data element's values to a visual representation).

(xv) Properties of the data (e.g., its dimensionality and number of elements) and the analysis tasks to be performed by the viewer.

Question for Discussion

1. What is importance of warning signals; factorial and graphic display?
2. What is Anthropometry? Explain the use of human anthropometry in work place layout.
3. Compare and contrast between static and dynamic muscular load.
4. Discuss the application of anthropometric data in design.

5. What are the principle of good illumination and discuss the various colour codes used along with the purpose in any industries.
6. Discuss various laws and principles of appearance. Also discuss how quality and reliability is incorporated in the design?
7. How the designs of displays play an important role in the product design? Give examples to support your answer.
8. Define anthropometry and show its importance. Explain the principles important *for* the 'application of anthropometric data to the design problem.
9. What do you meant by anthropometry ? How anthropometry data base is useful, discuss its limitations.
10. What are the functional aspects of control; how they' are classified into different categories, also discuss the disadvantages of different controls.
11. Discuss various types of displays and the principles used in the design of visual displays?
12. Explain the various Characteristics for selection of Display systems in an industry .
13. Explain about Safety Signs and Colour code used at Work-place for precautions and safety .
14. Discuss about Quantitative display and Qualitative Display.

References

1. W. Jürgens Hans, "International data on anthropometry", Occupational Safety and Health Series, No. 65, Geneva, ILO, 1990.
2. R.M. Currie, "Work Study", 4th edn, Pitman, London, 1977.
3. B.W Niebel and A. Freivalds, "Methods, Standards, and Work Design", New York: WCB/McGraw-Hill, 1998.
4. B Das and R.M. Grady, "Industrial workplace layout design", An application of engineering anthropometry. Ergonomics, Vol. 26, Pp. 433-447, 1983a.
5. W. Rohmert, "Anthropometric elements of ergonomics", Bulletin de l'AISS, Pp. 25-35, 1975.
6. E. Nowak, "CIS 78-2068 Determination of the spatial reach area of the arms for workplace design purposes", Ergonomics, Vol. 21, No. 7, Pp. 493-507, 1978.
7. ISSA. N.d. W.G. 3E, "Checklist for a Classification of Machines On the Basis of Ergonomic Principles", International Section 'Machine Safety' On the Prevention of Occupational Accidents and Diseases. Mannheim. ISSA.
8. A. Grieco and M. Masali, "Setting up a method for the measurement of anthropometric parameters with a view to the ergonomic design of workplaces", Med Lavoro, Vol. 62, No. 11, Pp. 505-531, 1971.

9. T.M. Fraser, "The Worker At Work. London: Taylor & Francis", 1989.
10. Niosh, "The occupational safety and health office of communication workers of America: a survey", Washington, DC: NIOSH, 1991.
11. E.N. Corlett and R.P. Bishop, "A technique for assessing postural discomfort", Ergonomics, Vol. 19, Pp. 175-182, 1976.
12. S. Pheasant, "Body space, Anthropometry", Ergonomics and Design. London: Taylor & Francis, 1986.

CHAPTER 7

WORKING CONDITION

7.0. Working Condition

Any industrial organization at very initial stage try to attempting to improve methods of work in a factory but the second basic objective will be on prime consideration that is working condition under which the workers have to perform their tasks without strain thereby his / her performance does not affected the overall productivity of manufacturing organization and achieving the profit with higher amount , the major responsibility is always depend upon the working atmosphere . It was the experience of this mission that productivity could often be increased simply by improving the working conditions, before method study techniques were applied. for example -- if lighting is so bad that operatives have to strain their eyes to see what they are doing or if the atmosphere is so hot, humid and fume full will not be feasible to the operator performance .thus we can say that bad working conditions are uneconomic for any manufacturing organization so therefore small improvements can produce marked increases in productivity.

One of the most important statement that was given by largest work study departments of the world **"Do not use a spoon when a steam shovel is needed."** In other words, it is no use improving the layout of an individual workplace or the methods of the worker, using refined techniques and saving seconds on an operation, when hours are being lost through bad working conditions throughout the shop or factory. In terms of increased productivity the "steam shovel" of improved working conditions may be worth many "spoonfuls" of detailed method study.

Physical working conditions depend on many factors: the site of the building, the type of construction, layout, ventilation, temperature, lighting, sanitation, the nature of the floors and stairs, equipment installed, etc.

since the worker is a human being first and foremost, and a producer second, that his working efficiency is so much affected by the environment in which he has / she to work. Unhealthy and unhappy workers, doing their jobs under conditions of physical or mental strain, are inefficient producers. Following are very important terms that should always be on prime consideration regarding the working condition or environment surrounding which the operator create the job / performing the task .

7.1. Cleanliness

One of the obvious benefits of keeping your workplace clean is the reduction of germs and filth in the vicinity. This means healthier employees, which in turn means fewer sick days and higher energy levels. When people think of sanitation, the bathroom is likely the first place to come to mind.

Here are Six Reasons Why a Clean Work Place Also Means a Safe Work-Place

(i) Clean, dry floors to prevent slips and falls. Maintaining clean, dry floors is essential for the prevention of slips and falls in the workplace.

(ii) Disinfectants prevent the spread of germs and illness, including the flu.

(iii) Proper air filtration lowers employee exposure to hazardous substances. You may not see them, but dusts and vapors are hazardous substances that can create an unsafe environment for employees. Building ventilation is one important factor in reducing airborne transmission of respiratory infections and maintaining the health and productivity of workers.

(iv) Clean light fixtures improve lighting efficiency. Dirty light fixtures can reduce essential light levels, making it difficult and unsafe for employees to complete their daily tasks. Clean light fixtures significantly improve lighting efficiency in the workplace.

(v) Green cleaning products are safer for both your employees and the environment.

(vi) Proper disposal of waste and recyclable materials keeps work areas clutter-free.

Spitting is especially dangerous to health. Notices expressly forbidding spitting on the floor, walls, or stairs should be posted in all parts of the factory. Where it is necessary to provide spittoons they should be sufficient in number, hygienic and properly cleaned and disinfected at least once every working shift Particular care must be taken to rid workrooms and shops of rats, insects and other vermin, as these are amongst the worst carriers of disease. In districts where malaria is endemic all workplaces should, as far as possible, be screened against mosquitoes, thus we can say that it is an important factor which will be responsible for productivity improvement amongs others.

7.2. Availability of Drinking Water

Adequate supplies of clean and cooled drinking water from a reliable and regularly tested source should be made freely available at convenient locations close to all workplaces. This is particularly important in areas where the quality of the water from piped supplies is suspect and on open worksites where workers may otherwise be tempted to refresh themselves from pools and streams liable to pollution. In such situations it is imperative to enforce strict discipline to ensure that only designated areas are used for latrines and sanitary purposes. During intensive studies on manual earth-moving operations carried out on a large dam site in India it was found that most of the workers, including those enjoying the best diets and feeding communally, were infested with intestinal parasites and had been apparently so infested for a considerable time. 2 When such conditions are present or suspected it will not be enough merely to provide clean drinking water: employers should consider it part of their responsibilities to make available suitable medicaments to combat the infestations.

Hygiene: - Protection against health hazards on the job Potential workplace health hazards include air contaminants and chemical, physical, biological, and ergonomic hazards.

(i) Air contaminants can include dusts, fumes, mists, aerosols, fibers, and vapors. When there are contaminants in large enough quantities in the air you breathe, they can make you sick. That's why we use ventilation and, in some situations, respirators to eliminate air contaminants or at least to reduce them to safe levels.

(ii) Chemical hazards can also make you sick. How sick you get depends on the amount of exposure, the length of exposure, and the toxicity of the chemical. We work hard to eliminate chemical hazards. When they can't be eliminated, we use safety procedures and personal protective equipment (PPE) to protect you.

(iii) Physical hazards include excessive levels of noise, vibration, illumination, and temperature. You can do your part to minimize these hazards by using available protective devices, following safety procedures, and always wearing PPE (personal protective equipment) .

(iv) Biological hazards include bacteria, viruses, fungi, and other living organisms that can cause infections by entering the body either directly or through breaks in the skin. Good housekeeping to maintain a clean and sanitary workplace can help reduce these hazards. So can frequent hand washing.

(v) Ergonomic hazards in the workplace, such as repetitive motion, awkward working positions, heavy lifting, and prolonged periods of sitting or standing, can cause health problems and can be controlled by redesigning jobs, processes, tools and workstations. When this approach alone doesn't solve the problem, we add safe work procedures to help protect your health.

7.3. House Keeping

A good housekeeping program plans and manages the orderly storage and movement of materials from point of entry to exit. It includes a material flow plan to ensure minimal handling. The plan also ensures that work areas are not used as storage areas by having workers move materials to and from work areas as needed. Part of the plan could include investing in extra bins and more frequent disposal. The costs of this investment could be offset by the elimination of repeated handling of the same material and more effective use of the workers' time. Often, ineffective or insufficient storage planning results in materials being handled and stored in hazardous ways. Knowing the plant layout and the movement of materials throughout the workplace can help plan work procedures. Worker training is an essential part of any good housekeeping program. Workers need to know how to work safely with the products they use. They also need to know how to protect other workers such as by posting signs. Housekeeping, when used in reference to a factory or workroom, is a general term embracing cleanliness, tidiness and general state of repair. Tidiness in the workshop is an aid to productivity as well as a means of reducing accidents. Poor housekeeping is responsible for accident in the work-place, some important point regarding the housekeeping must be remembered:-

(i) Tripping over loose objects on floors, stairs and platforms.
(ii) Being hit by falling objects.
(iii) Slipping on greasy, wet or dirty surfaces.
(iv) Striking against projecting, poorly stacked items or misplaced material.
(v) Cutting, puncturing, or tearing the skin of hands or other parts of the body on projecting nails, wire or steel strapping.
(vi) To avoid these hazards, a workplace must "maintain" order throughout a workday. Although this effort requires a great deal of management and planning, the benefits are many.

Benefits of Good House-Keeping

(i) Improved productivity (tools and materials will be easy to find).
(ii) Reduced property damage by improving preventive maintenance.
(iii) Reduced handling to ease the flow of materials.
(iv) Better control of tools and materials, including inventory and supplies.
(v) Fewer tripping and slipping accidents in clutter-free and spill-free work areas
(vi) Decreased fire hazards.
(vii) Lower worker exposures to hazardous substances (e.g. dusts, vapours).
(viii) More efficient equipment cleanup and maintenance.
(ix) Better hygienic conditions leading to improved health.

(x) More effective use of space.

(xi) Less janitorial work.

(xii) Improved morale.

Dust and Dirt Removal: - In some jobs, enclosures and exhaust ventilation systems may fail to collect dust, dirt and chips adequately. Vacuum cleaners are suitable for removing light dust and dirt. Industrial models have special fittings for cleaning walls, ceilings, ledges, machinery, and other hard-to-reach places where dust and dirt may accumulate.

Employee Facilities: - Employee facilities need to be adequate, clean and well maintained. Lockers are necessary for storing employees' personal belongings. Washroom facilities require cleaning once or more each shift. They also need to have a good supply of soap, towels plus disinfectants, if needed. If workers are using hazardous materials, employee facilities should provide special precautions such as showers, washing facilities and change rooms, Smoking, eating or drinking in the work area should be prohibited where toxic materials are handled.

Floors: - Poor floor conditions are a leading cause of accidents so cleaning up spilled oil and other liquids at once is important. Allowing chips, shavings and dust to accumulate can also cause accidents. Trapping chips, shavings and dust before they reach the floor or cleaning them up regularly can prevent their accumulation.

Walls: - Light-coloured walls reflect light while dirty or dark-coloured walls absorb light. Contrasting colours warn of physical hazards and mark obstructions such as pillars , Paint can highlight railings .

Aisles and Stairways: - Keeping aisles and stairways clean & clear is an important requirement because Aisle space allows for the movement of people, products and materials.

Tools and Equipment: - Tool housekeeping is very important, whether in the tool room, on the rack, in the yard, or on the bench. Tools require suitable fixtures with marked locations to provide orderly arrangement, both in the tool room and near the work bench. Returning them promptly after use reduces the chance of being misplaced or lost.

Maintenance: - The maintenance of buildings and equipment may be the most important element of good housekeeping. Maintenance involves keeping buildings, equipment and machinery in safe, efficient working order and in good repair.

Waste Disposal :- Placing scrap containers near where the waste is produced encourages orderly waste disposal and makes collection easier. All waste receptacles should be clearly labelled (e.g., recyclable glass, plastic, scrap metal, etc.).

Storage: - Good organization of stored materials is essential for overcoming material storage problems whether on a temporary or permanent basis , stored materials should not obstruct aisles, stairs, exits, fire equipment . Flammable, combustible, toxic and other hazardous materials should be stored in approved containers in designated areas that are appropriate for the different hazards that they pose.

7.4. Lighting Facility in Work-Place

The quality of lighting in a workplace can have a significant effect on productivity. With adequate lighting workers can produce more products with fewer mistakes, which can lead to a 10-50 % increase in productivity. Good lighting can decrease errors by 30-60 % as well asdecrease eye-strain and the headaches, nausea, and neck pain which often accompany eyestrain. Adequate lighting allows workers to concentrate better on their work which increases productivity.

Good Lighting in the Workplace Promotes

(i) A reduced risk of occupational accidents and health problems.

(ii) Better concentration and accuracy in work.

(iii) A brighter, cleaner workplace resulting in a more active, cheerful environment.

(iv) Improved work performance.

(v) Better visibility, improved accuracy and increased work speed enhancing production.

It can be easy to improve lighting without increasing the number of light bulbs, light fixtures or the electric bill. Improved lighting can be achieved by using more daylight, by changing the position of light sources or workstation layouts and by effectively using reflected light

Eliminate Shadows and Glare: -Shadows and direct or indirect glare reduce lighting, make it difficult to work and decrease productivity. Glare is usually caused by light from lamps, skylights and windows. Here are some low-cost suggestions to eliminate shadows and glare:

(i) Change the position of light sources.

(ii) Change the position of the workstation.

(iii) Use multiple light sources by mixing direct and reflected light.

(iv) Use lamp shades which reflect light upwards, since reflected light from ceilings provides the best visibility.

(v) Provide lamp shades with white inside and black inside-edges.

(vi) Use lamps with a movable stand or a flexible arm to easily change the direction of light

(vii) Use a magnetic lamp stand with a clip to easily change the position of the lamp.

(viii)Hang lamps high and use light sources with a relatively large surface area such as fluorescent lamps to obtain even general lighting with less glare.

(ix) Construct skylights and windows on the non-sunny side to obtain an evenly lit working area.

(x) Provide blinds, curtains, louvers, trees and vines to shade the building.

(xi) Use matte paint or darker colours for all surfaces (for example, tables, machines, tools).

(xii) Install screens, covers or partitions for shielding strong light which produces glare.

(xiii)Combine daylight from windows and skylights with ceiling and local lights to reduce sharp shadows and glares and to achieve optimum lighting conditions.

Maintenance of Lighting :-Periodic maintenance is important for productivity and cost-savings. Lamps give off less brightness when they are dirty but still consume the same amount of electricity. If lamps, skylights, windows, ceilings, walls and work areas are maintained in good condition, lighting can be improved without increasing the number of fixtures or light bulbs. In a dusty workplace, light coming from windows is reduced 30-40% after three months and 45-55% after 6 months. Therefore it is important to:

(i) Clean all the essential elements in workplace at regular intervals.

(ii) Replace light bulbs and tubes which have blown or are reduced in brightness.

(iii) Place light bulbs which have reduced in brightness at another workstation where less light is required.

(iv) Use open top type lamp shades to prevent dust from collecting on light bulbs and tubes.

7.5. Ventilation, Heating and Cooling

General ventilation is required for the health and comfort of the workers, and is thus a factor in their efficiency. Most of the factories and mills which they entered little had been done to mitigate the effects of heat by proper ventilation, and as a result of this a great deal of time was lost by workers having to gooutside in order to recover from the "unbearable working conditions". Excessively high or excessively low temperatures and inadequate ventilation reduce productivity through sickness, discomfort and lowered vitality of the workers. The main cause of discomfort are due to the greater concentration of carbon dioxide (CO2) in the air but the reduced rate of heat loss from the body to the surroundings , Ventilation may be natural or artificial or a combination of the two. Air conditioning is employed chiefly to combat extremes of temperature , All dust, fumes, gases, vapours or mists generated and released in industrial processes should as far as possible be removed at their points of origin and not be permitted to spread throughout the atmosphere of the workrooms .

The temperature and humidity of the air should be suitable for the work being performed;one of the most important instruments for daily use in factories where excessive heat and humidity are likely to occur is the wet kata thermometer, which gives a direct reading of the "cooling power" of the environment. Every work study man should understand its use.

The efficiency of natural ventilation depends largely on outside conditions, which often vary widely. It is often least efficient when ventilation is needed most. It is also difficult to control. Where natural ventilation is not enough, mechanical ventilation by means of fans must be employed. This may be an exhaust system (extraction of air from the shop); an induction system (blowing in of air), using plenum ducting to conduct the air to where it is wanted, or a combination of both .

Color Coding System for Shop Safety

(i) **Red** represents danger and identifies dangerous areas on machines; Fire equipment, safety switches, and other types of emergency equipment are red.

(ii) **Orange** signifies warning. The safety color code assigns orange to identify machine hazards, such as edges and openings. Electrical controls, switches, and levers have orange backgrounds.

(iii) **Yellow** reminds people to use caution. The yellow signal in a traffic light means caution; this association should be extended to machines in a shop. Levers and adjustment knobs are yellow.

(iv) **Blue** signifies information. Signs informing the user that a machine is “out of order” are blue. A blue sign on a machine informs the operator of something that requires extra attention.

(v) **Safety green** is a special shade that means safety. Safety equipment is identified by safety green. Safety green is also used for first aid, safety areas, and areas where medical treatment is given.

(vi) **Gray** is used on work area floors. It is relaxing and contrasts with other colors. Another use for gray is to color tabletops and the bodies of machines. When gray is used as a background, the contrast with other colors allows good visibility and easy color recognition.There are also a three colour whose function is to provide contrast for the safety colors. An added benefit of using focal colors is the pleasing and attractive appearance they create in a work area. **Ivory** is used to improve visibility because of the contrast between ivory and the other shopcolors. **Vista green** is sometimes used as the body color on machines instead of gray. Man stationary tools are painted vista green. It is one of the cool colors and considered a good colorfor work environments. **Aluminum** is used on waste containers.

7.6. Noise

The noise pollution is defined as the unwanted sound which is released into the environment. It disturbs the human being and cause an adverse effect on the mental and psychological well being. It is measured in the units of decibels and is denoted by the dB. The noise which is more than 115 dB is tolerant. The industrial limit of sound in the industries must be 75 dB according to the world health organization.

Noise Generators: - Following factors are responsible for noise generation in mechanical – manufacturing industries. Aerodynamic factors – unpleasant sound is generated by fans ,flow induced ,pneumatics ,nozzles ,exhausts ,combustion and Mechanical factors – unpleasant sound is generated by impacts presses, mechanical handling , rotating machines ,gears pumps , motors bearings , electrical forces , friction forces , cutting tools, brakes .

Noise Control Steps: - if we want to control the noise inside the manufacturing system, we have to control the following steps, these are

1. List all the potential noise sources on each piece of noisy plant.
2. Rank the sources.
3. Assess all the noise control options for the dominant source.
4. potential reduction in noise from this source.
5. operational, productivity, hygiene constraints.
6. operator acceptance and cost.
7. If engineering control is not practical for the dominant source, then you have proved that screening / enclosure etc are the only options.

The results are used to generate cost v noise reduction trade-offs for each item of noisy plant and to plan the most practical and cost effective noise control programme possible across the company .There are many others methods which help to control the noise pollution. The source of noise must be reduced. The path of transmission of sound must be stopped and the receiver of noise must be safe guarded. The amount of traffic must be reduced near the residential homes, educational institutes and hospitals. The machinery must be redesigned and the vehicles must be properly maintained. The acoustical furnishing must be done so that the sound can be absorbed. The voice must not be passed directly to the receiver and must be diverted upwards instead of downwards as in case of ear splitting jet exhaust noise. The industries must be built away from the residential areas and the legal laws must be established and observed to protect the humans from noise pollution. A general awareness programme to educate the people must be done. It is basically a public nuisance which affects the human beings in almost all aspects.The maximum tolerable intensity may be taken to be in the region of 90 decibels, but lower intensities may be objectionable at very high frequencies. Noise can be measured by instruments called audiometers, of which different types exist.

7.7. Working Space and Seating

It is obvious that no worker can be efficient unless he has sufficient space in which to work, to lay down his tools and materials, and to move without being hindered by fellow workers, other machines or stacks of material.

Although in certain conditions the health of the worker may be affected by overcrowding at the workplace, in general the question is more one of efficiency and will be dealt with more fully in the chapters on the application of method study. Prolonged standing at work is one of the most common sources of avoidable discomfort or fatigue, and seats should be provided to enable workers men and women to sit at their work whenever practicable or, where work cannot be done sitting, to rest at intervals. It is generally realized that the prevention of unnecessary fatigue promotes efficiency. Progressive firms are usually alive to the advantages of good seating but many employers do not pay enough attention to the subject. Workers are often made to stand continuously, either because of the mistaken idea that they will thus work harder and better, or simply because it has always been the tradition to do so.

In many parts of the world, including India, the natural position is squatting, and many workers prefer this position to sitting. Indeed, even when provided with seats some workers, especially the older ones, may be found squatting on them. Note must be taken of such traditional or local, preferences, and where squatting is regarded as more comfortable the workplace must, be studied so as to make it as convenient as possible for that position.

Designing a workplace that provides opportunities for the broadest potential workforce makes good business sense. This allows employers to select the most qualified people from the largest possible applicant pool. It may also improve work efficiency, employee productivity, workplace safety and the quality of work.

Sitting is generally less strenuous than standing, and allows easier control over hand and body movements. It is not surprising that the sitting workstation is the most common.

1. Use sitting workstations to facilitate hand movement accuracy when smaller hand manipulation forces are required and when less area needs to be covered by the employee.
2. Seated workstation table heights should be approximately the elbow height of the seated individual and therefore should be adjustable. Seated workstations should allow enough clearance for users of wheeled mobility devices.

7.8. Accidental Prevention at Work-Place

In any manufacturing industries , accidental prevention is always on prime consideration because , if any kind of accident occurs then industries suffers with heavy losses so therefore it is very necessary to control all those factors which are responsible for accidents . A workshop is a place where you learn to use tools and machines to make things; It can be a dangerous place, so you must learn the safety rules for the workshop. The safety rules tell you how to dress appropriately and how to behave whilst working with tools that may cause harm. You must never play in the workshop, run around or throw equipment to one another.

Personal safety: - The basic dress rules that you should always follow are:

1. **Proper clothing: -** You must not wear loose clothes that can be caught in moving machinery. You must wear tight fitting.
2. **Proper eye protection: -** You must always wear goggles to protect your eyes while you are working in the workshop.

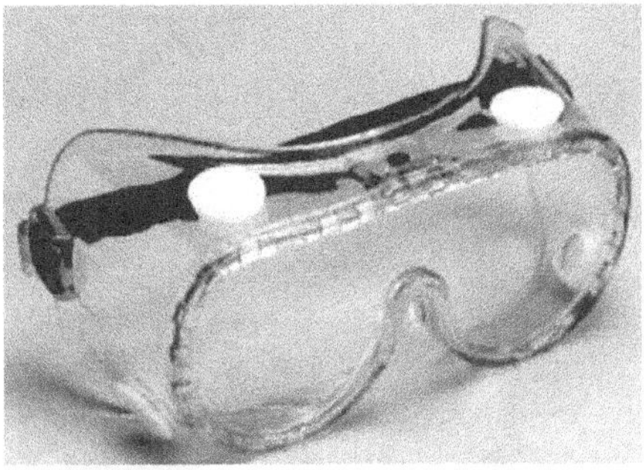

Goggles for Eye Protection

(i) **Remove tie and jewelry: -** Ties, watches, rings and other jewelers increase the chance of getting caught by moving machinery. You must remove them before entering the workshop.

(ii) **Cut or secure long hair:** - Long hair is also dangerous as it may be caught by the machine and pulls you into it.

(iii) **Proper shoes:** - You must not wear sandals or soft shoe inside the workshop as they will not protect your feet from falling objects. A safety shoes (steel-toe shoes) will protect your feet if you accidentally drop something. The way you dress in the workshop is very important for your safety, always be sure to wear properly and encourage your friends to do the same.

(iv) **Use of tools:** - You will learn to use many tools and machines in the workshop. Always ask for your teacher's advice before using a machine for the first time. You must listen to the teacher when he tells you how to use them.

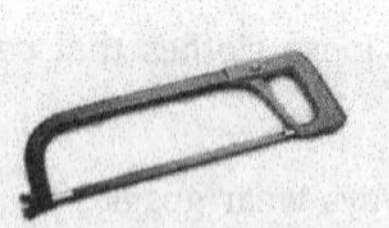

Hacksaw

Hand File

Vernier caliper

Hammer

Drilling machine

Bench grinder

Workshop cleanliness: - It is very important to keep the workshop clean and tidy. This will help save time and prevent accidents. When you have finished using tools, they must be put away and kept in the proper place. Do not leave tools lying around the benches or workshop; accidents may occur.

7.9. Fire Prevention

Accidental fires are likely to be very frequent in hot and dry countries, especially in certain industries. The prevention of fires is primarily a matter of the proper training of all concerned and the strict enforcement of fire prevention regulations, such as rules against smoking in places where it would be hazardous. Prevention is always better than cure, but the provision of adequate extinguishers and other appliances maintained in good condition is essential. It is also important that all members of the management and supervisors should be instructed on exactly what they are to do should a fire occur and that workers should also know what to do and the exits by which they should leave their shops.

Fire precautions :- Fire is a great danger in a workshop and care must be taken not to start a fire.

The Main Causes of Fire Are

(i) Careless storage of flammable materials.

(ii) Careless electrical maintenance.

(iii) People throwing away cigarettes.

Classification of Fire Classes

Class of Extinguisher	Class Symbols	Materials to Use On
Class A		Ordinary Combustibles: paper, wood, cloth, rubber, many plastic.
Class B		Flammable Liquids: oil, grease, petrol, some paints, solvents, etc.
Class C		Electrical: wiring, DB boards. Electrical equipment.
Class D		Combustible Metals: magnesium, sodium.

Diagram 7.1: Clour Code for Different Material

Fire Colour Code for Various Substances Are Given Below

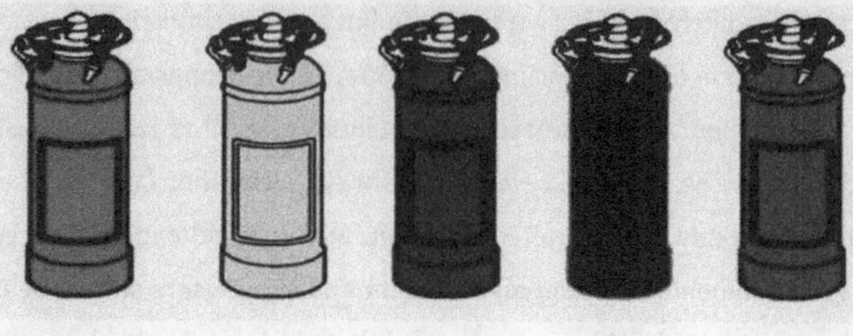

Diagram 7.2: Fire Code for Liquid

Following important points are must be consideredthroughout the work place for preventing accidents

(i) **Avoid Shortcuts :** -Many times accidents happen not because the safety precautions are unknown but because people take short cuts when they're familiar with their job or are in a hurry.

(ii) **Always be alert on the job** – Being awake and alert all the time while at work will not only prevent accidents from happening at work. It will also enhance the performance of the worker and can even earn him a promotion or a salary increase. Most of the people who become involved with accidents at work are those who feel sleepy while working.

(iii) **Wear the required uniform** – A person who works in a factory has a greater chance of being involved in an accident at work. Thus, he should be more vigilant about the wearing of proper uniforms and other protective garments when working

(iv) **Listen and actively participate during emergency drills** – Some work places conduct emergency drills to make sure their employees know what to do in cases of emergencies so as to avoid accidents.

(v) **Always asks your supervisor about the possible risks of doing a certain task** – There are workers who are exposed to danger just by doing their regular workloads. A worker should always check with his supervisor if he is unsure about the task at hand or if he feels that doing that task will expose him to great danger.

(vi) **Never take a high-risk job for which you have not been trained for** – It is stupid for anyone to take on a high-risk job especially if he has not been trained for the job. Imagine an untrained person doing the job of a fireman? Doing this will not only expose you to a great danger but will expose other people to danger as well.

(vii) Always follow the safety program of your workplace - Employers are responsible for the well-being of their workers while at work so they have installed safety programs which the employees should follow. Employees can avoid an accident at work if they follow the safety program mandated by the employer.

(viii) Always be on the lookout for possible cause of accidents and then report it to the management - There are workers who are already aware of a disaster waiting to happen but they refuse to see the warning and go on with their work. If you see telltale signs of a possible cause of accident at work like a ceiling that is about to give way or a stairway that's about to give way then you should report it to the proper authority immediately. Do not wait for the accident to happen as it might happen to you.

(ix) Owners of a company should always post signs or rules that should be observed by their employees in order to avoid a possible accident at work. These posters should be posted in areas that are highly visible to the workers so they will be able to read them.

(x) Formation of an emergency team among the workers - The team will be responsible for monitoring possible hazards at work that may cause an accident. They will also formulate rules and regulation to be observed in case of possible incidents that may cause harm to the workers.

(xi) Never risk the health and safety of the employees - Profit may be the main motivation of a company but they should earn their keep without exposing their workers to hazard. Company owners should always prioritize the health and safety of their workers no matter what.

Thus the prevention of accidents must be tackled first by eliminating possible causes, technical and human. The means of doing so are too varied to be dealt with here. They include compliance with technical regulations and standards, a high level of supervision and maintenance, the fostering of good industrial relations, care for the workers' health and welfare, training all members of the works to observe safety regulations and safe practices, posting warning notices and using distinctive colours to draw attention to objects likely to prove dangerous.

Question for Discussion

1. Discuss the various effects of heat on body and work performance.
2. What is meant by the term noise ? Discuss the physiological effects of noise .
3. Discuss the selection criteria of ventilation in an industries ? described about natural & mechanical ventilation .

4. How noise and heat affect the efficiency and performance of the operator ? what essential steps need to be undertaken for safeguard the health of the workers in an industries
5. Differentiate between natural ventilation & mechanical ventilation as related to the industrial safety .
6. Discuss the factors on which vibration depends, types of vibrations, their effects and methods of reducing those vibrations.
7. Write short note on the followings:
   a) Effect of heat on body and work performance.
   b) History, development and scope of ergonomics.
8. Discuss the heat exchange process of human body with the, environment?
9. What remedial measures you Will suggest to provide the thermal comfort in the working areas?
10. How do the high heat and high cold affect the human performance?
11. How does the human body respond to 'low frequency vibrations' and high frequency vibrations'? Discuss the effects of these vibrations on human performances.
12. Briefly discuss the methods of reducing vibrations.
13. Describe the effects of the following on the productivity.
    (i) Plant layout.
    (ii) Working conditions.
14. What is meant by the term noise ? Discuss the physiological effects of noise .
15. Define the term ergonomics and described about stress and fatigue ?
16. How do high heat and high cold affect the human performance? Discuss the heat exchange process of human body with the environment:
17. How does human body respond to vibration? Discuss the methods of protection against vibration.

References

1. "Protection of workers against noise and vibration in the working environment", (Geneva,3rd impression (with modifications), 1984.
2. "Management of working time in industrialised countries", (Geneva), 1978.
3. "Encyclopaedia of occupational health and safety", 2 vols. (Geneva, 3rd éd.,), 1983.
4. International Labour Office (ILO). Accident prevention: A workers' education manual (Geneva,2nd éd.,), 1983.
5. "Automation, work organization and occupational stress", (Geneva), 1984.
6. A.A. Evans, "Hours of work in industrialised countries", (Geneva, ILO), 1975.

7. J.E. Veney and A.D. Kaluzny, "Evaluation and Decision Making for Health Services Programs", Prentice Hall, Englewood Cliffs, New Jersey, 1984.
8. A. Pallin and R.P. Kittell, "Mercy hospital: Simulation techniques for ER processes", Industrial Engineering, Vol. 24, No. 2, Pp. 35–37, 1992.
9. J.M. Clerc, "Introduction to working conditions and environment", (Geneva, ILO), 1985.
10. D. Gold, "Safely in the use of chemicals at work", (Geneva, ILO, forthcoming).
11. R.N. Kramer and J.V. Douglas, "Planning for performance", Performance Improvement Through Information Management, M.J. Ball & J.V. Douglas (eds), Springer-Verlag, New York, 1999.
12. S. Konz,. Hand tools. In Work Design Columbus, Ohio: Grid. 1979
13. "International Labour Organization (ILO)", Protection of Workers against Noise and Vibration in the Working Environment, an ILO Code of Practice. Geneva: ILO, 1984.
14. H.M. Parsons, "Human factors/ergonomics in office design and automation", Proceedings: 1982 Meeting of the American Psychological Association, 1992.
15. A. Mehrabian and J. Russell, "An approach to environmental psychology", Cambridge, MA: MIT Press, 1974.
16. "Environmental psychology man and his physical setting", NewYork: Holt, Rinehart & Winston.
17. "In Fundamentals of Industrial Hygiene", edited by BA Ploog. Chicago: National Safety Council.
18. M. Ahonen, M. Launis and T. Kuorinka, "Ergonomic Workplace Analysis", Helsinki: Finnish Institute of Occupational Health, 1989.

CHAPTER 8

PLANT LAYOUT

8.0. Introduction

The concept of Plant layout is nothing but it is a arrangement of internal facilities that is required in manufacturing process .such as machinery, equipment, tools, furniture etc. In order to achieving easy & quick flow of raw material , finished product , semi finished product sometimes flow of smallest devices , equipment in association with the handling of the material from initial stage to finished product at lowest possible and examination aswellas full utilization of space positively .

Factors that influence the plant layout: - following are some important factors upon which plant layout is dependent.

(i) Location in notion of marketing area how it has far away from the plant layout, means the plant is located near the customers/markets; the cost of transportation is low.

(ii) This reduces the product cost.

(iii) Transportation facilities is another's factor upon which plant layout is dependents & considering the point regarding the various type of transportation facility like - road,rail, etc according to the situation of the plant.

(iv) Availability of power or electricity, that is required for running / operating the plant.

(v) Availability of labor is on prime consideration for that particular location, the labour may be a skilled or semi-skilled.

(vi) Availability of suitable land and how much land is needed, and it should not be close to the public area having a low-priced land.

(vii) Environmental impact and disposal facility of waste material.

(viii)Support of local community.

(ix) Climate condition is always considering for example -temperature,humidity, atmospheric pressure etc.

(x) Facility of raw material supply.

(xi) Political supports thats included Relaxed taxation policies, excise duty exemption and various other promotional efforts attract industrial activities in a region

(xii) Basic Amenities - The location site should have certain basic facilities like sewage system, piped water supply, security, and also availability of housing facilities, schools, colleges, banks, post office, hospitals, etc. are added advantages for locating a facility in an area.

(xiii) Construction costs of a plant may be low at a particular place due to cheap labour available there. The construction material may also be cheaper at another place. Such places are preferred for locating a plant.

8.1. Objective of Plant Layout

Plant layout is the method to plan and arrange materials and facilities so that a steady flow of production is ensured at minimum cost. A good plant layout always results in comfort and satisfaction of workmen and this automatically increases the production. A bad plant layout leads to accidents and unnecessary problems.

A good plant layout is designed to achieve the following objectives.

(i) Economic handling of materials and finished goods.

(ii) Fast and efficient quality production.

(iii) Enhanced utilization of available space.

(iv) Flexibility in change of plant design and possibility of expansion at a later date.

(v) Improvement in work condition leading to higher productivity.

(vi) Unidirectional/systematic flow of production operation.

(vii) Reduction in waiting time.

(viii) Reduction in manufacturing cost.

8.2. Advantage of Good Plant Layout

A good plant layout results in better production and lower costs. The advantages of a good plant layout are as follows:

(i) Well-organized work-space: - A good plant layout means well-organized workspace with adequate facilities provided for the machines as well as for the workmen. Proper arrangement of machineries and tools eliminates congestion.

(ii) Ergonomics aspects: - A good plant layout results in labour satisfaction due to improved and clean working conditions. A good plant layout ensures that the machine is properly placed, with adequate space in between so that there is no congestion and no danger of the workmen getting injured. This provides safety to the workmen and creates a good environment for work.

(iii) Minimization of material handling costs: - A good plant layout minimizes material handling costs. The machinery and equipment are placed in such a manner that there is no difficulty in transferring materials between workstations.

(iv) Minimization in damage and spoilage of material -: In a good plant layout, materials are handled properly which results in good quality of production. There is minimum damage and spoilage of materials. Minimizing waste also leads to increase in profits for a company

(v) Flexibility in changing production conditions: - The advantage of flexibility is that in future if the market conditions change, the firm can easily put up new machinery, etc. without having to dismantle the existing ones and with minimum hindrance to the daily schedule or work.

8.3. Classification of Layout

There are four types of lay-out generally used in production system.

- Layout by fixed position.
- Layout by process or function.
- Layout by product or line layout.
- Group layout.

(i) Layout by fixed position :- Such kind of arrangement is mostly used when the material to be processed does not passes through the various workstation plant but all the necessary equipment and machinery are brought to it for operation . This is the case when the product is bulky and heavy. For example – shipbuilding, aircraft construction, and the manufacture of diesel engines or large motors.

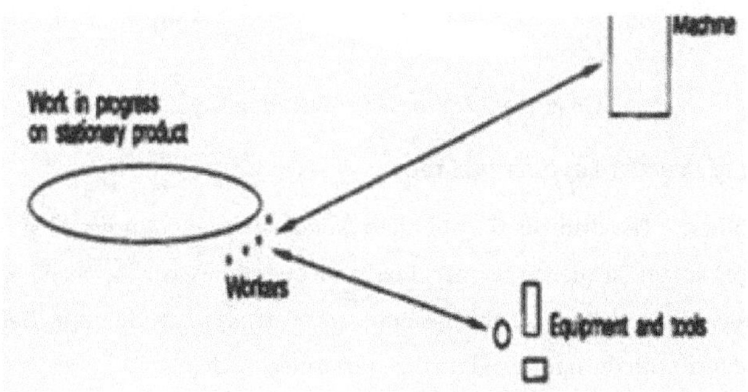

Diagram 8.1: Layout by Fixed Position

Advantages of Fixed Position Layout

- The investment on layout is very small.
- The layout is flexible as change in job design and operation sequence can be easily incorporated.
- Adjustments can be made to meet shortage of materials or absence of workers by changing the sequence of operations.

Disadvantages of Fixed Position Layout

- As the production period being very long so the capital investment is very high.
- Very large space is required for storage of material and equipment near the product.
- As several operations are often carried out simultaneously so there is possibility of confusion and conflicts among different workgroups.

(ii) Layout by process or function: -Process layout or function layout is one in which all the equipment/machineries performing similar tasks are grouped together is called the process layout or functional layout,dividingthe whole work place into small units helps in faster production and better utilization of the work place. The process layout can give a higher variety of products .for example textile spinning and weaving, maintenance of workshops , pressing machines , knitting machines , Grouped of grinding machines to form one department , Grouped of milling machines to form one department .

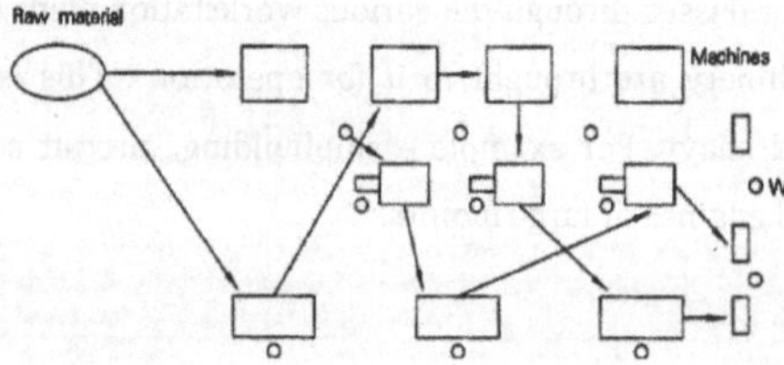

Diagram 8.2: Process or Function Layout

Advantages of Process Layouts Include

(i) Flexibility: - The firm has the ability to handle a variety of processing requirements. Changes in the sequence of machines and operations can be made without much difficulty. This is because the machines are arranged in different departments in accordance with the nature of function performed by them.

(ii) Cost: - Sometimes, the general-purpose equipment utilized may be less costly to purchase and less costly and easier to maintain than specialized equipment.

(iii) **Motivation:** - Employees in this type of layout will probably be able to perform a variety of tasks on multiple machines, as opposed to the boredom of performing a repetitive task on an assembly line. A process layout also allows the employer to use some type of individual incentive system.

(iv) **System protection:** - Since there are multiple machines available, process layouts are not particularly vulnerable to equipment failures.

(v) **Maximum utilization of machines:**-This method ensures fuller and effective utilization of machines and consequently investment in equipment and machines becomes economical.

(vi) **Greater flexibility:** -Changes in the sequence of machines and operations can be made without much difficulty. This is because the machines are arranged in different departments in accordance with the nature of functions performed by them.

(vii) **Scope for expansion:**- Production can be increased by installing additional machines without much difficulty.

(viii) **Specialization:** - As has already been pointed out that under this method, specialized machines are used for performing different production operations. This leads to specialization.

(ix) **Effective utilization of workers:** -Specialized workers are appointed to carry different type of work in different departments. This leads to effective and efficient use of their talent and capabilities.

(x) **More effective supervision:** -As the machines are arranged on the basis of functions, performed by them, the specialized and effective supervision is ensured by the specialized knowledge of supervisors. Each supervisor can perform his task of supervision effectively as he has to supervise limited number machines operating in his department.

(xi) **Lesser work stoppages:**-Unlike the product method, if a machine fails, it does not lead to complete work stoppage and production schedules are not seriously affected. Due to breakdown in one machine, the work can be easily transferred to the other machines.

Disadvantages of Process Layouts Include

(i) **Utilization:** - Equipment utilization rates in process layout are frequently very low, because machine usage is dependent upon a variety of output requirements.

(ii) **Cost:** - If batch processing is used, in-process inventory costs could be high. Lower volume means higher per-unit costs. More specialized attention is necessary for both products and customers. Setups are more frequent, hence higher setup costs. Material handling is slower and more inefficient. The span of supervision is small due to job complexities (routing, setups, etc.), so supervisory costs are higher. Additionally, in this type of layout accounting, inventory control, and purchasing usually are highly involved.

(iii) **Confusion:** - Constantly changing schedules and routings make juggling process requirements more difficult.

(iv) **Coverage of more floor area :-** Under this method, more floor space is needed for the same quantum of work as compared to product layout.

(v) **Higher cost of material handling :-** Material moves from one department to another under this method, leading to the higher cost of material handling. The mechanical devices of material handling cannot be conveniently employed under this method on account of functional division of work. Material has to be carried by applying other methods from one department to another, resulting into higher cost of material handling.

(vi) **Higher labour costs : -** As there is functional division of work, specialized workers are to be appointed in different departments for carrying specialized operations. The appointment of skilled worker leads to higher labour costs.

(vii) **Longer production time: -** Production takes longer time for completion under this method and this leads to higher inventories of work-in-progress.

(viii) **Difficulties in production, planning and control : -** Due to large variety of products and increased size of the plant, there are practical difficulties in bringing about proper coordination among various areas (departments) and processes of production. The process of production, planning and control becomes more complex and costly.

(ix) **Increased inspection costs :-** Under this type of layout more supervisors are needed and work is to be checked after every operation which makes the process of supervision costlier

(x) **Layout by product or line layout : -**This type of layout is also known as "mass production" layout . In this layout all the necessary machinery and equipment needed to make a given product are set out in the same area and in the sequence of the manufacturing process . for examples -soft drinks bottling, car assembly and some canning operations. For example - The automobile industry , soft drinks bottling , soft drinks bottling , canning operations .

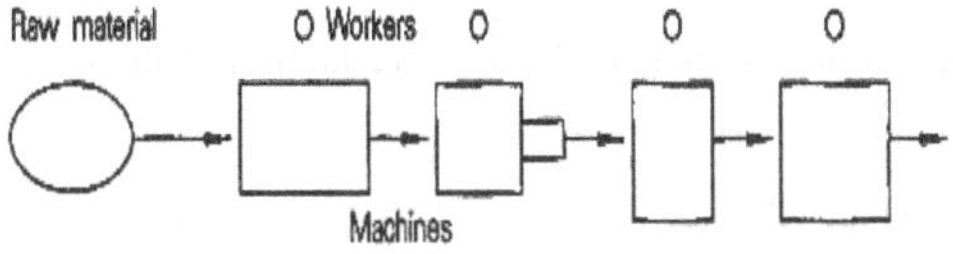

Diagram 8.3: Layout by Product or Line Layout

Advantages of Product Layouts Include

- **Output**: - Product layouts can generate a large volume of products in a short time.
- **Cost: - Unit** cost is low as a result of the high volume. Labor specialization results in reduced training time and cost. A wider span of supervision also reduces labor costs. Accounting, purchasing, and inventory control are routine. Because routing is fixed, less attention is required.
- **Utilization: -**There is a high degree of labor and equipment utilization.
- Product layout ensures unrestricted and continuous production thereby minimizing bottlenecks in the process of production, this is because work stoppages are minimum under this method.
- **Economies in material handling: -** Under this method there are direct channels for the flow of materials requiring lesser time which considerably eliminate back-tracking of materials. On account of this, cost of material handling is considerably reduced. This is greatly helpful in achieving desired quality of the end product.
- **Lesser manufacturing time**:-Under this method (as already pointed), backward and forward handling of materials is not involved; it leads to considerable saving in manufacturing time.
- **Lesser work in progress: -**On account of continuous uninterrupted mass production, there is lesser accumulation of work in progress or semi-finished goods.
- **Proper use of floor space:**-This method facilitates proper and optimum use of available floor space. This is due to non- accumulation of work in progress and overstocking of raw materials.
- **Economy in inspection: -**Inspection can be easily and conveniently undertaken under this method and any defect in production operations can be easily located in production operations. The need for inspection under this method is much less and can be confined at some crucial points only.

- **Lesser manufacturing cost:-**On account of lesser material handling, inspection costs and fullest utilization of available space, production costs are considerably reduced under this method.
- **Lesser labour costs:-** Due to specialization and simplification of operations and use of automatic simple machines, employment of unskilled and semi-skilled workers can carry on the work. The workers are required to carry routine tasks under this method. This leads to lesser labour costs.

8.4. Introduction of Effective Production Control

Effective production control on account of simple operation of this method can be employed successfully. Production control refers to the adoption of measures to achieve production planning

Disadvantages of Product Layouts Include

- **Motivation: -** The system's inherent division of labor can result in dull, repetitive jobs that can prove to be quite stressful. Also, assembly-line layouts make it very hard to administer individual incentive plans.
- **Flexibility: -** Product layouts are inflexible and cannot easily respond to required system changes—especially changes in product or process design.
- **System protection:-** The system is at risk from equipment breakdown, absenteeism, and downtime due to preventive maintenance.
- **Lesser flexibility:-**As work is carried in sequence and process arranged in a line, it is very difficult to make adjustments in production of operations. Sometimes, certain changes under this method become very costly and impractical. On account of this drawback, this method is not suitable in the production of goods which are subject to quick style and design changes.
- **Large investment:-**Under this method, machines are not arranged in accordance with functions as such similar type of machines and equipment is fixed at various lines of production. This leads to unavoidable machinery duplication resulting in idle capacity and large capital investment on the part of the entrepreneur.
- **Higher overhead charges:-**Higher capital investment leads to higher overheads (fixed overheads) under this method. This leads to excessive financial burden.
- **Interruption due to breakdown:-**If one machine in the sequence stops on account of breakdown, other machines cannot operate and work will be stopped. The work

stoppage may also take place on account of irregular supply of material, poor production scheduling and employee absenteeism etc.

- **Lack of specialization in supervision:-**Supervision of different production jobs becomes difficult under this method as there is absence of specialized supervision as the work is carried on in one line having different processes and not on the basis of different departments for different specialized jobs. Under this method a supervisor is supposed to have detailed knowledge of all the machines and processes which leads to absence of specialization in the process of supervision.
- **Under-utilization of machines**:- As has already been pointed out, separate set of one type of machines is fixed at different lines of production. Usually, these machines are not properly and fully utilized and there remains idle capacity in the form of underutilized equipment.

Group layout: -In modern age , several enterprises have arranged their operations in a new way, with a group of workers working together on a given product or on a part of a product and having at hand all the machinery and equipment needed to complete their work. With these kinds of layout, we may now analyze the flow of materials in a working area In some. Situations, rapid changes in output may be realized by switching from one type of layout to another but it is a very costly process

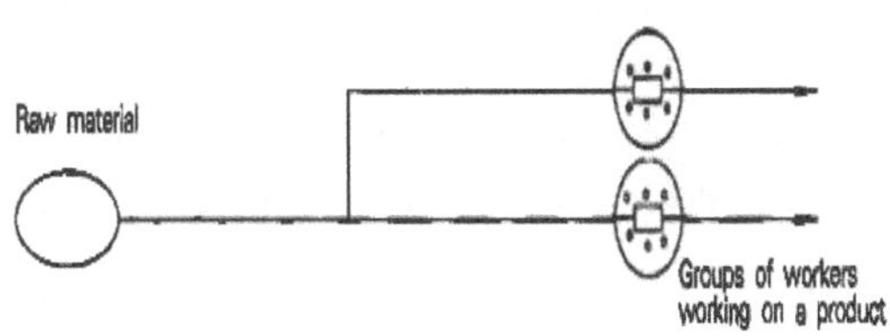

Diagram 8.4: Group Layout or Combination of Layout

Advantage of Group Layout

- Material movement is reduced, because the production process is located in one location for its entire manufacturing, it can be minimize damage or cost of moving.
- Provides job enrichment opportunities.
- Promotes pride and quality because an individual can complete the "whole job".
- Promotes pride and quality because an individual can complete the "whole job". Production volume.

Limitation of Group Layout

- Personal and equipment movement is increased, because it is necessary many equipment to brought the material from storage to the process location and moved to the warehouse. Movement of people and equipment to and from the work site may be expensive
- Requires greater skill forpersonnel because the same workers areinvolved in more operations, skilledand versatile workers are required. The necessary combinations of skills may be difficult to find andhigh pay levels may be necessary.
- Requires greater increased space and greater work-in process.skill for personnel.
- May result.

8.5. Material Handling Technique used in Manufacturing Industries

Material handling is the movement, protection, storage and control of materials and products throughout manufacturing, warehousing, distribution, consumption and disposal. In the field of engineering and technology, the term materials handling is used with reference to industrial activity. In any industry, be it big or small, involving manufacturing or construction type work, materials have to be handled as raw materials, intermediate goods or finished products from the point of receipt and storage of raw materials, through production processes and up to finished goods storage and dispatch points. the important point in favour of materials handling is that it helps production that is depending on the weight, volume and throughput of materials, mechanical handling of materials may become unavoidable. In many cases, mechanical handling reduces the cost of manual handling of materials.

According to American Materials Handling Society material handling is defined as

Materials handling is the art and science involving the moving, packaging and storing of substances or product in any form.

Some others Important Definition Regarding the Title is Given Below

- Materials handling is the movement and storage of materials at the lowest possible cost through the use of proper methods and equipment.
- Materials handling is the moving of materials or product by any means, including storage, and all movements except processing operations and inspection.
- Materials handling is the art and science of conveying, elevating, positioning, transporting, packaging and storing of materials.

In one of the definitions, it is define very clearly,

All the functions of materials handling have been referred, to which are conveying, elevating, positioning, transporting, packaging and storing. Storage or warehousing is very much a part of materials handling. Materials handling uses different equipment and mechanisms called Materials Handling Equipment.

Purpose of a Good Materials Handling System

(i) Efficient and safe movement of materials to the desired place.

(ii) Timely movement of the materials when needed.

(iii) Supply of materials at the desired rate.

(iv) Storing of materials utilizing minimum space.

(v) Lowest cost solution to the materials handling activities.

Often materials handling extends beyond the boundary of the industry in the form of movement of raw materials from the sources to the plant or in the form of finished goods from the plant to the points of consumption. These long distance movements of materials are generally termed as transportation of materials through various modes of transport like, road, rail, ship or air. Transportation is generally excluded from the scope of materials handling.

Example of Materials Handling System

(i) The feeding mechanism in an automatic machine.

(ii) Coiler and de-coiler in a strip rolling mill.

(iii) Paper feeding and transportation arrangement in a multi-station printing machine.

These are special material handling devices, but when integrated with specific production machines, they become specialized parts of those machines. Such special devices and their functions are generally not considered to be within the scope of materials handling.

8.6. Principle of Material Handling System

The designing of material handling system consisting "10" principles, material handling system does not a production process but it is supporting to the production process.

(i) Planning :-A material handing plan defines the material (**what**)and the moves (**when and where**); together, they define the method (**how and who**). Five key aspects must be considered in developing a plan.

- The plan should be developed in consultation between the planner(s) and all who will use and benefit from the equipment to be employed

- Success in planning on large-scale material handling ,involving suppliers ,consultant aswellas end-user from management, engineering, computer and information systems, finance, and operations
- The plan should be a analyses of existing methods or related problems then innovation of method / development of method , physical and economic constraints, and future requirements and goals.
- The plan should promote concurrent engineering of product, process design, process layout, and material handling methods.

(ii) Standardization:-Material handling methods, equipment, controls, and software should be standard-ized within the limits of achieving overall performance objectives , without affecting the desired flexibility, modularity, and throughput , there are three key aspects of achieving standardization.

- The planner should select methods and equipment that can perform a variety of tasks under a variety of operating conditions.
- Standardization applies to sizes of containers and other load-forming components, as well as operating procedures and equipment.
- Standardization, flexibility, and modularity must not be incompatible.

(iii) Work: -The measure of work is material handling flow or flow of volume, weight, or count / unit of time multiplied by the distance moved. Material handling work should be minimized without sacrificing productivity. **Thereare five important key points are considered in optimizing the work.**

- Simplifying processes by reducing, combining, shortening, or eliminating unnecessary moves that will reduces the work.
- Placing material in and out of storage—as distinct moves and components of the distance moved.
- Process methods, operation sequences, and process/equipment layouts should be prepared that support the work minimization objective.
- Where possible, power of gravity should be used to move materials respecting consideration of safety and the potential for product damage.
- The shortest distance between two points is a straight line , avoid the zigzag motion.

(iv) Ergonomics:-Ergonomics is a branch of science that aims to learn about human abilities and limitations, and then apply this learning to improve people's interaction with products, systems and environments.Ergonomics aims to create safe, comfortable and productive workspaces by bringing human abilities and limitations into the design of a workspace, including the individual's body size, strength, skill, speed, sensory abilities (vision, hearing), and even attitudes.

Ergonomics deals following branch of science which are described as,

Anthropometry	Body sizes, shapes; populations and variations
Biomechanics	Muscles, levers, forces, strength
Environmental physics	Noise, light, heat, cold, radiation, vibration body systems: hearing, vision, sensations
Applied psychology	Skill, learning, errors, differences
Social psychology	Groups, communication, learning, behaviours

The key points in the ergonomic principles is given below :-

- Equipment should be selected that eliminates repetitive and strenuousmanual labor and that effectively interacts with human operators and users. The ergonomic principle embraces both physical and mental tasks .
- The material handling workplace and the equipment employed to assist in that work must be designed so they are safe for people

(v) Unit Load:-A unit load is one that can be stored or moved as a single entity at one time, such as a pallet, container, or tote, regardless of the number of individual items that make up the load . When unit load is used in material flow, six key aspects deserve attention.

- Less effort and work are required to collect and move many individualitems as a single load than to move many items one at a time.
- Load size and composition may change as material and products movethrough stages of manufacturing and the resulting distribution channels.
- Large unit loads are common both pre- and postmanufacturing in the formof raw materials and finished goods.
- During manufacturing, smaller unit loads, including as few as one item,yield less in-process inventory and shorter item throughput times.
- Smaller unit loads are consistent with manufacturing strategies thatembrace operating objectives such as flexibility, continuous flow, andjust-in-time delivery.
- Unit loads composed of a mix of different items are consistent withjust-in-time and/or customized supply strategies as long as item selectivityis not compromised.

(vi) Space Utilization : Space in material handling is three-dimensional and therefore is counted as Cubicspace.

- Eliminate cluttered and unorganized spaces and blocked aisles in work areas.
- In storage areas, balance the objective of maximizing storage density against accessibility and selectivity. If items are going to be in the ware-house for a long time,

storage density is an important consideration. Avoid honeycombing loss . If items enter and leave the warehousefrequently, their accessibility and selectivity are important. If the storage density is too high to access or select the stored product, high storage density may not be beneficial

- Consider the use of overhead space when transporting loads within afacility. Cube per order index (COI) storage policy is often used in awarehouse.

(vii) System :- Asystem is a collection of interacting or interdependent entities that form a unified whole, system consists of material movement and storage activities, receiving, inspection, storage, production, assembly, packaging, unitizing, order selection, shipping, transportation, and the handling of returns.

Here are five key aspects of the system principle:-

- Systems integration should encompass the entire supply chain, includingreverse logistics. It should include suppliers, manufacturers, distributors,and customers.
- Inventory levels should be minimized at all stages of production anddistribution, while respecting considerations of process variability andcustomer service.
- Information flow and physical material flow should be integrated andtreated as concurrent activities.
- Methods should be provided for easily identifying materials and products,for determining their location and status within facilities and within thesupply chain, and for controlling their movement. For instance, bar codingis the traditional method used for product identification. Radio frequencyidentification (RFID) uses radio waves to automatically identify objects as they move through the supply chain .
- Customer requirements and expectations regarding quantity, quality, and on-time delivery should be met without exception.
- **Automation :-** Automation is concerned with the application of electro-mechanical devices, electronics, and computer-based systems to operate and control production and service activities. It suggests the linking of multiple mechanical operations to create a system that can be controlled by programmed instructions. Material handling operations should be mechanized and/or automated where feasible to improve operational efficiency, increase responsiveness, improve consistency and predictability, decrease operating costs and eliminate repetitive or potentially unsafe manual labor.

There are Four Key Points in Automation

- Preexisting processes and methods should be simplified and/or reengineered before any efforts at installing mechanized or automated systems.
- Computerized material handling systems should be considered where appropriate for effective integration of material flow and information management.
- All items expected to be handled automatically must have features that accommodate mechanized and automated handling.
- All interface issues should be treated as critical to successful automation, including equipment to equipment, equipment to load, equipment to operator, and control communications.

(viii) Environment : Environmental consciousness stems from a desire not to waste natural resources and to predict and eliminate the possible negative effects of our daily actions on the environment. Here are the three key points:

- Containers, pallets, and other products used to form and protect unit loadsshould be designed for reusability when possible and/or bio-degradabilityas appropriate.
- Systems design should accommodate the handling of spent dunnage,empty containers, and other byproducts of material handling.
- Materials specified as hazardous have special needs with regard to spill protection, combustibility, and other risks.

(ix) Life Cycle: Life-cycle costs include all cash flows that will occur between the time the first dollar is spent to plan or procure a new piece of equipment, or to put in place a new method, until that method and/or equipment is totally replaced. There are four key aspects:

- Life-cycle costs include capital investment, installation, setup and equipment programming, training, system testing and acceptance, operating(labor, utilities, etc.), maintenance and repair, reuse value, and ultimate disposal.
- A plan for preventive and predictive maintenance should be prepared for the equipment, and the estimated cost of maintenance and spare parts should be included in the economic analysis.
- A long-range plan for replacement of the equipment when it becomes obsolete should be prepared.
- Although measurable cost is a primary factor, it is certainly not the only factor in selecting among alternatives. Other factors of a strategic nature to the organization that

form the basis for competition in the market place should be considered and quantified whenever possible.

These ten principles are vital to material handling system design and operation. Most are qualitative in nature and require the industrial engineer to employ these principles when designing, analyzing, and operating material handling systems

8.7. The Purpose of Material Handling

A more realistic aim would be to move material by the most appropriate methods and equipment at the lowest possible cost and with regard to safety. This aim may be met by:

- Eliminating or reducing handling.
- Improving the efficiency of handling.
- Making the correct choice of material-handling equipment.

Eliminating or reducing handling: -There is often ample scope for eliminating or reducing handling. In practice, it becomes obvious that there is a need to improve an existing situation when certain symptoms are observed, e.g. too much loading and unloading, repeated manual handling of heavy weights, material travelling considerable distances , non-uniform flow of work with congestion in certain areas, frequent damage or breakage resulting from handling, and so on.

The approach to be followed here is similar to the traditional method study approach, using outline and flow process charts and flow diagrams and asking the same questions as to "where, when, by whom, how" and, above all, "why" this handling is done.

Improving the efficiency of handling: - Improving the efficiency of material handling system have been consisting the following items.

(i) Increase the size or number of units being handled at any one time. If necessary, review product design and packaging to see if you can achieve this result more readily.

(ii) Increase the speed of handling if this is possible and economical.

(iii) Let gravity work for you as much as possible.

(iv) Have enough containers, pallets, platforms, boxes, etc., available in order to make transportation easier.

(v) Give preference in most cases to material-handling equipment that lends itself to a variety of uses and applications.

(vi) Try to ensure that materials move in straight lines as much as possible, and ensure that aisles are kept clear.

8.8. Type of Material-Handling Equipment

There are various kind of material handling equipment are available that is generally used in manufacturing industries, Amongs them following are important from subject point of view .

(i) **Conveyors:-**Conveyors are mechanical devices or assemblies used to move items or packages with minimal effort between two fixed workstations, either continuously or intermittently. They are mainly used for continuous or mass production operations indeed, they are suitable for most operations where the flow is more or less steady.. They usually consist of frames that support rollers, wheels, or belts and may be motor powered or manual devices. They are used to move bulk material too, such as gravel or aggregate. They include belt conveyors that use moving belts, bucket and vertical conveyors that lift material, vibrating conveyors that use vibratory motion to move material, and overhead conveyors from which items hang during transport. Other types include screw conveyors for moving bulk material, chute conveyors which rely on smooth surfaces and gravity, and drag or tow conveyors that use cables to drag objects along.

Type of conveyor : - Following type of converter are generally used in manufacturing industries.

(ii) **Apron conveyor :-** Apron Conveyors are material handling systems that use steel, wood, or other materials typically mounted on roller chains to convey product. Apron conveyors are used primarily in material handling applications for moving large, heavy objects including crates, drums, or pallets in heavy-industry settings such as foundries and steel mills. The slats allow for heavy duty use with less wear and tear.

Apron Conveyor

(iii) **Ball Transfer :** -Ball Transfer Conveyors are material handling systems that use a series of mounted ball casters to allow for unpowered, multi-directional conveyance of the product, Ball transfer conveyors are used in material handling applications such as assembly lines and packaging lines, among others. Ball transfer conveyors are operated by external forces to move the product.

Ball Transfer

(iv) Belt conveyor :-Belt Conveyors are material handling systems that use continuous belts to convey products or material. Belt conveyors are used material handling applications such as food services, baggage handling, packaging, scrap handling, and postal service, among many others. **Various** belt materials are available including metals and non-metals depending on the application requirements. Belt conveyors are typically powered and can be operated at various speeds depending on the throughput required.

Belt Conveyor

(v) Beltless Magnetic :-Beltless Magnetic Conveyors are material handling systems that use moving magnets beneath **stationary** plates, tables, etc. to move magnetic (ferrous) materials usually in the form of machining scrap, etc

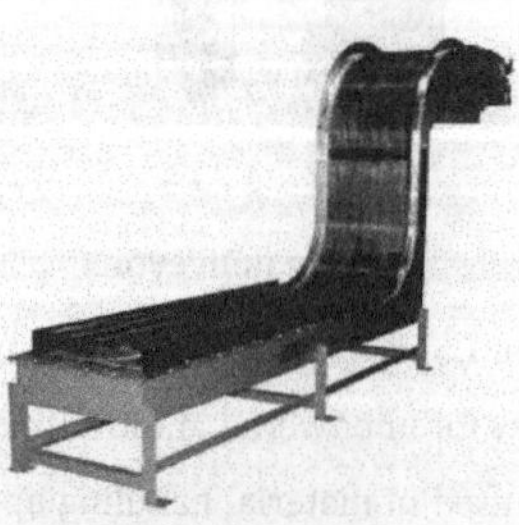

Beltless Magnetic

(vi) **Bucket Conveyors :-**Bucket Conveyors are material handling systems that use multi-sided containers attached to cables or chains to convey products or materials. Bucket conveyors are used in applications such as parts, bulk material, or food handling. This conveyed material can be liquid or **dry** such as sludge, sand, manure, sugar, etc.

Bucket Conveyor

(vii) **Chute Conveyors : -**Chute Conveyors are material handling systems that use gravity to convey product along smooth surfaces from one level to another. Chute conveyors are used for scrap handling, packaging, postal service, etc. Their chutes are generally smooth allowing the product to slide easily and can be straight or curved depending on the application.

Chute Conveyors

(viii) **Pneumatic/Vacuum conveyor :-** Pneumatic / Vacuum Conveyors are material handling systems that use air pressure or vacuum to transport materials or items in or through closed tubes or ducts or along surfaces. Pneumatic/vacuum conveyors are used primarily in materials handling applications such as dust collection, paper handling, ticket delivery, etc. and in processes such as chemical, mineral, scrap, and food. Materials for the conveyors can be metallic or non-metallic depending on the media being conveyed.

(ix) Powered Roller conveyor :- Powered Roller Conveyors are material handling systems that use powered rollers mounted in frames to convey products , Powered roller conveyors are used primarily in material handling applications that require powered conveyance of the product. Various drive types include belts, chains/sprockets, and motorized rollers.

Powered Roller Conveyor

(x) Roller conveyor :-Roller Conveyors are material handling systems that use rollers mounted in frames to convey product either by gravity or manually. Roller Conveyors are used primarily in material handling applications such as on loading docks, for baggage handling, or on assembly lines among many others. The rollers are not powered and use gravity, if inclined, to move the product, or manually if mounted horizontally.

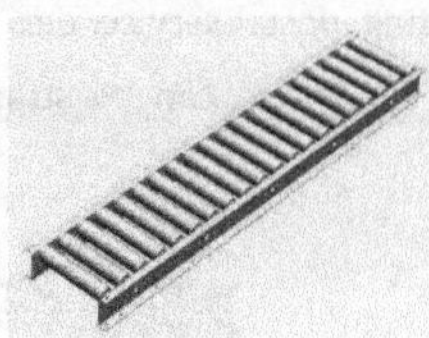

Roller Conveyor

(xi) Vibrating conveyor : -Vibrating Conveyors are material handling systems that use rotary or linear vibration to move material along their system beds. Key specifications include the intended application, load capacity, and throughput, along with the physical dimensions and electrical requirements. Vibrating conveyors are used to move dry, bulk materials such as aggregate, gravel, coal, etc. The bed of the conveyor vibrates to move the material along its length.

Vibrating conveyor

(xii) Walking Beam conveyor :- Walking Beam Conveyor Walking Beam Conveyors are material handling systems that use a combination of static supports and moving supports to index work pieces through manufacturing cells.

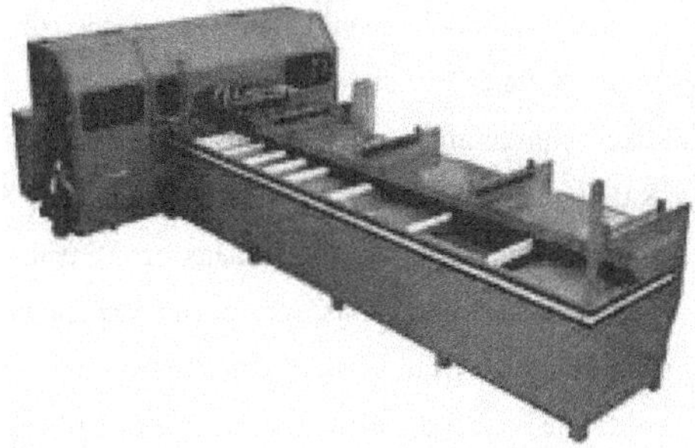

Walking Beam Conveyor

Walking beam conveyors are used on automation and assembly lines, where items need to be placed in specific locations in a sequential manner.

The item being supported at a stationary position is picked up by the walking beam and indexed to its next position, where another manufacturing step takes place. This is repeated over the length of the conveyor.

(xiii) Wheel conveyor ; -Wheel Conveyors are material handling systems that use unpowered wheels to move objects along their lengths by gravity or manual power. Wheel conveyors are used for item or package handling, and are commonly employed for loading and unloading trucks and for moving packages, pallets, etc. through facilities or along assembly lines.

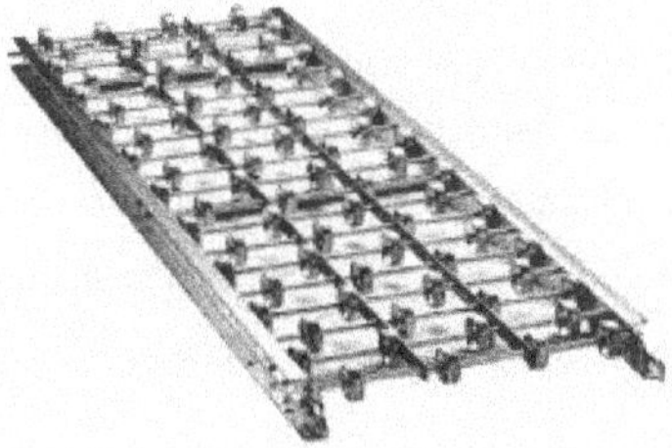

Wheel Conveyor

(xiv) Selection of Conveyor: -Conveyor selection depends on product type, throughput or speed, elevation change, and in some cases industry focus. Belt conveyors, for instance, come in variety of sizes and can range from foot-long units used on packaging lines to mile-long systems used for mining operations. Conveyors can be hand powered where products are moved along manually over rollers or wheels or air powered, engine powered, etc. In general, though, they are driven by ac and dc motors either directly or through reduction gears, chains and sprockets, etc. Products generally ride on the top surfaces of conveyors though some exceptions exist. Overhead conveyors, as found at dry cleaners, slaughterhouses, etc., suspend their loads from trollies which ride along overhead tracks. Other conveyors such as screw and pneumatic types convey their products through semi-enclosed troughs or tubes. These conveyors often handle dry products and powders. Some conveyors are designed to locate products precisely between manufacturing operations. Walking beam conveyors are an example of this type. Other conveyors transport hard to handle products such as cosmetic bottles through fillers, labelers, etc. by holding each container in individual pucks or pallets. Conveyors are sometimes custom designed from modular components such as straight-aways, curves, transitions, merges, dividers, etc. Manufacturers of such components often provide design expertise and installation assistance. Other conveyors are stand-alone systems, complete with drives and controls. An example of this is a take-away conveyor used at the discharge of an extrusion line, pictured at right. Manual type roller and wheel conveyors can often be purchased in discrete sections and bolted up to form material handling systems of nearly any length. In general, powered conveyors employ head shafts and tail shafts where the head end provides the drive arrangement and the tail end provides for adjustment of the chain or belt tension.

(xv) Industrial trucks: -Industrial trucks are more flexible in use than conveyors since they can move between various points and are not permanently fixed in one place. They are therefore most suitable for intermittent production and for handling various , sizes and shapes of material. There are many types of truck petrol-driven, electric, hand-powered, and so on. Their greatest advantage lies in the wide range of attachments available; these increase the trucks' ability to handle various types and shapes of material. Following are important trucks that are generally used in industries.

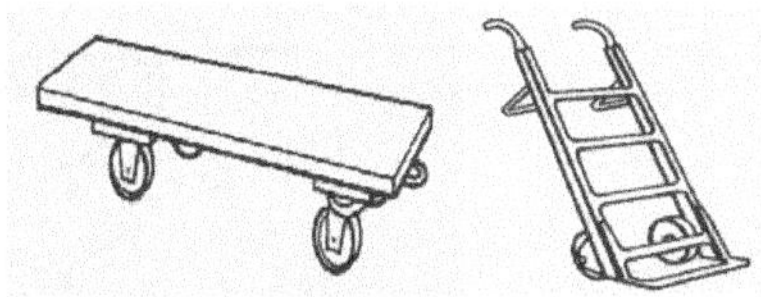

(i) dolly (ii) Two-Wheeled Hand Truck

 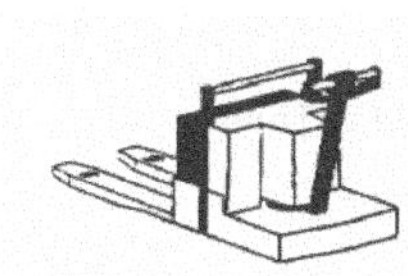

(iv). Floor Hand Truck (v) Pallet Truck

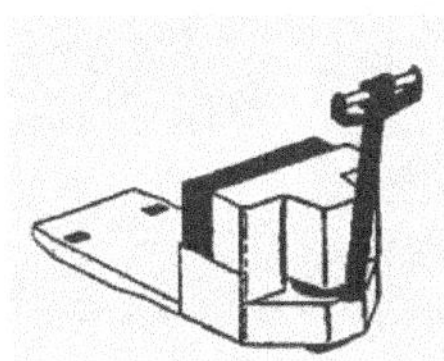 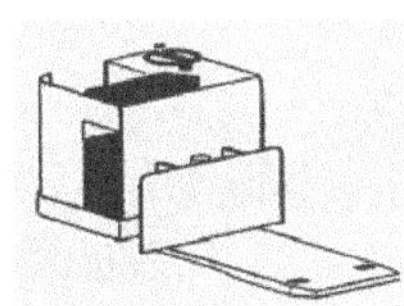

(vi). Walkie Platform Truck (vii) Rider Platform Truck

 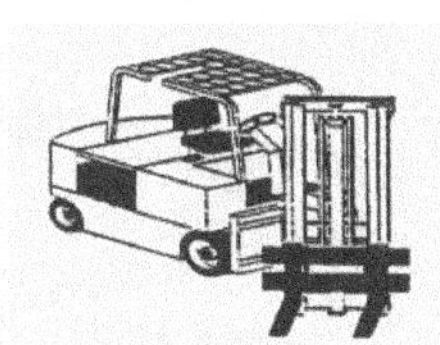

(viii) Sit-Down Counterbalanced Lift Truck (ix) Operator-Down Turret Truck

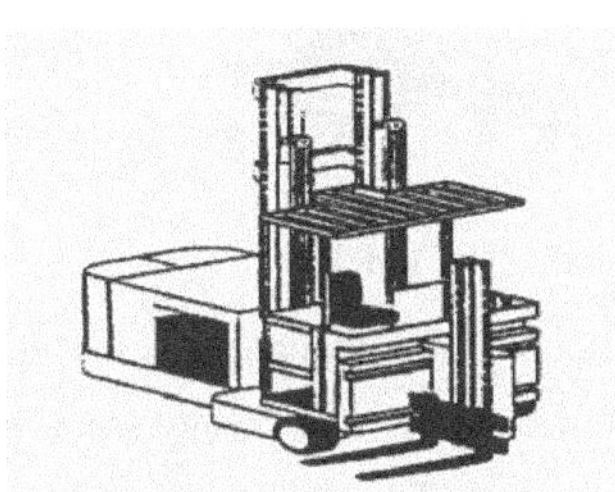 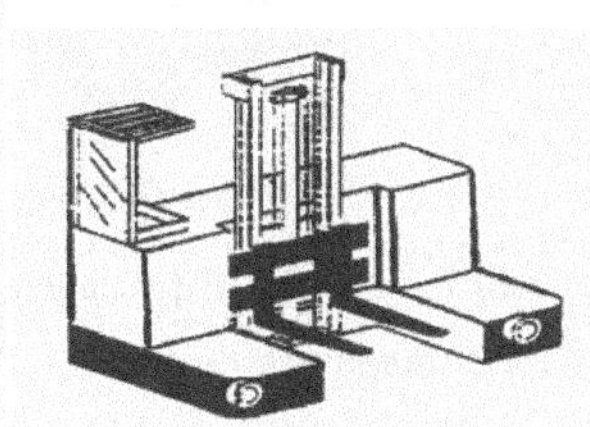

(x) Operator- up Turret Truck (xi) Side loader

Diagram 8.5: Industrial Truck

(xii) Tractor-Trailer

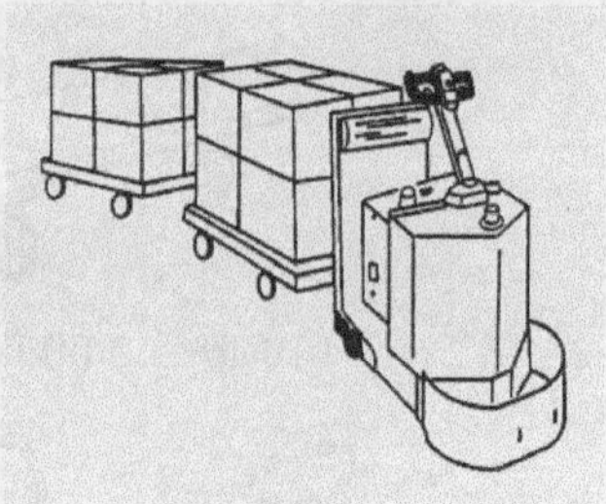

(xiii) Tow AGV

Power Operated Industrial Truck

(xvi) Cranes and hoists:-The major advantage of cranes and hoists is that they can move heavy material through overhead space. However, they can usually serve only a limited area. Here again, there are several types of crane and hoist, and within each type there are various loading capacities. Cranes and hoists may be used both for intermittent and for continuous production.

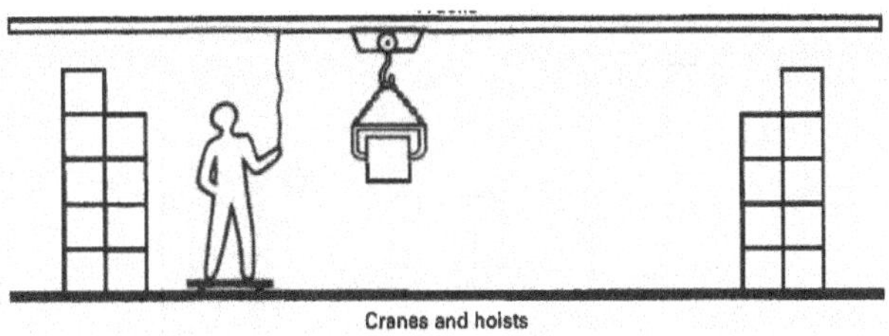

Diagram 8.6: Cranes and Hoists

(xvii) Containers:-These are either "dead" containers (e.g. cartons, barrels, skids, pallets) which hold the material to be transported but do not move themselves, or "live" containers (e.g. wagons, wheelbarrows or **computer** self-driven containers). Handling equipment of this kind can both contain and move the material, and is usually operated manually.

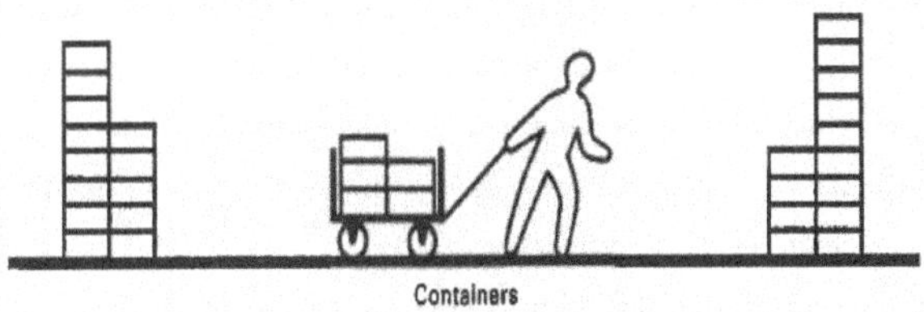

(xviii) Robots: - Many types of robot exist. They vary in size, and in function and man euverability . While many robots are used for handling andtransporting material, others are used to perform operations such as welding or spray painting. An advantage of robots is that they can perform in a hostileenvironment such as unhealthy conditions or carry on arduous tasks such as therepetitive movement of heavy materials.

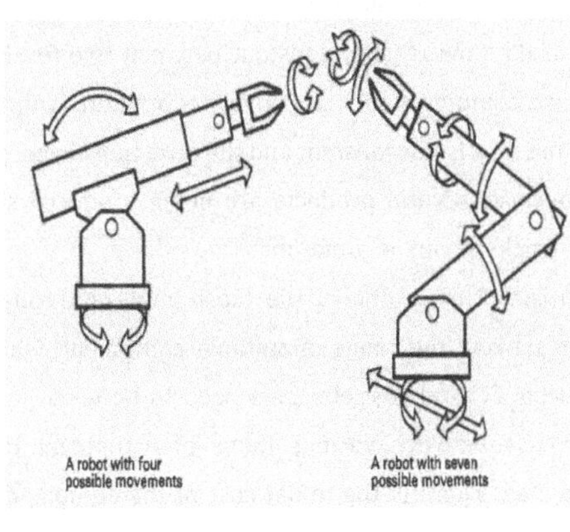

Diagram 8.7: Industrial Robots

8.9. Decieding Factor for Selection of Material-Handling Equipment

(i) **Properties of the material :-**Whether it is solid, liquid or gas, and in what size, shape and weight it is to be moved, are important considerations and can already lead to a preliminary elimination from the range of available equipment under review. Similarly, if a material is fragile, corrosive or toxic this will imply that certain handling methods and containers will be preferable to others.

(ii) **Layout and characteristics of the building: -** Another restricting factor is the availability of space for handling. Low-level ceilings may preclude the use of hoists or cranes, and the presence of supporting columns in awkward places can limit the size of the material-handling equipment. If the building is multi-storeyed, chutes, or ramps for industrial trucks, may be used. Finally, the layout itself will indicate the type of production operation (continuous, intermittent, fixed position or group) and can already indicate some items of equipment that will be more suitable than others.

(iii) **Production flow:-**If the flow is fairly constant between two fixed positions that are not likely to change, fixed equipment such as conveyors or chutes can be successfully used. If, on the other hand, the flow is not constant and the direction changes occasionally from one point to another because several products are being produced simultaneously, moving equipment such as trucks would be preferable.

(iv) **Cost considerations: -** This is one of the most important considerations. The above factors can help to narrow the range of suitable equipment, while costing can help in taking a final decision. Several cost elements need to be taken into consideration when comparisons are made between various items of equipment that are all capable of handling the same load. There is the initial cost of the equipment, from which one can derive the investment cost in terms of interest payment (i.e. if the company has to borrow money to buy the equipment) or opportunity costs (i.e. if the company possesses the funds and does not have to borrow, but the purchase of the equipment would deprive it of an opportunity to invest the funds at a certain rate of return). From the cost of the equipment one can also calculate the depreciation charges per year, to which will be added other charges such as insurance, taxes and additional overheads. Apart from these fixed charges, there are also operating costs, such as the cost of operating personnel, power, maintenance and supervision. By calculating and comparing the total cost for each of the items of equipment under consideration, a more rational decision can be reached on the most appropriate choice.

8.10. Importance of Material Handling

The first & foremost function of materials handling is to enhancing the productivity and thereby increases profitability of an industry. Many enterprises go out of business because of inefficient materials handling With the help of well designed materials handling system, we can obtained following.

(i) Improve efficiency of a production system by ensuring the right quantity of materials delivered at the right place at the right time most economically.

(ii) Reduced indirect labour cost of production.

(iii) Reduce damage of materials during material handling like storage , movement.

(iv) Maximize space utilization by proper storage of materials and thereby reduce storage and handling cost.

(v) Risk of accident is reduces during materials handling.

(vi) Reduce overall cost by improving materials handling.

(vii) Improve customer services and satisfaction .

(viii)Increase efficiency of plant and equipment with integral materials handling features.

All those positive features are obtained by utilizing the material handling technique in manufacturing industries . But every system have also some draw back these are given below

(i) Additional capital investment is associated due to materials handling system

(ii) Flexibility is reduced

(iii) If any kind of failure is occurs in material handling system then it increasing the downtime of the production system.

(iv) Maintenance cost of the material handling system is increased.

Some production equipment are fitted with facilities for handling of the materials being processed. Such materials handling equipment are generally considered to be an integral part of the production equipment. A few typical examples are:

(i) The feeding mechanism in an automatic machine,

(ii) Coiler andde-coiler in a strip rolling mill or

(iii) Paper feeding and transportation arrangement in a multi-station printing machine .

In an industry, materials handling is a subsystem (or part) of the production system. Materials handling itself can also be considered to be a system whose subsystems are

(i) Design or method to be adopted

(ii) Types of materials handling equipment to be used

(iii) Different operations like packing /unpacking,movement and storage involved

(iv) Maintenance required for the equipment employed

(v) Mode oftransportation by the raw materials suppliers, distributors / customers, waste / scrap collectors etc

(vi) The common objective by which the different subsystems are guided is the lowest cost solution of the materials handling system for that industry.

Application of material handling system: -material handling technique having a widely used in process industries, heavy manufacturing industries, construction industries, mining industries, shipbuilding or aircraft industries etc.

8.11. Site Selection

Site selection is very critical and complex in nature, that is depends upon type of plant which is being established, there are so many type of plant like - Solar Power Plant Site Selection, Thermal power plant site selection, Selection of the site of the waste water treatment plant, site selection of nuclear power plant , Site Selection for bio-gas power plant . the basic concept of site selection is to find out the the optimal location that is cost effective, efficient, and meets your operating requirements , here we will discussed about the site selection criteria for manufacturing plant .

The Site Selection Process : - An efficient site selection process is really more of a site "elimination" process.

Location screening and scoring criteria, beginning with high-level, large-geography considerations and then evolving into site-specific considerations, will guide you through a progressively shorter and shorter list of potential locations. The following methodology provides a basic foundation for site selection process .

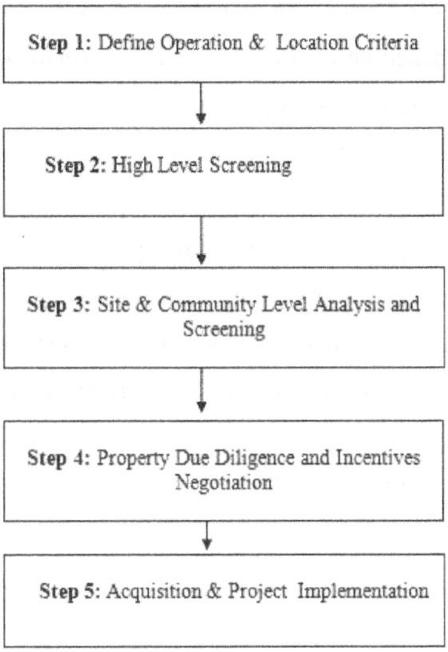

Diagram 8.8: The Site Selection Process

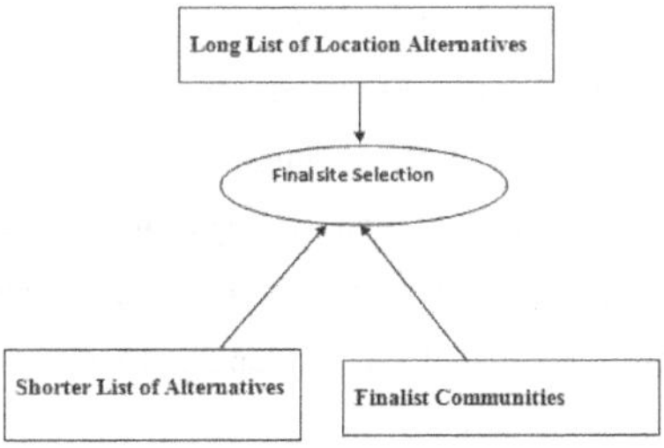

Diagram 8.9: Site Selection Process

The Site Selection Process

Step 1: Define your facility and operational requirements

Before beginning a site selection, it is critical to first define the proposed manufacturing facility and its operational requirements in order to evaluate location opportunities. Below is a list of criteria to guide this process.

- **Inbound Shipments** - Identify raw material sources, locations, annual volumes, number of shipments, modes of transportation.
- **Outbound Shipments** - Identify customers, locations, annual volumes, number of shipments, modes of transportation.

- **Labor Requirements** – Number and types of jobs, desired skill sets, wages and benefits, shifts and schedules, work ethic, and labor/management relationships.
- **Building Requirements** – Size, configuration, breakdown ofmanufacturing manufacturing/warehouse/office space needs, clear heights, truck docks, and otherfacility needs.
- **Site Requirements** – Size, configuration to support vehicle circulation, on-sitestorage, employee and truck parking, and other project needs.
- **Electric Power Requirements** – Electric power usage (demand and consumptionby month), line size required, service type preference (primary vs secondary),service redundancy needs.
- **Natural Gas Requirements** – Natural gas usage (consumption by month), line sizerequired, service redundancy needs.
- **Water Requirements** – Water usage (volume by day and month, peak usage byhour), line size required, redundancy needs, fire service needs
- **Wastewater Requirements** – Discharge volume (by day and month, peak volumeby hour), line size required, wastewater effluent characteristics.
- **Telecommunications Requirements** – Service requirements.
- **Air Emissions** – Description and estimated volumes of air emissions by type.
- **Community Preferences** – Community characteristics and amenities preferred,training resources available, highway proximity and infrastructure, communityculture, sustainability aspects, other requirements.
- **Project Investment** – Estimated value of land and building, equipment purchasesand installation, other start-up costs.
- **Incentives** – Desired types of incentives programs including tax credits, loans grants, and other financing assistance.
- **Project timeline** – Target real estate acquisition date, staffing ramp-up schedule, utility ramp-up schedule, equipment commissioning, production start date, etc.
- **Other Factors** – Other location criteria factors that are unique to the operation

Step 2: High Level Location Screening :-Once requirements for the manufacturing plant are defined, the next step is to begin using the criteria to narrow the search, beginning with those that can be measured at a high (state- or regional-) level .

The following factors are typically reviewed at this stage in the process and weighted as appropriate.

High Level Site Selection Screening Factors	
Transportation and Logistics	• Inbound transportation costs to ship raw materials into the plant. • Outbound transportation costs to ship finished products tocustomers. • Transportation Service and Infrastructure Availability Airport Accessibility.
Labor	• Labor Costs. • Labor / Management Relations. • Mandated Benefits. • Population/Demographics. • Concentration of Specificed Labor Skills. • Labor Training Programs.
Utilities	• Electric Power availability and cost. • Natural Gas availability and cost. • Water availability and cost. • Wastewater availability and cost. • Telecommunications availability and cost.
Taxes	• State Corporate Taxes • Property Taxes – Real and Personal • Inventory Taxes
Real Estate	• Availability of Sites and/or Buildings • Cost of Real Estate
Regulatory Environment	• Business Climate • State Regulations • Environmental Regulations

Step 3: Site and Community level Analysis and Screening: -The next step is to conduct a detailed evaluation of specific properties and communities to select the optimal location for your specific operational needs. The following factors should be evaluated and considered as part of this process:

Community and Property Specific Site Selection Factors	
Property Issues : Evaluate Each Factor Relative to Each Specific Property	• Property Characteristics • Site Image • Transportation Infrastructure and Characteristics • Electric Power Service Characteristics • Water Service • Waste Water Service • Fire Protection • Natural Gas Service • Telecommunications Service • Employee Amenities • Zoning • Site Security • Environment Manmade Hazards • Environment - Natural Hazards
Community Issues : Evaluate Each Relative to Each Specific Community	• Labor Force, Including Availability, Skilled Labor Relations • Water and Waste Water Infrastructure • Community Development Trends • Industrial Support Services • Local Economy • Regulation • Sustainability • Community Support
State and Local Incentives : Evaluate For Each Short Listed Location	• State Incentives • Local Incentives • Utility Service Incentives • Other Programs

Step 4: Property Due Diligence and Incentives Negotiation:-Once the finalist locations are selected, it is prudent to conduct property due diligence to obtain commitments on utility services and other needs, as well as to ensure there are no surprises once the first shovel goes into the ground for development. At this phase, it may be appropriate to engage an engineering firm to assist with the technical review of the site attributes.If considering a Greenfield site, you may need to conduct a number of site studies – e.g. Environmental Study, Geotechnical Analysis, Hydrologic Analysis, Site Survey, Archaeological Study, Endangered Species Analysis, and other technical site reviews pertaining to the development of the site, this is also the time to initiate formal incentives negotiation for your finalist locations

In this step, we will prepare a detailed financial analysis for each finalist site and negotiate incentive programs that will help "close the deal" by providing upfront and/or ongoing cost savings and risk management.

Step 5: Acquisition and Project Implementation: - The final step is to acquire the property and begin detailed engineering of the facility and its implementation.

Site Selection Time line: - The timeline for completing a site selection effort for a new manufacturing plant will vary based on the size and the specifics of the project, but most can be completed through all the steps above between three and six month or it may be 12 month to 18 month.

(i) State and Local Development Incentives: - Economic development incentive programs are varied across the country at the state level and local levels. In general, programs can be grouped into the following categories:-

Grants	Speed and markets	Operating costs
• Low cost loan • Infrastructure improvements • Bonds financing • Free land / building • Relocation assistance	• Expedited permitting • Recruitment assistance • Training assistance • Incentive approval process • Ready site	• Tax credits • Property tax • Utility riders • Sales tax exemption • Incumbent training

(ii) Positioning for optimal incentives :- Access to many state incentives programs throughout the country that is governed by what is known as a "but for" test, meaning the incentives status require documentation and/or attestation by the company that the incentives program had to be offered as a condition for the company to locate the project in that state .

(iii) Timing ;- It is important to understand the incentives approval process and timeline , Every state has a different process and schedule for approving incentives offerings, that is depends upon the type of plant .

(iv) **Program Terms:** - Although economic development agencies are excellent resources for incentives programs, it is important to understand that they are advocating for the state or community in which they represent. A company must exercise due diligence when reviewing incentive offers and agreements. A multitude of requirements will accompany incentive benefits, including job creation and investment obligations, as well as on-going commitments to continue operations at the new location.

(v) **Site Selection Resources:** - The following resources may be very helpful to you during your site selection process.

Economic Development Agencies:-There may be so many Economic Development Agencies are working with in the country at different level like - state, local and regional levels. They may be a governmental organizations, public-private or even private entities that have been given responsibility to help attract and retain business investment and jobs to their respective states and communities. They may represent an entire state, a county, a municipality, or a region comprised of multiple countries and municipalities .They can be excellent initial resources of preliminary information, at no cost, including .

- **Available properties** - Most states and local communities maintain databases of available properties in their areas, many of which are available on their respective websites. However, these databases may not be complete and/or may not represent the most up-to-date information. Hence we have to hire appropriate economic development agency for assurance.
- **Workforce** – Local economic development agencies can typically provide data regarding the local labor force in their region or area, including types of skills available. In addition, they can also offer information on workforce training programs and resources within their communities.
- **Utilities** – Economic developers can connect you with appropriate utility service providers in the community for information on costs and capacities .
- **State and local tax rates** – Economic developers can provide specific rates for property taxes, income taxes, and other burdens in their states and local communities.
- **Available incentives** – Initially, an economic developer can provide you with an overview of available programs at the state and local level for their respective area.
- **Site Selection and Incentives Consultants** – A professional site selection consultant functions as a specialized management consultant, serving as the company's advocate and advisor, and assisting companies in formulating their site selection process. Consultants also manage the execution of that strategy including data collection,

analysis and field work. Most site selection consultants work closely with state and local economic development agencies as well as utilizing internal and subscription-based data services and other tools through an established methodology.

Question for Discussion

1. Discuss the relative advantages and disadvantages of process and product type layouts.
2. What are the main objectives of a good plant layout.
3. Discuss the various factors to be considered for finalizing plant's location.
4. Explain with examples the significance of the term standardisation.
5. "Materials handling does not add value to the product but adds to the 'cost". Comment.
6. Write the. difference between product layout and process layout.
7. What are the objectives of material handling ?
8. Discuss the relationship between plant layout and material handling equipments.
9. Explain.
   (a) Material handling in process layout.
   (b) material handling in product layout.
10. Explain the working principle of conveying equipment. How do you classify conveyors?
11. With a neat diagram explain the working of an apron conveyor.
12. What types of powered trucks are self loaders? Give their operating details.
13. What types of powered trucks are best suited to the movement of long length items such as pipe, bar stock or lumbers?
14. With the help of a sketch explain the working of belt feeders.
15. Explain how chutes can be used for delivering different types of goods to different stations?
16. What care should be taken while designing the feeding equipment to avoid spillage of materials?
17. Name the various hoists. What are their features?
18. Explain different principles of material handling.
19. Explain principles of plant layout.
20. What are the different types of plants layout? Explain any one with its advantages & disadvantages with its application.
21. What are the factors considered while selecting location for an foundry plant ?
22. Discuss about the Driverless trains and wire-guided pallet trucks.
23. Give the types of belt cleaners in belt conveyor. With sketches describe them
24. How roller conveyor can be used to handle unit loads?

25. What are the differences between roller conveyor and belt conveyor.
26. Discuss the main features of four wheeled powered drives such as fork lift, side loading trucks and straddle trucks with aid of sketches.
27. What are the uses of chutes in feeding equipment?
28. Compare the features of various types of hoists.
29. Name 6 types of surface handling equipment with a typical application for each type.
30. Draw a sketch for horizontal and ascending conveyor showing various components of belt conveyor.
31. Name various types of feeders used in a pneumatic handling system.
32. Explain the steps in selecting the plant site.
33. Explain the advantages and disadvantages while selecting a site in rural areas.
34. Explain objectives of good plant layout.
35. Explain the advantages and disadvantage while selecting the site in urban area.
36. Write various types of plant layout and explain any two with neat sketch.
37. Define plant lay out. discuss following types of plant layout.
    (a) Process layout.
    (b) Group technolog.
    (c) Fixed position layout.
38. What is material handling? Explain its major principles in brief.
39. A company is to decide on the location of a new plant. It has narrowed down the choice to 3 locations A, B, and C; data in respect of which is furnished below: Use suitable criterion and advise the company on the best choice.

Data	Location		
	A	B	C
Data wages & salaries	20000	20000	20000
Power, water supply expenses	20000	25000	30000
Raw material & other supplies	80000	75000	60000
Total initial investment	200000	300000	25000
Distribution expenses	50000	40000	60000
Miscellaneous expenses	40000	25000	30000
Expected sales per year	225000	250000	225000

40. What are the various Principle of good layout.
41. Differentiate -Process layout plan v/s product layout plan.
42. Discuss the factors to be considered for the selection of the site for Thermal Power Plant.
43. List the factors which should be considered for the selection of site for the automobile industry.

44. Discuss the factors to be considered for the selection of the site for textile industry.
45. Explain the types of plant lay outs with their applications.
46. Discuss the various factors must be considered for the site selection.

References

1. D. Van Cott and R. Kinkade, "Human engineering guide to equipment desigm", New York: McGraw-Hill, 1972.
2. S. Konz, "Workstation organization and design", Ergonomics, Vol. 32, Pp. 795-811, 1990.
3. R.W. Mallick and A.T. Gaudreau, "Plant Layout Planning and Practice", New York. John Wiley and Sons, 1951.
4. J.W. Hendry, "A Manual of Time and Motion Study", London. Sir Isaac Pitman and Sons. Third Edition, 1950.
5. "Idem: Time study Fundamentals for Foremen", New York. McGraw-Hill. Second Edition, 1951.
6. "Idem: Motion and Time Study Applications", New York. John Wiley and Sons. Fourth Edition, 1961.
7. M. Bannes Ralph, "Motion and Time Study: Design and Measurement of Work", New York and London. John Wiley and Sons. Fifth Edition, 1966.
8. J. Sandier Henry and Luckiewicz Edward, "Practical process engineering, a working approach to plant design", (New York, McGraw-Hill), 1987.
9. N.F.T. Saunders, "Factory organization and management", (London, Pitman, 5th éd.,), 1973.
10. M. Apple James, "Plant layout and materials handling", (London, John Wiley, 3rd éd.,), 1977.
11. J.A. Alie, "Who designs work?", Technology in Society (Newark, New Jersey), Vol. 12, Pp. 301-312, 1990.
12. "British Standards Institution", Glossary of terms used in work management services, BS 3138, London, 1991.
13. S. Parker and T. Hall, "Job and Work Design: Organizing Work to Promote Well-Being and Effectiveness", Thousand Oaks, CA: Sage, 1998.
14. S. Pheasant, "Body space, Anthropometry, Ergonomics and Design", London: Taylor & Francis, 1986.
15. E. Eggleton, "Ergonomic design for people at work, Belmont, CA: Lifetime Learning", Vol. 1, 1983.

16. S. Sauter and L. Schlufir, "Work posture, workstation design, and musculoskeletal discomfort in a VD data entry task", 1991.
17. H.A Bolz and G.E Hagemann, "Materials Handling Handbook", Ronald Press, 1991.
18. IS 8005:1976, Classification of Unit Loads, Bureau of Indian Standards.
19. IS 8730:1997, Classification and Codification of Bulk Materials for Continuous Material Handling Equipment, Bureau of Indian Standards.
20. J.A. Apple, "Material Handling System Design", John Wiley & Sons.
21. T.H Allegri, "Materials Handling: Principles and Practice", CBS Publishers & Distributors, N. Delhi.
22. International Material Management Society Staff and American Society of Mechnical Engineering Staff. Materials handling handbook (London, John Wiley, 2nd éd.,), 1985.
23. Eastman. Materials handling (New York, Dekker, Marcel Inc.,), 1987.

16. S. Sauter and [illegible], "Work posture, workstation design, and musculoskeletal discomfort in a VDT data entry task", [illegible].

17. R.C. Bolz and G.E. Hagemann, "Materials Handling Handbook", Ronald Press, [illegible].

18. IS 8005-1976 Classification of Unit Load, Bureau of Indian Standards.

19. IS 8730-1997, Classification and Codification of Bulk Materials for Continuous Material Handling Equipment, Bureau of Indian Standards.

20. J.M. Apple, "Material Handling System Design", John Wiley & Sons.

21. T.H. Allegri, Materials Handling: Principles and Practice, CBS Publishers & Distributors, New Delhi.

22. International Material Management Society SME and American Society of Mechanical Engineers, Bulk Materials Handling Handbook (London: John Wiley, 2nd ed.), [illegible].

23. Eastman, Materials Handling (New York: Dekker, Marcel Inc.), 1987.

CHAPTER 9

QUALITY & RELIABILITY

9.0. Introduction

The history of quality is as old as civilization. The Harappa's of the ancient Indus Valley civilization (3000 BC) achieved high precision in the measurement of length, mass, and time. The smallest division, which is marked on an ivory scale from Lothal, was approximately 1.704 millimeters, recorded in the Bronze Age. The 0 Dimensions of the pyramids, built around 2500 BC, show a high degree of accuracy. However, the use of tolerancing systems for the specification of quality and statistical principles to monitor quality are of recent origin. The quality movement may be traced back to medieval Europe. Craftsmen began organizing into unions called guilds in the late thirteenth century. Manufacturing in the industrialized world followed the craftsmanship model throughout the eighteenth century. The factory system, with its emphasis on product inspection, started in Great Britain in the mid- 1750s and grew into the Industrial Revolution in the early nineteenth century. In 1798 Eli Whitney introduced the concept of producing interchangeable parts to simplify assemble.

Frederick W. Taylor introduced the principles of scientific management around 1900 and emphasized the division of labor with a focus on productivity. There was a significant rise in productivity but it had a negative effect on quality.

Henry Ford's moving automobile assembly line was introduced in 1913. This required that consistently good-quality parts were available so that the production assembly line would not be forced to slow down

In 1924 **Walter A. Shewhart** introduced the basic ideas of the statistical process controlchart, which signaled the beginning of the era of statistical quality control. By the mid-1930s, statistical quality control methods were widely used at Western Electric, a manufacturing arm of the Bell system. World War II brought increased recognition of quality in manufacturing industries and military applications. The American Society for Quality Control was formed in 1946. A quality revolution in Japan followed World War II , Quality stalwarts W. Edwards Deming and Joseph M. Juran lectured extensively in Japan. As aResult, the Japanese became leaders in quality by the 1970s.The ISO 9000 quality system standards were published in 1987. The Baldrige NationalQuality Program and the Malcolm Baldrige National Quality Award were established by the U.S. Congress in the same year.

Quality : There was so many definitions were introduced by the various quality guru's , all these famous quality gurus , recognize as Crosby , Juran , W. Edwards Deming and the famous organization American Society for Quality . All these definitions are summarize as given below

According to Crossby – it is defined as,

Quality is nothing but it is a conformance to requirements

The definition considers the specification and requirement of products , which conform the requirement.

According to Juran – it is defined as,

Quality is fitness for use.

Jurans definition is leads to customer and how much he is satisfied with that products, when using.

According to W. Edwards Deming – it is defined as,

Good quality means a predictable degree of uniformity and dependability with a quality standard suited to the customer .

Sometimes it may also be defined as,

Quality is the degree to which performance meets expectations.

One of the most important definitions of the quality that is by the American Society for Quality is given below (ASQ).

Quality denotes an excellence in goods and services, especially to the degree they conform to requirements and satisfy customers.

9.1. Quality Control

It is defined asan aggregate of activities (such as design analysis and inspection for defects) designed to ensure adequate quality especially in manufactured products. Or it may also be defined as,

Quality Control (QC) is a system of routine technical activities, to measure and control the quality of the inventory as it is being developed.

The QC system is designed to:

(i) Provide routine and consistent checks to ensure data integrity, correctness, and completeness.

(ii) Identify and address errors and omissions.

(iii) Document and archive inventory material and record all QC activities.

QC activities include general methods such as accuracy checks on data acquisition andcalculations and the use of approved standardized procedures for emission calculations, Measurements, estimating uncertainties, archiving information and reporting. Higher tier QC activities include technical reviews of source categories, activity and emission factor data, and methods.

In order to implement an effective QC program, an enterprise must first decide which specific standard the product or service must meet. Then the extent of QC actions must be determined, then real-world data must be collected and the results reported to management personnel. After this, corrective action must be decided upon and taken. If too many unit failures or instances of poor service occur, a plan must be devised to improve the production or service process and then that plan must be put into action.

Finally, the QC process must be ongoing to ensure that remedial efforts, if required, have produced satisfactory results and to immediately detect recurrences or new instances of trouble. The basic concept of Quality control associated with the inspection of product which is the process of measuring, examining, and testing to gauge one or more characteristics of a product or service and the comparison of these with specified requirements to determine conformity. Products, processes, and various other results can be inspected to make sure that the object coming off a production line, or the service being provided, is correct and meets specifications.

General Quality Control Procedure: - The focus of general QC techniques is on the processing, handling, documenting, archiving and reporting procedures that are common to all the inventory source categories.

The procedure is summarize as,

Table 9.1: Quality Control Procedure

GENERAL INVENTORY LEVEL QC PROCEDURES	
Quality Control Activity	**Procedures**
Check that assumptions and criteria for the selection of activity data and emission factors are documented.	• Cross-check descriptions of activity data and emission factors with information on source categories and ensure that these are properly recorded and archived.
Check for transcription errors in data input and reference	• Confirm that bibliographical data references are properly cited in the internal documentation. • Cross-check a sample of input data from each source category (either measurements or parameters used in calculations) for transcription errors.
Check that emissions are calculated correctly.	• Reproduce a representative sample of emissions calculations. • Selectively mimic complex model calculations with abbreviated calculations to judge relative accuracy.
Check that parameter and emission units are correctly recorded and that appropriate conversion factors are used	• Check that units are properly labeled in calculation sheets. • Check that units are correctly carried through from beginning to end of calculations. • Check that conversion factors are correct. • Check that temporal and spatial adjustment factors are used correctly
Check the integrity of database files.	• Confirm that the appropriate data processing steps are correctly represented in the database. • Confirm that data relationships are correctly represented in the database. • Ensure that data fields are properly labelled and have thecorrect design specifications. • Ensure that adequate documentation of database and model structure and operation are archived
Check for consistency in data between source categories	• Identify parameters (e.g. activity data, constants) that are common to multiple source categories and confirm that there is consistency in the values used for these parameters in the emissions calculations.
Check that the movement of inventory data among processing steps is correct.	• Check that emissions data are correctly aggregated from lower reporting levels to higher reporting levels when preparing summaries. • Check that emissions data are correctly transcribed between different intermediate products.
Check that uncertainties in emissions and removals are estimated or calculated correctly.	• Check that qualifications of individuals providing expert judgement for uncertainty estimates are appropriate. • Check that qualifications, assumptions and expert judgments are recorded. Check that calculated uncertainties are complete and calculated correctly. • If necessary, duplicate error calculations or a small sample of the probability distributions used by Monte Carlo analyses
Under take review of internal documentation	• Check that there is detailed internal documentation to support the estimates and enable duplication of the emission and uncertainty estimates. • Check that inventory data, supporting data, and inventory records are archived and stored to facilitate detailed review. • Check integrity of any data archiving arrangements of outside organisations involved in inventory preparation
Check methodological and data changes resulting in re-calculations.	• Check for temporal consistency in time series input data for each source category. • Check for consistency in the algorithm/method used for calculations throughout the time series. • Confirm that estimates are reported for all source categories and for all years from the appropriate base year to the period of the current inventory. • Check that known data gaps that result in incomplete source category emissions estimates are documented
Compare estimates to previous estimates	• For each source category, current inventory estimates should be compared to previous estimates. If there are significant changes or departures from expected trends, re-check estimates and explain any difference

9.2. Fifteen Quality Improvement Tools

following 15th steps will leads to improve the quality control of the products.

(i) **Top management's visible commitment and explicit involvement**: - Without top management's visible commitment and explicit involvement, quality improvement efforts will never be successful. Top management holds the key to quality improvement as it determines the various systems in which people work. According to W. Edwards Deming, management is responsible for as much as 94% of a company's quality problems. Joseph M. Juran puts the figure as 85% or more. Moreover, the actions of employees greatly depend on top management's attitude towards quality.

(ii) **Strategic Quality Planning**: - A Strategic Quality Plan is absolutely vital for an organization to develop competitive advantage and effectively manage organizational quality improvement efforts. It sets the organizational mission statement, specifies key quality goals, promotes collective action and determines the strategies of how the organization is going to attain its mission and quality goals.

(iii) **Organizational quality awareness: -** Organizational quality awareness is necessary to gain employee support for quality improvement efforts and reducing potential resistance to change. All employees must fully understand the need for quality improvement, how quality relates to their jobs and how it can be measured and improved. Many quality improvement efforts have failed in the past due to neglect of organizational quality awareness programmes. Numerous organizations made the mistake of plunging directly into quality improvement efforts.

(iv) **Quality through people :-** Quality begins and ends with the individual. Quality people do quality work. People manage processes and make the systems work; processes don't do work, people do. In short, quality is the expression of human excellence.6 Quality improvement efforts should focus more on people rather than on techniques and machines. Management must have faith in the ability of employees to produce quality work. Employees must be given the opportunity to realize their full potential through training, participation and empowerment.

(v) **A Culture of Quality** :- Successful organizations have a culture that creates and sustains a work environment that is conducive to long-lasting quality improvement. Quality is deeply embedded in virtually every aspect of organizational life. Culture is important as it influences employee behaviour and action towards work, customers and one another. Satisfied employees help to create satisfied external customers. Some of the core values of a Culture of Quality are customer focus, employee autonomy, teamwork, open and honest communication, basing rewards on quality work and Doing the right things right the first time, every time.

(vi) Quality management structure :- A suitable organizational structure is necessary to ensure effective and efficient management of quality improvement efforts. The quality management structure would depend upon the size of the organization and the nature of its business. However, most successful organizations have a Corporate Quality Council consisting of the top management team, a TQM Manager and Divisional Quality Councils. The major responsibilities of the Corporate Quality Council are to set an overall strategic quality improvement plan for the company, review and approve divisional quality improvement plans and cooperation and monitor quality improvement efforts. The TQM Manager assists the Corporate Quality Council in planning and implementing organization-wide quality improvement efforts. The Divisional Quality Councils set specific quality goals, formulate plans of action and implement quality improvement efforts at the divisional level.

(vii) Customer-driven :- Quality is defined and judged by the customers. The mission of quality organizations is centered around customer satisfaction. Organizational processes and procedures are designed to meet the requirements of both the external and internal customers. Market research must be undertaken to determine the requirements of external customers, how well is the organization meeting them and what are the areas for improvement as perceived by the customers. Customer's complaints should be tracked and acted upon promptly. With regard to the internal customers, employee opinion surveys should be undertaken to determine, inter alia, their views on how the organization is run and how quality can be improved.

(viii) Process Management :- Long-lasting quality improvement is attained through preventive management (building quality into the work processes, particularly key processes). All work ranging from administrative to engineering is a process. A process is essentially a sequence of tasks or activities that transforms inputs into desired outputs. At very initial stage define a process as "the sequential integration of people, materials, methods and machines in an environment to produce value-added outputs for customers."

(ix) Teamwork :- Teamwork is vital as the success of quality improvement efforts is greatly dependent on close cooperation between managers and employees and among departments. Teamwork generally enhances the quality of decision-making; helps to break down departmental barriers; and creates a sense of ownership and commitment. Teamwork can be encouraged by creating cross-functional Quality Improvement Teams.

(x) **Management by Fact :-** TQM involves systematic and fact-based decision making. Facts or reliable data and not opinions or hearsay form the basis of making intelligent decisions or solving problems systematically.

(xi) **Training: -** A TQM organization is a learning organization. Training is viewed as a valuable investment and not as an unnecessary expense. Companies noted for world-class quality typically devote 40-80 hours per year, per person to training.8 Training sessions must start with top management and cascade down the organization. Training should be related to actual work performed by employees. Managers should train their own work groups. Quality training should cover quality awareness, basic quality concepts and tools, process management, problem solving, and team building.

(xii) **Reward and recognition systems: -** The reward and recognition systems must be aligned with quality improvement efforts. Recognition and rewards must be based on quality work and not on seniority or other non-merit factors. Examples of recognition and rewards are bonuses, merit certificates and pay increase. Quality success stories should be published.

(xiii) **Quality standards :-** Quality standards are important as they help to clarify work roles, communicate priorities and form the benchmark for assessing and rewarding employee performance. In short, what gets measured gets done. Quality standards can be set in terms of reliability, responsiveness and courtesy of frontline employees. Quality standards should be customer-oriented, specific and verifiable, realistic and challenging, and mutually established and agreed upon by management and employees.

(xiv) **Measurement :-** Measurement is a prerequisite to any quality improvement effort. Quality is defined as "meeting or exceeding customers' requirements consistently". These requirements have to be quantified and measured to determine the success of quality improvement efforts. Measurement provides a data base for decision-making, establishing customers' requirements, identifying opportunities for quality improvement, and assessing performance.

(xv) **Continuous improvement :-** Quality is a moving target. There is no one best or optimum level of quality. Companies have to continuously improve the quality of their products and services to stay ahead in an increasingly competitive business world. Technological changes often render current "quality" products as obsolete. Customer expectations often change over time. Hence, quality improvement is a never-ending journey. In this regard, the organizational work climate should promote employee creativity, and problems viewed as opportunities for improvement.

9.3. Customer Orientation

Customer orientation is defined as an approach to sales and customer-relations in which staff focus on helping customers to meet their long-term needs and wants. Here, management and employees align their individual and team objectives around satisfying and retaining customers.

Examples of customer orientation include providing quality products or services, responding quickly and thoroughly to customer feedback and complaints and being sensitive to the community's needs.

(i) Market Orientation and Customer Orientation

The principal element underlined by the current marketing implementation is the customer aspect. That describes the evolution of marketing and the influence of customers. marketing has moved from a focus on the mass market to a focus on market segments to a focus on one-to-one customer relations. He thinks that current marketing is moving from a transaction-orientation to a customer-relationship-building orientation. "There is only one valid definition of business purpose: to create a customer ... It is the customer who determines what the business is "In a simple definition, the importance of customers in the role of marketing: "Marketing is making sure you make the stuff customers want" On market places, companies have also realized that adapting the products to the customers' needs or tastes was more profitable than attempting to change the customers to fit the firm's purposes . Today, customers are becoming better organized, well informed and more demanding. Those changes are the reason why companies may develop customer-orientation. In actually , it is sometimes really hard to make the difference between "customer-oriented" and "market-driven" or "customer-driven" and "market-oriented". The goals of those four concepts are to create or adapt products to the customers . foreign markets. Customer-orientation dimension. Hence, we need to differentiate it from the other terms. Customer-orientation is a part of market-orientation but some authors defined it more precisely. Customer-orientation is "the dissemination of information about customers throughout an organization, formulation of strategies and tactics to satisfy market needs inter-functionally and achievement of a sense of company-wide commitment to these plans . Customer-orientation is here presented as an organizational culture that is devoted to satisfy customers' needs on their market. That culture is developed through communication or "dissemination of information" throughout the company. Customer orientation also "represents the degree to which customer information is both collected and used by the business unit". These emphasize the need to "collect" the information of customer needs or tastes and also introduce the term of "business unit" instead of "organization" to be closer to the customers customer-orientation is "the set of beliefs that puts the customer's interest first, while not excluding those of all other stakeholders such as owners, managers, employees, in order to develop a long-term profitable enterprise" Thus, customer orientation is a crucial component of an organizational culture and attention to information about customers' needs should be considered as a basic value of the firm. In other words, "customer orientation is a concept comprising customer understanding orientation and customer satisfaction focus" . This is the reason why customer orientation involves cultural attitudes such as developing and enhancing value to customers. Customer orientation is all about implementation and since culture is "the way things are done" customer orientation should permeate the overall activities of a company.

(ii) Necessity of Customer Orientation?

Market-orientation, and especially customer-orientation, has been indicated as having positive effects on firms' performances. Some researchers argue that profit is a consequence of customer orientation and some others assert that the best way to beat the competition is through customer orientation rather than competitor preoccupation. Moreover, there are a growing number of contexts wherein market orientation is found to be appropriate. However, many managers do not use this strategy to run their companies.

(iii) Organizational Culture

Several marketing researchers tried to look for the reasons and the possible barriers to market- or customer-orientation , Some argue that the principal barrier to market orientation is organizational culture , for instance, explains that the development of market-orientation depends from the ability of an organization to overcome its cultural barriers because firm's level of customer orientation is positively related to its performance and that there is a positive link between the extent of adoption of innovation orientation and the degree of customer orientation in small and medium enterprises ,indicate that high levels of market dynamism and competitive intensity do not tend to result in an increased emphasis on customer-oriented strategies amongst SMEs (Small and Medium Enterprises)."Although some firms are likely to engage in increased customer-oriented activities in highly dynamic or competitive conditions, others may adopt cost or price cutting strategies in an effort to counter or respond to the competitive pressures"

(iv)Management Influence

A lot of other studies explain that an organization has or does not have a customer or market-orientation depending from its management. It is the first to identify and discuss four potential pitfalls to market orientation: lack of inter-functional integration; political manoeuvring; weak management skills; and executive inexperience describes that these behavioural obstacles can severely restrict market orientation development and thus organisational performance.

In the case of shop floor employee behaviour and market orientation with the aim of exploring operational level behavioural barriers.Thus we can say management is a source of barriers to customer-orientation, following seven different barriers at the shop floor level are achieved , including: instrumentalism; short-termism; and weak management support. Sometimes managers and executives often share the notion of improving market orientation levels but behave in a different manner. Thus, the "illogical" behaviour is caused by different cultural beliefs and concludes that "the difficulty often is not so much in getting management to accept this vision, but rather in overcoming the inertia bred of individual corporate cultures. mostly management behaviour which is formalised, conflictual or politically motivated is negatively associated with the extent of market orientation.

(v) Market and Customer Types

Since the level of market- or customer-orientation depends upon the type of markets and customers that the organisation targets. Means , companies have the possibility to choose between a market-, and thus customer-orientation or a competitor-orientation. By adopting a competitor-orientation strategy, companies choose to follow the strategy of the competitors and to try to influence them in order to get the large market share as well as the competitive advantage.

Thus, when market demand is predictable, the competitive structure is concentrated and stable, and there are a few powerful customers, the emphasis is necessarily on competitors.

In other words, even if the competitor-orientation seems to be the better strategy to gain market shares and competitive advantage, customer- or market-orientation may also be a great opportunity for differentiation and so, for a better adaptation to customers' needs or tastes. Furthermore, the small or medium firms are less likely to have a market- or customer-orientation.

But a small firm with a high level of market orientation may secure a strong competitive advantage in a commodity product environment because most large firms respond to the very high level of price-oriented competitive intensity with an emphasis on cost-oriented activities. A market-oriented small firm might be able to achieve high customer satisfaction levels in a commodity market.

9.4. Customer Orientation and Global Corporate Strategy

Basically "**Customer is king**", as a motto, lead the company to succeeding on the crowded and fiercely competitive market. From market orientation perspective, the company should learn how to adapt to customer and to changing consumer lifestyle and behaviour. Studying and understanding consumer behaviour, which provides insights into product, pricing, retail, advertising, and communication, is prerequisite to build up and strengthen the market place of a company. This is described as the marketing concept—the process of planning and executing the conception, pricing, promotion, and distribution of ideas, goods, and services to create exchanges that satisfy individual and organisational objectives . Traditionally, most of the studies emphasise on "why people buy" or "what makes them buy". Lately, why and how people consume comes into consideration. Customers only want to buy some valuable things that can meet their satisfaction. To create value is the primary task of a firm. there are many variables that affect the consuming process and organizational influences finally determine what customer can buy from manufacture to retail. When we look at the large global corporation, most of the strategies, such as market segmentation, a focus on customer service, monitoring consumer trends, database management, and global marketing, regard customers as the business centre to extend their enterprise.

Steps in Corporate Strategy: -following steps must be taken for global corporate strategy.

(i) Developing a shared vision about the market and how it is expected to changein the future.

(ii) Selecting avenues for delivering superior value to customers.

(iii) Positioning the organization and its brands in the marketplace using distinctive Competencies.

(iv) Recognizing the potential value of collaborative relationships with customers, Suppliers, distribution channel members, internal functions, and evencompetitors.

(v) Reinventing organizational designs to implement and manage future strategies.

Here are ten steps generally used by the industrial organization that helps in customer orientation.

- **Create a customer value proposition (CVP) :-** Decide at board level the sum total of benefits which you as a vendor promise to deliver to your customer, in return for the customer's loyalty.
- **Recruit customer-friendly people: -** Hire for attitude, train for skills is the mantra here. Empathy, good communication and problem solving abilities are the qualities to look for.
- **Treat your employees well: -** How your employees feel at work has a much bigger impact on how they deal with customers than any training will.
- **Train your team: -** Frontline as well as back-office staff must gain a full understanding of the customer, product and industry they support. Coaching must also focus on the soft skills; communication and teamwork.
- **Walk the talk :-** Leaders must fully embrace the customer service orientation process and take to the frontline from time to time. Companies that value a culture of servant leadership will excel in this area.
- **Implement a CORE program :-** CORE is an acronym for: Customer Orientation, Referencing & Engagement. In this program cross-functional teams focus on the needs and behaviors of customers as well as internal procedures. An orientation is built around customer needs and the customer profile forms a reference point for the business.
- **Listen to the Voice of the Customer (VOC) :-** An honest appraisal of how you are doing in the eyes of your customer is critical. This can be achieved by conducting a formal customer satisfaction survey or by gathering and recording customer comments.
- **Define your standards :-** By analyzing feedback obtained via your VOC program you will be able to fine tune your service levels according to customer needs and expectations. Create a Balanced Scorecard to measure your performance and rectify any shortcomings.
- **Empower your staff:-** Make sure your team has the authority to resolve most customer complaints without further escalation to a supervisor.
- **Co-ordinate functions :-** Avoid silo mentality by encouraging different departments and functions to work closely together.

9.5. External & Internal Customer

An external customer is someone who uses your company's products or services but is not part of your organization. If you own a retail store, for example, an external customer is an individual who enters your store and buys merchandise. An internal customer is any member of your organization who relies on assistance from another to fulfill her job duties, such as a sales representative who needs assistance from a customer service representative to place an order. External customers are essential to the success of any business, as they provide the revenue stream through their purchases that the enterprise needs to survive. Satisfied external customers often make repeat purchases as well as refer your business to other people they know. A customer who suffers through a negative experience with a business, such as being treated rudely by an employee, can also hinder a business by dissuading others from patronizing it.

Internal customers :-Internal customers include current employees, distributors, vendors or departments. These type of customers are integral in varying points of the product line and depend on their employer to provide the overall finished goods to external customers. Unlike external customers, **internal** customers don't generally have a choice in certain business aspects. For example, if your payroll department doesn't agree with guidelines created by your human resources division, the payroll department can't fire human resources. The success of your business may depend on improving your internal customer satisfaction. If this is the case, you can take steps to bring about improvement.

(i) **Set clear expectations for your internal customers :-** These may include following daily departmental protocol, adhering to specific policies or guidelines, or accomplishing certain goals every business quarter. Inform internal customers of timelines and deadlines to better guide them toward completing objectives.

(ii) **Hold individual, all-staff or inter-departmental meetings where you can present your priorities** to specific people or groups. Following your meetings, send out an email to your internal customers reiterating what your key priorities are so there is no confusion.

(iii) **Anticipate internal customer expectations, such as merit raises or promotions,** title advancements when projects are accomplished or better performance guidance by upper-level management. Monitor daily performance and encourage customers to come to you with specific needs and concerns so you will be better prepared to deal with them going forward.

(iv) Clear communication about what priorities are to internal customers is essential for a business to function efficiently. This communication will ensure that everyone is on the same page when it comes to daily tasks, responsibilities, concerns and objectives. Understanding your internal customers' expectations can result in better work flow, and can clear up potential misunderstandings or muddled priorities.

9.6. Life Cycle Approach to Quality Costs – Prevention

Cost of Quality (COQ) is the sum of the costs incurred by a company in preventing poor quality, the costs incurred to ensure and evaluate that the quality requirements are being met, and any other costs incurred as a result of poor quality being produced. Poor quality is defined as non-value added activities, waste, errors or failure to meet customer needs , and requirements. These COQ costs can be broken down into the three categories of prevention, appraisal and failure costs. The cost of quality is also defined as "Cost of Poor Quality". Sometimes poor quality costs refer only to the "failure" costs. Crosby refers to the COQ costs as the "price of conformance" – the prevention and appraisal costs and the "price of nonconformance" – the failure costs .

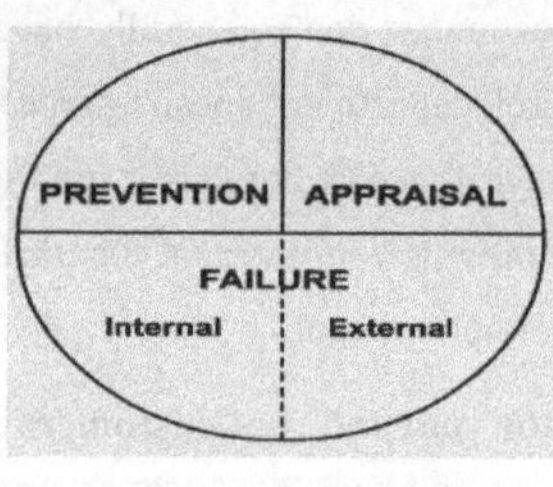

Diagram 9.1: Quality Cost

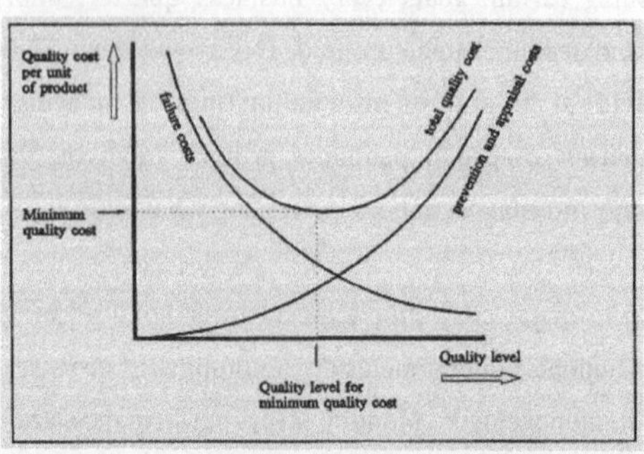

Diagram 9.2: The Relationship Amongst Prevention Costs, Appraisal Costs and Failure Costs

9.6.1. *Prevention Costs*

The planned costs incurred by an organization to ensure that errors are not made at any of the various stages during the delivery process of that product or service to a customer. The delivery process may include design, development, production and shipping.

Examples of prevention costs include education and training, continuous improvement efforts, quality administration staff, process control, market research, field testing and preventive maintenance.

9.6.2. *Appraisal Costs*

The costs of verifying, checking or evaluating a product or service at the various stages during the delivery process of that product or service to the customer.

Examples of appraisal costs include receiving or incoming inspection, internal product audit, inspection activities, inventory counts, quality administration salaries, supplier evaluation and audit reports.

9.6.3. *Failure Costs*

The costs incurred by a company because the product or service did not meet the requirements and the product had to be fixed or replaced or the service had to be repeated. These failure costs can be further subdivided into two groups – internal or external failures. Internal failures include all the costs resulting from the failures that are found before the product or service reaches the customer.

Examples include scrap, rework, extra inventory, repair stations, re-design, salvage, corrective action reports and overtime due to nonconforming product or service. External failures are all the costs incurred by the company resulting when the customer finds the failure. These external failure costs do not include any of the customer's personal costs.

Examples of these costs include warranty, customer complaint administration, replacement product, recalls, and shipping costs, analysis of warranty data, customer follow-up and field service departments. Many of the COQ costs are hidden and very difficult to identify by formal COQ measurement systems. Many of these costs, if identified, would be considered as the cost of doing business. Three major groups of hidden costs that are not considered in most COQ systems include customer incurred costs, lost reputation costs and customer dissatisfaction costs.

9.7. Quality Philosophies

Several individuals made significant contributions to quality control and improvement. We take a closer look at the approach and philosophies of W. Edwards Deming, Joseph M. Juran, Philip B. Crosby, and Armand V. Feigenbaum Contribution of quality improvements by the various quality gurus are given below.

W. Edwards Deming: W. Edwards Deming is perhaps the best-known quality expert in the world. He was instrumental in the post-war industrial revival of Japan. Subsequently his ideas were increasingly adopted in industry in the United States and other countries. Dem- ing received his electrical engineering degree from the University of Wyoming and his Ph.D. in mathematical physics , Deming firmly believed that quality is the responsibility of the management. The Deming philosophy is summarized in the following fourteen points, which were included in his monumental work Out of the Crisis

The fourteen points apply to both small and large organizations, to the service Industry as well as to manufacturing. They also apply to a division within a company. The fourteen points are presented here.

- **(i) Create constancy of purpose for improvement of product and service :-** The point stresses the need for a mission statement which must be understood by all employees, suppliers, and customers. The strategic plan should look for the long- term payback.
- **(ii) Adopt the new philosophy: -** Management must learn the responsibilities and take on leadership for change. Poor workmanship, defective products, or bad service are not acceptable.
- **(iii)Cease dependence on mass inspection: -** Eliminate the need for inspection on a mass basis by building quality into the product in the first place. Statistical methods of quality control are more efficient.
- **(iv) End the practice of awarding business on the basis of price tag alone**: - Instead, minimize the total cost. The aim in the purchase of new tools and other equipment should be to minimize the net cost per hour of operation or per piece produced. Move toward a single supplier for any one item, on a long-term relationship of loyalty and trust.
- **(v) Improve constantly and forever the system of production and service: -** The improvement of product and service is an ongoing process. The Deming cycle involves the four-step process of plan, do, check, act. At the plan stage, the opportunities for improvement are identified. The theory is tested on a small scale at the do stage, the

results of the test are analyzed at the check stage, and the results are implemented in the act stage.

(vi) **Institute training:** -On-the-job training must be provided for all employees. Employees must be encouraged to implement the knowledge developed through training.

(vii) **Adopt and institute leadership:** - The aim of supervision should be to help people to do a better job using machines. Supervision must create an environment where the workers take leadership roles in accomplishing their work.

(viii) **Drive out fear :-** Management must create an environment where workers are encouraged to ask questions and make suggestions. A climate of innovation leads to progress.

(ix) **Break down barriers between departments:** -People in research, design, material procurement, sales, and production must work as a team. They must understand the requirements and specifications. Teamwork leads to improvements in quality and productivity.

(x) **Eliminate slogans, exhortations, and targets for the work force:** - Exhortations such as asking for zero defects and new levels of productivity only create adversarial relationships. The bulk of the causes of low quality and low productivity belong to the system and thus lie beyond the power of the work force.

(xi) **Eliminate numerical quotas for the work force** and eliminate numerical goals for people in management. Quotas lead to the deterioration of quality. Learn the capabilities of processes and methods to improve them.

(xii) **Remove barriers that rob people of pride of workmanship**: - Quality is achieved in the company when all employees are satisfied and motivated. Management must create an environment where the workers take pride in thei

(xiii) **Encourage education and self-improvement for everyone:** - An organization needs people that are improving with education.

(xiv) **Take action to accomplish the transformation.**

Joseph M. Juran :- Joseph M. Juran is the founder of the Juran Institute, which offers consulting and management training in quality. Juran obtained his degree in electrical engineering from the University of Minnesota in 1924 and then worked with Shewhart atWestern Electric. He introduce first statistical process control techniques for factories. He published his Quality Control Handbook in 1951. In 1954 he was invited to Japan, where he conducted training courses in quality management. He defined quality as **fitness for use**.

Juran proposed the quality tribology :- Quality planning, Quality control, and quality improvement to develop a universal thought process for quality. Quality planning is the process for preparing to meet the company's goals. Both internal and external customers are identified and their needs are determined. Products and services are developed to fulfill these needs. Quality control is the process for meeting company goals during operations, and statistical process control techniques are the primary tools of control. Juran stated categorically that waste must be identified and eliminated. Juran conceptualized thePareto principle, which helps in identifying thevital fewout of the trivial many .This is commonly referred to as the 80–20 principle – 80% of the problems are createdby 20% of the causes.

Philip B. Crosby: - Philip B. Crosby is a businessman and author who influenced quality improvement through his writings and lectures. He started the Crosby Associates, which provides consulting and training in quality management.

He stated that quality is Free; Crosby provides a detailed quality management grid which provides various stages of management understanding and attitude, quality organization status, problem handling, cost of quality in relation to sales, qualityimprovement actions, and company quality posture.he also introduce a slogan **"doing it right the first time."**

He gives a four major principles these are –

- Quality is **"conformance to requirements,"**
- The management system is prevention.
- The performance standard is zero defects.
- The measurement system is the cost of nonconformance.

The concept of zero defects has led to the creation and development of Six Sigma, pioneered by Motorola, which has since been adopted worldwide by many other organizations.

Armand V. Feigenbaum :- Armand V. Feigenbaum is a pivotal figure in the history of quality. He received his Ph.D. from MIT. In 1951 he published his book Quality Control: Principles, Practice and Administration.

He proposed a three-step process for quality improvement, these are

- Quality leadership.
- Quality technology,
- Organizational commitment.

Total quality control is an effective system for integrating the quality development, quality maintenance, and quality improvement efforts of the various groups in an organization, enabling production and service to operate at the most economical level to achieve full customer satisfaction.

Thus all **GURU'S Philosophies** stressed the importance of quality and the pivotal role management must play in the implementation of quality improvement.

9.8. Quality Control Tools

The 7 QC Tools are simple statistical tools used for problem solving. These tools were either developed in Japan or introduced to Japan by the Quality Gurus such as Deming and Juran. In terms of importance, these are the most useful. Kaoru Ishikawa has stated that these 7 tools can be used to solve 95 percent of all problems. These tools have been the foundation of Japan's astonishing industrial resurgence after the Second World War. For solving quality problems seven QC tools used are **Pareto Diagram, Cause & Effect Diagram, Histogram, Control Charts, Scatter Diagrams, Graphs and Check Sheets**. All these tools are widely used in manufacturing process to monitor the overall operation and for continuous improvement. All these tools detected root causes and eliminates them, thus the manufacturing process can be improved.

For Solving Quality Problems Following Seven QC Tools are Described Below:

(i) Pareto Charts: - Alfredo Pareto was an economist who noted that a few people controlled most of a nation's wealth. "Pareto's Law" has also been applied to many other areas, including defects, where a few causes are responsible for most of the problems. Separating the "vital few" from the "trivial many" can be done using a diagram known as a Pareto chart. Pareto principle also known as the 80/20 rule is used in the field of materials management for ABC analysis. 20% of the items purchased by a company account for 80% of the value.

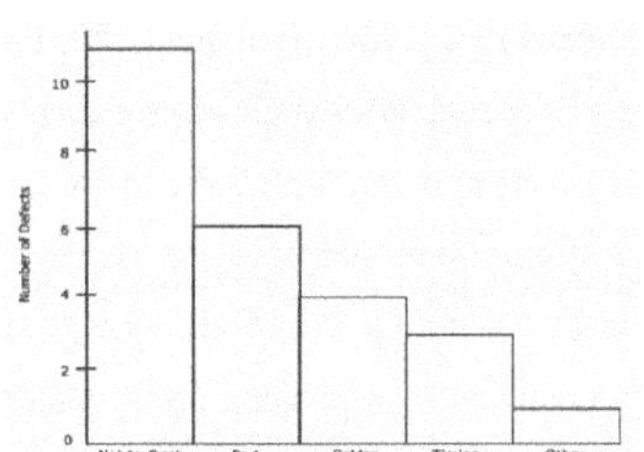

Diagram 9.3: Pareto Charts

Construction of Pareto Charts: -Following steps must be taken for constructing the Pareto Charts.

- From the available data calculate the contribution of each individual item.
- Arrange the items in descending order of their individual contributions. If there are too many items contributing a small percentage of the contribution, group them together as "others". It is obvious that "others" will contribute more than a few single individual items. Still it is kept last in the new order of items.
- Tabulate the items, their contributions in absolute number as well as in percent of total and cumulative contribution of the items.
- Draw X and Y axes. Various items are represented on the X-axis. Unlike other graphs Pareto Diagrams have two Y-axes - one on the left representing numbers and the one on right representing the percent contributions. The scale for X-axis is selected in such a manner that all the items including others are accommodated between the two Y-axes. The scales for the Y-axes are so selected that the total number of items on the left side and 100% on the right side occupy the same height.
- Draw bars representing the contributions of each item.
- Plot points for cumulative contributions at the end of each item. A simple way to do this is to draw the bars for the second and each subsequent item at their normal place on the X-axis as well as at a level where the previous bar ends. This bar at the higher level is drawn in dotted lines. Drawing the second bar is not normally recommended in the texts.
- Connect the points. If additional bars as suggested in step 6 are drawn this becomes simple. All one needs to do is - connect the diagonals of the bars to the origin.
- The chart is now ready for interpretation. The slope of the chart suddenly changes at some point. This point separates the 'vital few' from the 'useful many' like the A, B and C class items in materials management.

(ii) Ishikawa Diagrams: - Ishikawa diagrams are named after their inventor, Kaoru Ishikawa. They are also called fishbone charts, after their appearance, or cause and effect diagrams after their function. Their function is to identify the factors that are causing an undesired effect (e.g., defects) for improvement action, or to identify the factors needed to bring about a desired result (e.g., a winning proposal). The factors are identified by people familiar with the process involved. As a starting point, major factors could be designated using the "four M's": Method, Manpower, Material, and Machinery; or the "four P's": Policies, Procedures, People, and Plant. Factors can be subdivided, if useful, and the identification of significant factors is often a prelude to the statistical design of experiments. Ishikawa diagram attempting to identify potential causes of defects in a wave solder process.

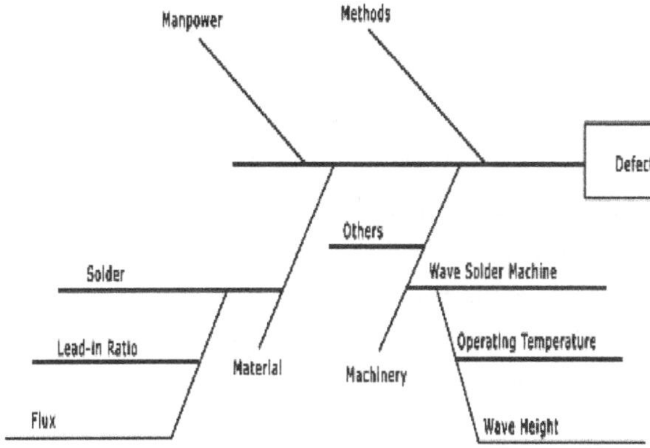

Diagram 9.4: Ishikawa or Fish-Bone Diagrams

Construction of Ishikawa Diagrams: - Following steps must be taken for constructing the Pareto Charts.

- Determine the main groups or categories of causes. Place them in boxes and connect them through large bones to the backbone. Brainstorm to find possible causes and subsidiary causes under each of the main groups.
- Make sure that the route from the cause to the effect is correctly depicted. The path must start from a root cause and end in the effect.
- After completing all the main groups, brainstorm for more causes that may have escaped earlier.
- Once the diagram is complete, discuss relative importance of the causes. Short list the important root causes.

(iii) Histograms: Histograms are another form of bar chart in which measurements are grouped into bins; in this case each bin representing a range of values of some parameter. For example, in given Figure, X could represent the length of a rod in inches. The figure shows that most rods measure between 0.9 and 1.1 inches. If the target value is 1.0 inches, this could be good news. However, the chart also shows a wide variance, with the measured values falling between 0.5 and 1.5 inches. This wide a range is generally a most unsatisfactory situation.

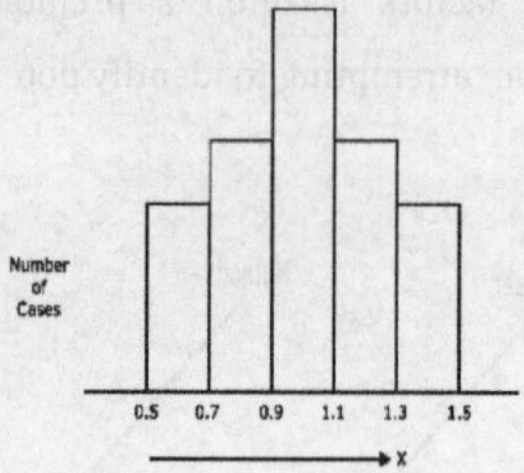

Diagram 9.5: Histograms

Besides the central tendency and spread of the data, the shape of the histogram can also be of interest. For example, the given histogram shows a bi-modal distribution. This indicates that the measurements are not from a homogeneous process, since there are two peaks indicating two central tendencies. There are two (or more) factors that are not in harmony. These could be two machines, two shifts, or the mixed outputs of two suppliers. Since at least one of the peaks must be off target, there is evidence here that improvements can be made.

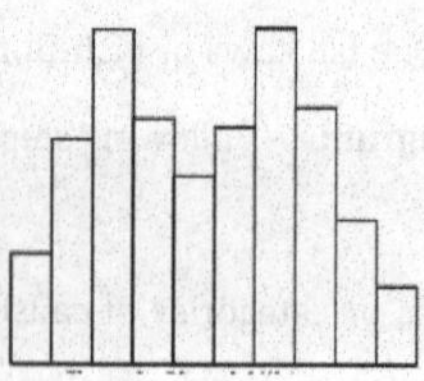

Diagram 9.6: Bio Model Histogram

- Collect data (preferably 50 or more observations of an item).
- Arrange all values in an ascending order.
- Divide the entire range of values into a convenient number of groups each representing an equal class interval. It is customary to have number of groups equal to or less than the square root of the number of observations. However one should not be too rigid about this. The reason for this cautionary note will be obvious when we see some examples.
- Note the number of observations or frequency in each group. 1
- Draw X-axis and Y-axis and decide appropriate scales for the groups on X-axis and the number of observations or the frequency on Y-axis.
- Draw bars representing the frequency for each of the groups.
- Provide a suitable title to the Histogram. 1. Study the pattern of distribution and draw conclusion.

In contrast, the Figure given below shows a situation in which the spread of measurements is lower on one side of the central tendency than on the other. These could be measurements of miles per gallon attained by an automobile. There are many situations that decrease fuel economy, such as engine settings, tire condition, bad weather, traffic jams, etc., but few situations that can significantly improve it. The wider variance can be attacked by optimizing any of the controllable factors such as tuning the engine, replacing the tires used, etc. Moving the central tendency in the direction of the smaller variance is unlikely unless the process is radically changed (e.g., reducing the weight of the vehicle, installing a new engine, etc.).

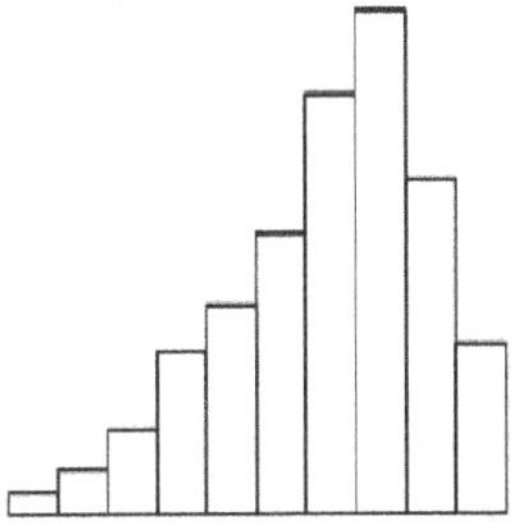

Diagram 9.7: Skewed Histogram

(iv) Scatter diagrams :- Scatter diagrams are a graphical, rather than statistical, means of examining whether or not two parameters are related to each other. It is simply the plotting of each point of data on a chart with one parameter as the x-axis and the other as the y-axis. If the points form a narrow "cloud" the parameters are closely related and one may be used as a predictor of the other. A wide "cloud" indicates poor correlation. The above Figure shows a plot of defect rate vs. temperature with a strong positive correlation, while Figure given below shows a weak negative correlation.

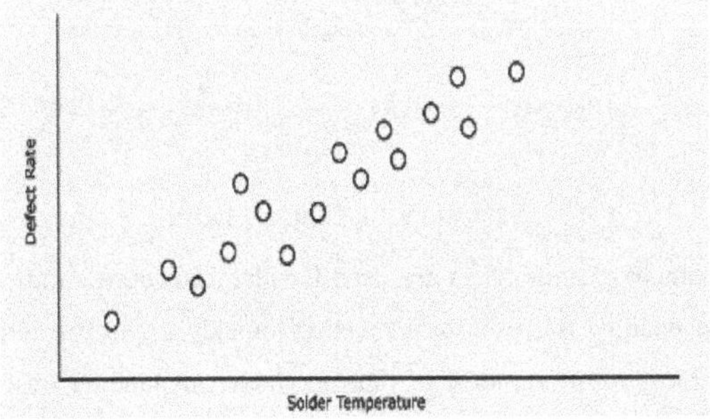

Diagram 9.8: Scatter Diagram Showing Strong Correlation

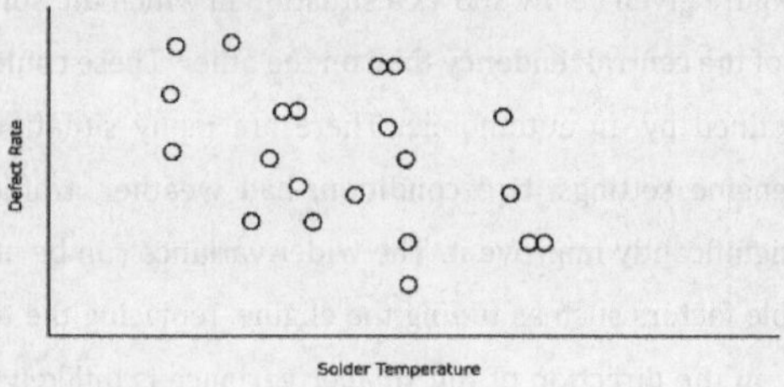

Diagram 9.9: Scatter Diagram Showing Weak Correlation

(v) Control Charts : - Control charts are the most complicated of the seven basic tools of TQM, but are based on simple principles. The charts are made by plotting in sequence the measured values of samples taken from a process. For example, the mean length of a sample of rods from a production line, the **number** of defects in a sample of a product, the miles per gallon of automobiles tested sequentially in a model year, etc. These measurements are expected to vary randomly about some mean with a known variance. From the mean and variance, control limits can be established. Control limits are values that sample measurements are not expected to exceed unless some special cause changes the process. A sample measurement outside the control limits therefore indicates that the process is no longer stable, and is usually reason for corrective action. Other causes for corrective action are non-random behavior of the measurements within the control limits. Control limits are established by statistical methods depending on whether the measurements are of a parameter, attribute or rate. For example control chart is given below:

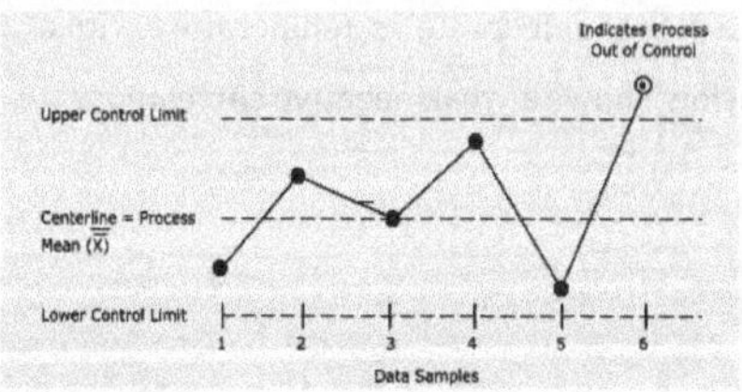

Diagram 9.10: Control Charts

(vi) Graphs: - Graphs of various types are used for pictorial representation of data. Pictorial representation enables the user or viewer to quickly grasp the meaning of the data. Different graphical representation of data is chosen depending on the purpose of the analysis and preference of the audience. The different types of graphs used are as given below:

S no	Type of graph	purpose
1	Bar Graph	To compare sizes of data
2	Line Graph	To represent changes of data
3	**Gantt chart**	**To plan and schedule**
4	**Radar chart**	**To represent changes in data (before and after)**
5	**Band Graph**	**Same as above**

(vii) Check Sheets: - Check sheets are tools for collecting data. They are designed specific to the type of data to be collected. Check sheets aid in systematic collection of data. Some examples of check sheets are daily maintenance check sheets, attendance records, production log books, etc. Data collected using check sheets needs to be meaningfully classified. Such classification helps gaining a preliminary understanding of relevance and dispersion of the data so that further analysis can be planned to obtain a meaningful output. Meaningful classification of data is called stratification. Stratification may be by group, location, type, origin, symptom, etc.

(viii) Quality control by PDCA-CYCLE: - we have 7- quality control tools, which is helpful in controlling the **quality** in all respect, except above we have another tool called PDCA-CYCLE, this cycle is described in short as given below.

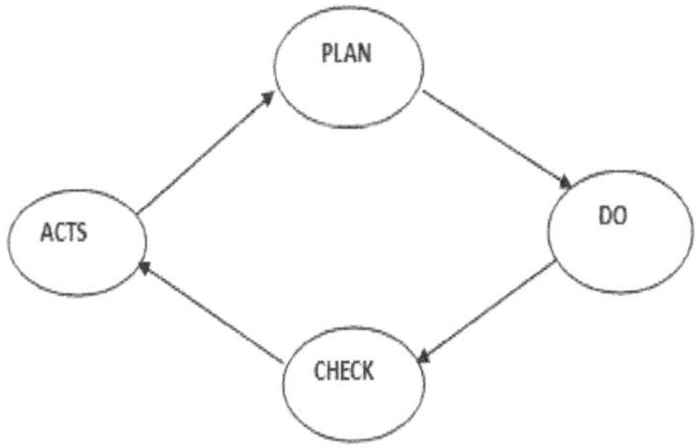

Diagram 9.11: PDCA - Cycle

1. **Plan -** Analysis of what needs to be improved by taking into consideration areas that hold opportunities - for change. Decision on what should be changed.
2. **Do -** Implementation of the changes that are decided on in the Plan step.
3. **Check -** Control and measurement of processes and products in accordance to changes made in previous steps and in accordance with policy, goals and requirements on products. Report on results.
4. **Act -** Adoption or reaction to the changes or running the PDCA-cycle through again. Keeping improvement on-going.

9.9. Process Capability Concepts

Process capability analysis (PCA) and Six Sigma methodology plays an important role in quality and process improvement initiatives. As a fundamental technique in any production, quality and process improvement efforts, PCA is used to improve processes, products or services to achieve higher levels of customer satisfaction. In order to measure process capability numerically, process capability indices (PCIs) have been developed. Six-Sigma is widely recognized as a systematic methodology that employs statistical and non statistical tools and techniques for continuous quality and process improvement and for managing operational excellence because it challenges to maximize an organization's return on investment (ROI) through the elimination of nonconforming units or mistakes in the processes.

Process: Process is defined as a combination of materials, methods, equipments and people engaged in producing a measurable output. As a matter of fact, all processes have inherent statistical variability, which can be identified, evaluated and reduced by statistical methods. The source and amount of variability should always be considered by organizations. In order to satisfy customer requirements, organizations must improve the quality by reducing variance in production processes. The less variation the system has, the better quality it provides. Thereby, the variability of critical-to-quality characteristics (CTQs) is a measure of the uniformity of outputs. When the variation is large, the numbers of products that are nonconforming are large. Nonconforming (NC) is the failure of meeting specification limits whereas specifications are the desired measurements for a quality characteristic.

Process capability: In particular, process capability deals with the uniformity of the process. Variability of CTQs in the process is a measure of the uniformity of outputs. Here, variability can be thought in two ways: one is inherent variability in a CTQ at a specified time, and the other is variability in a CTQ over time. It should be considered that process capability study frequently measures functional parameters or CTQs on the product. It does not measure the process itself. Process capability compares inherent variability in a process with the specifications that are determined according to the customer requirements. In other words, process capability is the proportion of actual process spread to the allowable process spread, which is measured by six process standard deviation units. Process capability compares the output of a process that is an in-control state to the specification limits by using PCIs. To sum up, a capable process is the one where almost all the measurements fall inside the specification limits and process capability study can be conducted to indicate the extent to which the process can meet these specifications.

In a true process capability study, when there is direct observation of the process, inferences can be made about the stability of the process over time by directly controlling or monitoring data collection activity and understanding the time sequence of the data. However, when there is no direct observation of the process, only sample units of product are known, in this case, the study is called product characterization.

In a product characterization study, distribution of the product quality characteristic or the fraction that conforms to specifications, which is referred to as process yield, can only be estimated, notably information about stability or dynamic behavior of the process cannot be given .

Process Stability and Process Capability : Process Stability and Process Capability are both extremely important aspects of any manufacturing process , Process Stability refers to the consistency of the process with respect to important process characteristics such as the average value of a key dimension or the variation in that key dimension.

If the process behaves consistently over time, then we say that the process is stable or in control. The graphic below illustrates a stable process. The process distribution remains consistent over time.

Diagram 9.12: Controlable Stable Process

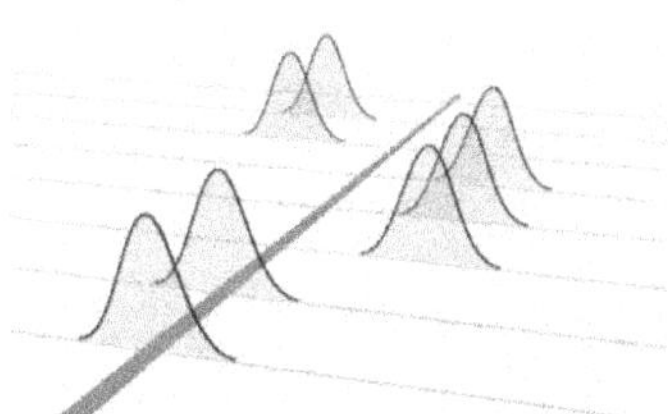

Diagram 9.13: Unstable Process

Statistical Process Control Charts are utilized to determine if the process is stable or not. Some charts are used to assess the stability of the process location (e.g. x-bar charts which monitor the process average), other charts are used to assess the stability of the process variation.

Process Capability is a measure of the ability of the process to meet specifications. It tells us how good the individual parts are. Several methods exist to measure process capability including an estimation of the ppm (defective parts per million). Capability indices such as Cp, Cpk, Pp, Ppk are very popular. Process capability is an important concept for industrial managers to understand. The challenge in today's competitive markets is to be on the leading edge of producing high quality products at minimum costs. In this context , a very important question arises ie. Why is process capability so important? Because it allows one to quantify how well a process can produce acceptable product. As a result, a manager or engineer can prioritize needed process improvements and identify those processes that do not need immediate process improvements. Process capability studies indicate if a process is capable of producing virtually all conforming product. If the process is capable, then statistical process controls can be used to monitor the process and conventional acceptance efforts can be reduced or eliminated entirely. This not only yields great cost savings in eliminating non-value added inspections but also eliminates scrap, rework and increases customer satisfaction. The benefits of performing process capability studies are certainly worth the effort in the long run. After a process capability study has been performed, a process will be classified as either capable or incapable. When the process is not capable of producing virtually all conforming product, the process is said to be incapable and acceptance sampling procedures (or 100%inspection) must remain part of the process. Figure shows the process output when compared to the specification limits for a typical incapable process.Processes may also start out as capable but changes over time to have more variability. In addition, the process mean may also shift placing the process too close to one of the specification limits. Both increases in process variability and shifting of the mean may result in once capable processes becoming incapable.

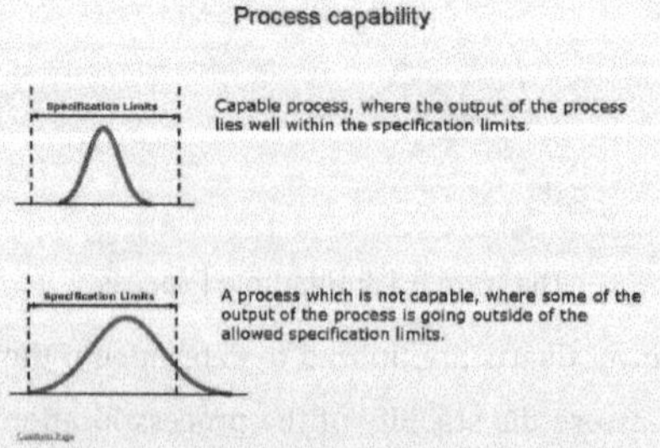

Diagram 9.14: Process Capability

Process Capability Study: A process capability study is a systematic procedure for determining the capability of a process. The process capability study may include studies to improve the process and in turn the capability of the process. Process capability studies are usually performed as part of a process certification effort or a process optimization effort. Process Capability Study Steps:

(i) **Select Critical Parameters: -** Critical parameters need to be selected before the study begins. Critical parameters may be established from drawings, contracts, inspection instructions, work instructions, etc. Critical parameters are usually correlated to product fit and/or function.

(ii) **Collect Data: -** A data collection system needs to be established to assure that the appropriate data is collected. It is preferable to collect at least 60 data values for each critical parameter. If this is not possible, corrections can be made to adjust for the error that is introduced when less than 60 data values are collected .Significant digits for each data should be the number of significant digits required per the specification limits plus one extra significant digit to assure that process stability can be evaluated.

(iii) **Establish Control over the Process: -** A distinction between product and process should be made at this point. The product is the end result from the process. The product may be a physical item (Example: fabricated part) or a service (example: typing a report). One may control the process by measuring and controlling parameters of the product directly or measuring and controlling the inputs to the process (once correlation between the process inputs and product critical parameters has been established). It is ultimately desirable to establish control over the process by controlling the process inputs. On the other hand, process capability indices are always performed using the critical parameters of the product , Calculation of predictable process capability indices is dependent on the statistical control of the process. If the process is not in statistical control, then the results of the study are subject to fluctuate unpredictably. The statistical control of the process can be studied using control charts (usually X-bar-R charts).

(iv) **Analyze Process Data: -** To calculate the process capability indices, estimates of the process average and dispersion (standard deviation) must be obtained from the process data. In addition, the formulas for process capability indices assume that the process data came from a normal statistical distribution. It is important that one prove that the data is normally distributed prior to reporting the process capability indices because errors in misjudgment can lead to the same undesirable effects as listed in Step 1. Methods for handling non-normal data and formulas for several process capability indices will be addressed in separate sections

(v) Analyze Sources of Variation: - Study of the component sources of variation and their magnitudes may range from simple statistical tests to complex experimental designs carried out over a long period of time. If possible, tests should be kept simple. Analyzing sources of variation involves determining what process factors affect the natural process spread (process variation) and the process centering. With this knowledge, it may be possible to improve the process' capability. Analyzing sources of variation always involve careful planning and data collection .

(vi) Establish Process Monitoring System :- Once the process capability indices indicate a capable process, a routine process control technique should be employed to assure that the process remains stable. This may be done by a variety of methods such as establishing a statistical process (SPC) program. The process capability indices should also be periodically recalculated to assure the process mean and spread has not significantly changed.

9.10. Process Capability Indices (PCIs)

Process capability indices (PCIs) are also called process capability ratios (PCRs). PCIs are used as tools for characterizing the process quality. In order to measure the process capability numerically, PCIs have been developed. PCIs are also defined as the quantitative indicators that compare the behavior of process or product characteristic to the specifications. In other words, PCIs are used to determine how well the process performs with respect to specifications and they express the ability of the process to meet these specifications, as a unique value quantitatively. There are several statistics that can be used to measure the capability of a process, In practice, Cp, Cpk (Cpl, Cpu), Cpm are some of the widely used PCIs. Process Capability index Cp In the literature, Cp index is also called process potential index, or process capability ratio, or inherent capability index, and two-sided PCI for two-sided specifications, that is, process is having both lower and upper specification limits. Cp is frequently used in industrial environment in order to express process capability in a simple quantitative way.

When the parameters are known, that is, in that case, when process standard deviation is known, PCI Cp is computed as follows:

$$Cp = \frac{USL-LSL}{6\sigma}$$

Where LSL and USL are lower and upper specification limits, respectively. The percentage of the speciation band used up by the process can be calculated in the following way:

$$P = \left(\frac{1}{Cp}\right) \cdot 100$$

Where LSL and USL are lower and upper specification limits, respectively. A Cp value less than 1 indicates that the process variation exceeds the specifications and a significant number of defects are made. A Cp value equal to 1 indicates that the process is exactly meeting the specifications. A Cp value greater than 1 indicates that the process variation is less than the specifications.A Cp value greater than 1.67 indicates that the process is highly capable.

9.11. Process Capability Index Cpk

Cpk is defined as one-sided PCI for specification limit nearest to the process mean. When the parameters are known, that is, in that case, when process mean μ and process standard deviation are known, PCI Cpk is computed as follows.

$$Cpk = \frac{1}{3\sigma}\min(USL-\mu,\ \mu-LSL) = \min(Cpu,\ Cpl)$$

Where LSL and USL are lower and upper specification limits, respectively. In practice, it is often impossible to know parameters. Generally, it is suitable to use sample mean x to estimate process mean μ and sample standard deviation s to estimate process standard deviation .When the parameters are unknown, that is, in that case, when process mean μ and process standard deviation are unknown, by replacing sample mean x and sample standard deviation s to estimate process mean μ and process standard deviation, respectively, the formula used for estimating Cpk is given below: Where LSL and USL are lower and upper specification limits, respectively.

$$\widehat{Cpk} = \frac{1}{3s} \cdot \min(USL-\bar{x},\ \bar{x}-LSL) = \min(Cpu,Cpl)$$

1. If Cp=Cpk, the process is centered at the midpoint of the specification limits.
2. k<Cp, the process if off-centered. This can be accepted as lower capability than the case that the process is centered. The reason is that it is not operating at the midpoint of the interval between the specification limits.
3. Cpk=0, the process mean is exactly equal to one of the specification limits.
4. If Cpk<0, the process mean lies outside the specification limits, that is for μ>USL or μ<LSL, Cpk<0.

5. If Cpk<-1, the entire process lies outside the specification limits. It should be noted that some authors define Cpk to be nonnegative so that values less than zero are defined as zero.1<Cpk<1.33 means that the process is barely capable. Automotive industry uses Cpk=1.33 as a benchmark in accessing the capability of a process.

9.12. Process Capability Index CPM

Cpm is defined as the ability of the process to be clustered around the target or nominal value, which is the measurement that meets to exact desired value for the quality characteristic. Actually, Cpm was developed because Cpk is observed to be inadequate measure of process centering although Cpk was developed to deal with the case of a process with mean μ that is not centered between the specification limits whereas Cp is inadequate in process centering. As a matter of fact, when μ is in the interval of the specification limits, LSL and USL, Cpk depends inversely on process standard deviation and becomes large as process standard deviation gets closer to zero. PCI Cpm is computed as follows.

$$Cpm=\frac{USL-LSL}{6\tau}$$

Where τ is the square root of expected squared deviation from target T. The target value T, which is the measurement that meets to exact desired value for the quality characteristic, is known to be the midpoint of the specification interval. Target T is evaluated as follows:

$$T=\frac{1}{2}(LSL+USL)$$

The formula for process variation around desired process target is given below:

$$\tau^2=E\left[(x-T)^2\right]=E\left[(x-\mu)^2\right]+(\mu-T)^2=\sigma^2+(\mu-T)^2$$

Computation of Cpm can also be performed with the following way:

$$Cpm=\frac{USL-LSL}{6\sqrt{\sigma^2+(\mu-T)^2}}=\frac{Cp}{\sqrt{1+\left(\frac{\mu-T}{\sigma}\right)^2}}$$

Process Capability index Cpkm :- The motivation of Cpkm is increased sensitivity to departures of the process mean μ from the desired target value T. Cpkm is known as a third generation PCI, since it is derived from the second generation PCIs Cpk and Cpm, in the same way that the PCIs, Cpk and Cpm are derived from the first generation PCI Cp. Computation of Cpkm is as follows:

$$Cpkm=\frac{Cpk}{\sqrt{1+\left(\frac{\mu-T}{\sigma}\right)^2}}$$

Reliability

9.13. Introduction

In the present scenario of global competition and liberalization, it is imperative that Indian industries become fully conscious of the need to produce reliable products meeting international standards. Even though the "Reliability Engineering" has taken birth during World War II with a significant contribution by defense personnel, today it has taken a new shape by blending itself in all phases of the product life cycle from proposal to manufacturing. It is a known fact that reliability program increases the initial cost of every device, instrument or system and also it is true that the reliability decreases when the complexity of the system increases. In this type of complex situation, reliability of a product or service is best assured when it is designed by the design engineer and built in by production engineer, rather than conducting externally an experiment by a reliability engineer. Once the product is accepted by the buyer and put in operation, either by itself or as a part of a larger assembly, the quality of performance would be judged by how long the product gives useful service; this is indicated by the word "Reliability". Hence reliability is defined as

The reliability is nothing but it is the probability of product / system/ items that will performed its intended function under specified conditions and environment for a certain period of time. the reliability is combination of two things probability of success (R_s) and probability of failure (Q_f) , the sum of both must be equal to 1 , the effect of reliability is measured by the MTBF (mean time between failure).

$$(R_s) + (Q_f) = 1$$

(R_s) = Reliability of success of an items / product / system.

(Q_f) = Reliability of failure of an items / product / system.

$$(Q_f) = 1 - (R_s)$$

$$Rs = \frac{Number\ \ of\ \ survivors\ \ at\ \ time\ \ t}{Numberofitemsputonte\ \ \ stattime\ \ t{=}0}$$

At time t = 0, the number of survivors is equal to number of items put on test. Therefore, the reliability at t = 0 isR(0) = 1 = 100%

Failure Rs The failure rate function λ(t) is analytically related to the time failure probability distribution, $\mathbf{Q_f}$.

$$R_S = 1 - Q_f$$

Now, let Δ(t) be a small time increment fromt. Then, consider the fractional probability with failure that occurs within Δ(t), this fraction of failure could occur only if the product has survived the time period from0 to t. Hence, the probability for the product that fails within Δt is a conditional one, denoted by: $\{Q_f \Delta t\}/Rs$. The time rate of changeof that probability is the failure rate at time t, denoted by λ (t);

$$\lambda(t) = \{\frac{Q\,f\,\Delta t}{R\,s}\} / \Delta t = \{\frac{Q\,f}{R\,s}\} \text{ ....Equ -1}$$

Thus

$$\lambda(t) = \{\frac{Q\,f\,\Delta t}{R\,s}\} / \Delta t = \{\frac{Q\,f}{R\,s}\}$$

Eq. -1 is a formal relation between λ (t) and Q_f. In general, one wishes to obtain Q_f when $\lambda(t)$ is known.

To this end, we note tha $t_f = dF(t)/dt = d[1\text{-}Rs]/dt = -dR(t) / dt$ and Eq. -1 can then be written in the form:

$$\lambda(t) = -[dRs]/dt]/Rs$$

or

$$\lambda(t)dt = -[dRs/Rs]$$

Integration of the above from0 to t on both sides and noting that R (0) =1, we obtain Rsin explicit termof $\lambda(t)$:

$$Rs = \exp[\,-\int_0^t \lambda(\tau)d\tau\,] \text{ ....Equ -2}$$

$$Q_f = \lambda(t)\exp[\,-\int_0^t \lambda(\tau)d\tau] \text{ ....Equ -3}$$

We can readily verify the following, noting that R (τ)=0:

$$\mu = MTTF = \int_0^\infty t\,Qfdt = \int_0^\infty Rs\,dt \text{ ....Equ -4}$$

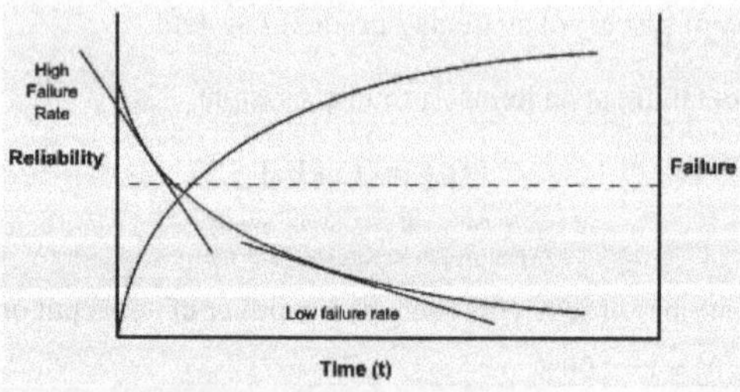

Diagram 9.15: Reliability Curve

Curve shows the reliability of success & failure of the system / product / items, the curve line that is indicated by blue line representing about system failure (Q_f) and reliability of success is represented by the faint red curve (R_S).

These curve showing the exponential behavior thus we have two equation of reliability for failure and success.

$$R_S = e^{-\lambda t} \quad \ldots\ldots\ldots\ldots\ldots\ldots\ldots\ldots 5)$$

$$Q_f = 1 - e^{-\lambda t} \quad \ldots\ldots\ldots\ldots\ldots\ldots\ldots\ldots 6)$$

9.14. Bath Tub Curve

A failure is the partial or total loss or change in the properties of a device in such a way that its functioning is seriously affected or totally stopped. The concept of failures and their details help in the evaluation of quantitative reliability of a device. In general, some components have well defined failures; others do not. In the beginning, when the item or component is installed, the item fails with high frequency, which is known as initial failure or infant mortality. These are generally due to manufacturing defects. They are very high at initial stages and gradually decreases and stabilize over a longer period of time. Stable or constant failures due to chance can be observed on an item for a longer period. These types of failures are known as random failures and characterized by constant number of failures per unit of time. Due to wear and tear with the usage, the item gradually deteriorates and frequency of failures again increases. These types of failures are called as wear-out failures. At this stage failure rate seems to be very high due to deterioration. Bath tub curve representing three stages of failure of product /system / items, these are **Infant Mortality, Useful life and Wear-Out. The diagrammatic structure is given below.**

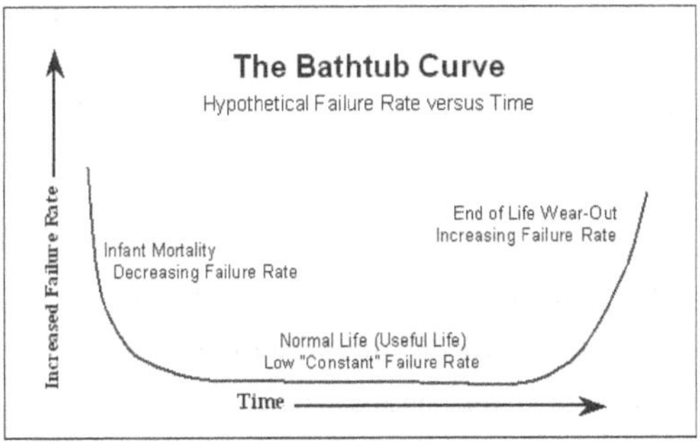

Diagram 9.16: Bath Tub Curve

(i) Infant Mortality or Early failure :- This stage is also called early failure or debugging stage. The failure rate is high but decreases gradually with time. During this period, failures occur due to manufacturing defect in the products. Therefore the failure rate at the beginning of infant mortality stage is high and then It decreases with time after early failures are removed by burn-in or other stress screening methods. For example poor welds, poor connections, contamination on surface in materials, incorrect positioning of parts, etc.

(ii) Useful life: - This is the middle stage of the bath-tub curve. This stage is characterized by a constant failure rate. This period is usually given the most consideration during design stage and is the most significant period for reliability prediction and evaluation activities. Product or component reliability with a constant failure rate can be predicted by the exponential distribution. it is also known as constant failure rate or normal life of product .

(iii) Wear-out stage: - This is the final stage where the failure rate increases as the products begin to wear out because of age or lack of maintenance. When the failure rate becomes high, repair, replacement of parts etc., should be done.

Objective of reliability: - following are some important Objective of reliability of the system.

- Higher customer satisfaction.
- Increased sales.
- Improved safety.
- Decreased warranty costs.
- Decreased maintenance costs, etc.

9.15. Design for Reliability

The objective of design for reliability is to design a given product that meets its requirements under the specified environmental conditions. To achieve this good sound engineering design rules should be followed. However there are a few general principles that should observed, these include:

- Component selection – well-established and known components should be used (company usually have their own approved components list). If this is not he case then analysis must be done to check the componentisfitforpurpose.
- Consider the load-strength relationship and ensure there is an adequate safety margin.
- Minimum complexity.
- Diversity–avoidscommonmodefailures.
- Analyses failure modes and their effects (FMEA).
- Identify any single point failures and either mitigate or design them out.
- Use lessons learned from previous products to design out any known weaknesses.

Ultimately the aim is to maximize reliability during service life by:

- Measurement & control of manufacturing quality/screening.
- Optimized design & build process to improve intrinsic reliability.
- Assure no systematic faults present in product.
- Provide sufficient margin to meet life requirements.

9.16. Availability&Maintainability

Availability of an item/system is the probability that this item/system will be in a state to perform a required function under given conditions, at a given instant in time or over a time interval, assuming that the given external resources are provided. Availability is defined by,

$$A = MUT/ (MUT + MDT)$$

Where **MUT** refers to Mean Uptime

MDTto Mean Downtime.

Depending on what is included in MDT, the availability measured is called Intrinsic Availability (A i), TechnicalAvailability (A t) or OperationalAvailability (A o) . Alternatively Availability is the measure of the percentage of time that an item or system is available to perform its designated function. The simplest expression for calculating availability is:

Availability (A_t) = **(uptime / totaltime)** for a continuously operating system. This can also be expressed as:

Availability (A_t) = Total time - Downtime (Scheduled + Unscheduled)/Total time

Or in terms of reliability (MTBF) and maintainability (MTTR) measures as:

$$\textbf{Availability } (A_t) = MTBF/ (MTBF + MTTR)$$

Maintainability is the probability that a given active maintenance action, for an item under given conditions of use can be carried out within a stated time interval when the maintenance is performed under stated conditions and using stated procedures and resources. Alternatively Maintainability is a characteristic of design and is essentially a measure of the ease with which the item can be maintained. A more formal definition is 'maintainability is a characteristic of design and installation, expressed as the probability that an item will be restored to operating condition, within a given period of time, using prescribed procedures and resources. The most commonly used measure of maintainability is the mean time to repair (MTTR). Where the logistical delays aren't considered and the time of interest is the time actively spent in repairing the system maintainability can be measured as mean active repair time (MART).

9.17. Redundancy

In general it is impossible to produce components with high reliability due to number of constraints, such as cost, non-availability of production facilities etc., In such cases redundancy comes handy to the reliability engineer. In simple words, redundancy is the existence of more than one means for carrying out a given function. The following are the methods for introducing redundancy in to a system for improving reliability.

(i) Element redundancy.

(ii) Unit redundancy.

(i) Element Redundancy: - Let *C1* and *C2* be the two elements with reliabilities *R1(t)* and *R2(t)* respectively connected in parallel as shown in the Figure . In this arrangement of elements, the reliability of the system will be much better due to the presence of redundant element and proper operation of one element is sufficient for the successful operation of the say

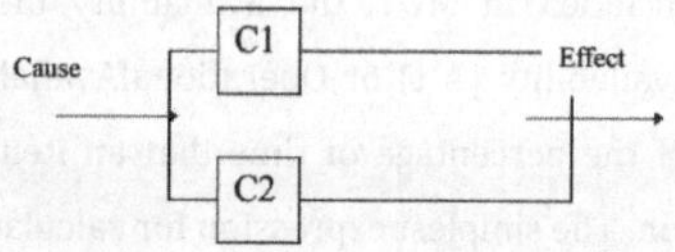

Diagram 9.17: Element Redundancy

(ii) Unit Redundancy: -To improve the reliability of the system, another similar system is connected either in series or parallel to the existing one is called the concept of unit redundancy. Consider a system with two elements C1 and C2 as shown below. For improving the reliability of the system, a similar system in parallel is added to the existing system.

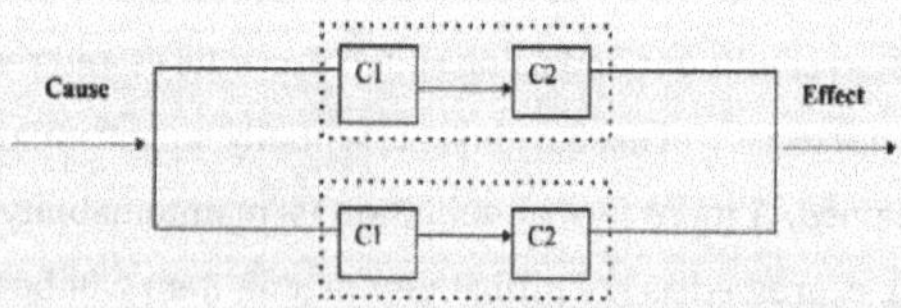

Diagram 9.18 Unit Redundancy

Unit redundancy is further classified into two types.

- Active redundancy.
- Stand by redundancy.

Active Redundancy: Redundant system consisting of two or more components connected in parallel and both components were operating simultaneously is called active redundancy. In active redundancy all the redundant units are operated simultaneously instead of switching on only when need arises. The schematic diagram representing the active redundancy.

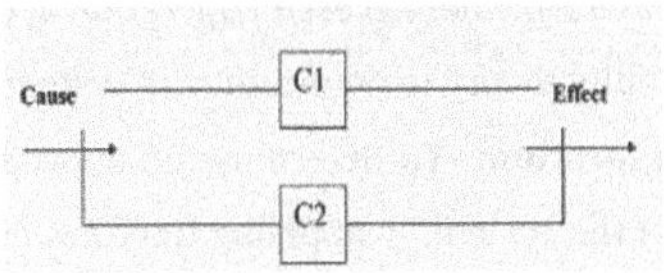

Diagram 9.19: Active Redundancy

Stand By Redundancy: In case of stand by redundancy the alternate means of performing the function is not operated until it is needed. The alternative means is switched on only when the primary means of performing the function fails. Standby redundancy is more appropriate for mechanical devices such as motors and pumps etc . The schematic diagram representing the standby redundancy.

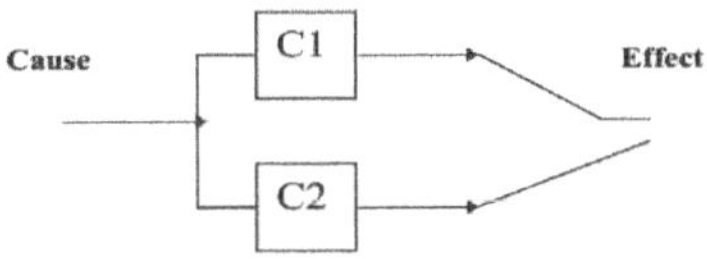

Diagram 9.20: Stand By Redundancy

Advantages of the Redundancy

(i) Desired level of reliability can be achieved under the resource flexibility.

(ii) Improvement in reliability per unit of resource is optimum when compared to any other approach.

(iii) Redundancy requires less skill on the part of the design engineer.

(iv) It is a quick method of solution.

(v) This method can be the best choice for improvement of reliability in case of failure of all other approaches.

9.18. Types of System Reliability Models

The system can be analyzed by decomposing it into smaller sub systems and estimating reliability of each subsystem to assess the total system reliability. The procedure to determine the system reliability is as follows.

(i) Identify the sub systems and elements of the given system.

(ii) Identify corresponding individual reliabilities of the sub systems and elements.

(iii) Draw a block diagram to represent the logical manner in which these units are connected.

(iv) Determine the constraints for the successful operation of the system.

(v) Apply rules of probability theory to determine the system reliability.

System reliability configuration: -To determine an appropriate reliability or reliability model for each component of the system by applying the rules of the probability according to the configuration of the components within the system is also known as system reliability. Several methods exist to improve the system reliability like using large safety factors, reducing the complexity of the system, increasing the reliability of the components etc. There are several types of configurations available, such as

- Series configuration.
- Parallel configuration.
- Mixed configuration.
- Series-parallel configuration.
- Parallel-series configuration.
- Non-series parallel configuration.
- Complex configuration.
- Coherent system.

(i) Series Configuration: - In series configuration all components must be connected in series in order to make the system to perform continuously. In this system all components are considered critical in that sense that their function must be performed in order to make the system to operate successfully. Under this concept if any one component connected serially fails, the System will fail. The reliability block diagram as shown in Figure.

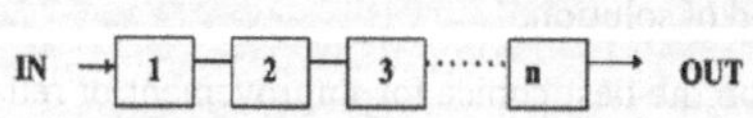

Diagram 9.21: Series Configuration

(ii) Parallel Configuration: - A system can have several components to perform the same operation and the satisfactory performance of any one of these components is sufficient to ensure the successful operation of the system. The elements for such a system are also said to be connected in parallel configuration as shown in Figure.

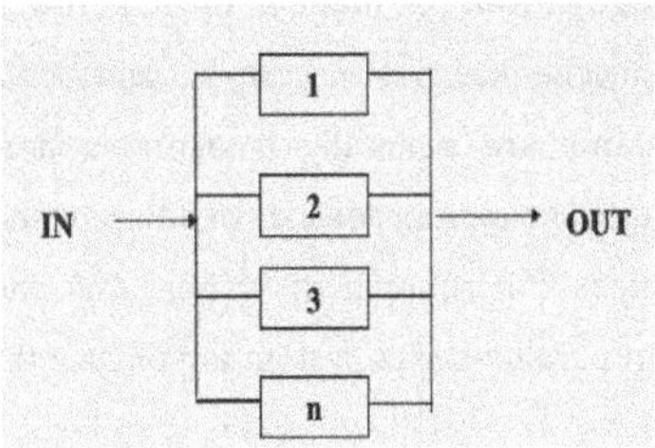

Diagram 9.22: Parallel Configuration

(iii) Mixed Configuration: In mixed configuration, the elements are connected in series and parallel arrangement to perform a required system operation. In mixed configuration, to compute the system reliability, the network is broken into series or parallel subsystems. The reliability of each sub system is found and then the system reliability may be obtained on the basis of the relationship among the sub systems. The schematic diagram representing the mixed configuration is shown in the Figure.

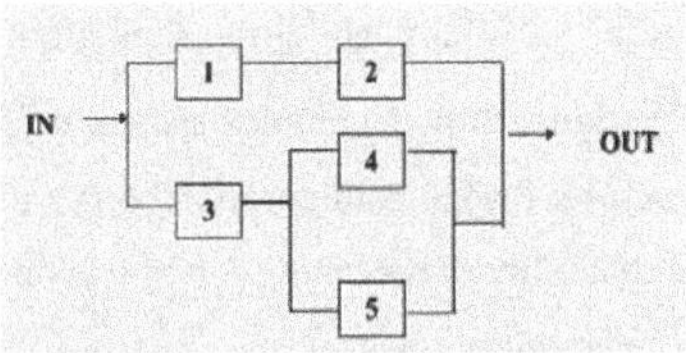

Diagram 9.23: Mixed Configuration

(iv) Series - Parallel Configuration: The System consists of different stages connected in series as shown figure,Each stage contains number of redundant elements. In the given figure , step -1 consists of three redundancy which is in parallel , step -2 consists of five redundancy which is again having parallel Configuration , further step -3 consists of two redundancy which is also in parallel . And finally all these three steps are connected in series. The diagrammatic arrangement is given below.

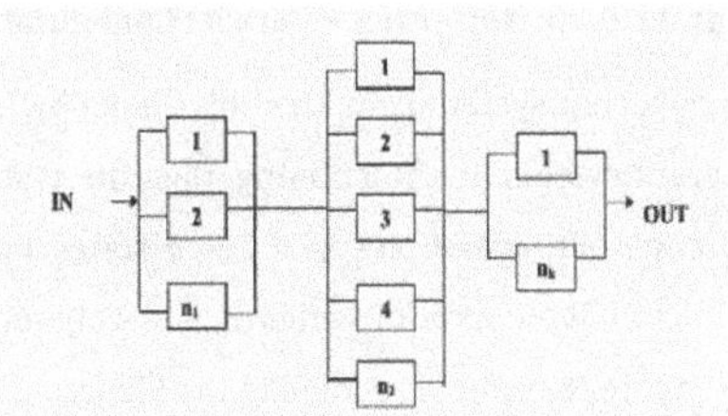

Diagram 9.24: Series - Parallel Configuration

(v) Parallel – Series Configuration: -A number of activities such as industrial process, economic systems and mechanical devices can be represented as multi stage systems. These multi stage systems are generally categorized according to the component configuration at each stage. The system consists of different stages connected in parallel as shown in Figure For successful running of system, one path is sufficient. Redundant components are added in parallel-series system to enhance the reliability of the stages as well as the system.

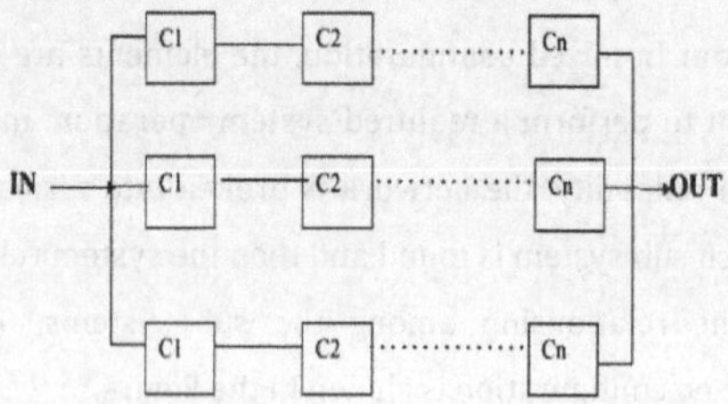

Diagram 9.25: Parallel – Series Configuration

(vi) Non Series – Parallel Configuration: A complex system on simplification can produce a non – series parallel structure. Bridge configuration comes under this category, which is shown in the Figure. The reliability of the below configuration can be evaluated using logic diagram and appropriate probability rules.

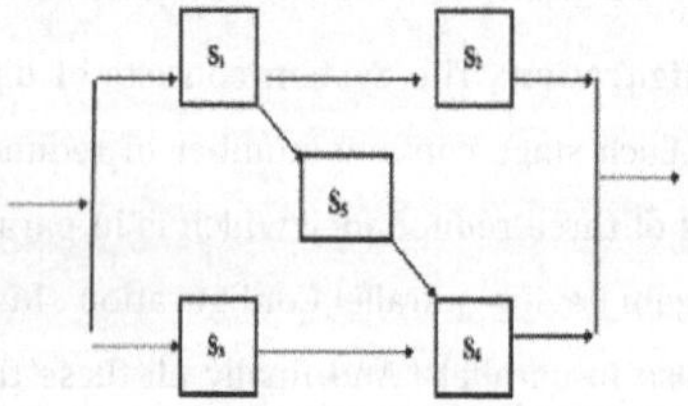

Diagram 9.26: Non Series – Parallel Configuration

(vii) Coherent System: A coherent system is defined by the following conditions: 1. If all the components in a coherent system are functioning, then the system is functioning. 2. If all the components in a coherent system are in a failed state, then the system is in failed state. If a system fulfills above conditions then it is a coherent system. The schematic representation of the coherent system is shown in the Figure.

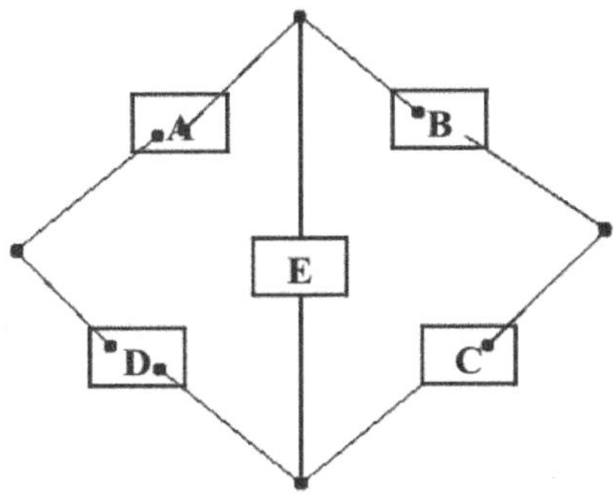

Diagram 9.27: Coherent System

Question for Discussion

1. Discuss the need, advantages and composition of quality circles.
2. What are the customer needs and expectations ? Explain.
3. Define the term Quality. Explain different dimensions of quality.
4. Explain the concept of Quality Circles. Discuss advantages and disadvantages of Quality Circles.
5. Define Prevention, Appraisal, Internal failure and External failure terms related to the cost of quality.
6. Explain PDCA cycle in details with suitable example.
7. Explain each phase of 5S technique. List the advantages offered by 5S technique
8. Define the following terms:
    (a) Reliability.
    (b) Failure Density.
    (c) Mean Time To Failure(MTTF).
    (d) Mean Time Between Failure(MTBF).
    (e) Mean Time To Repair (MTTR).
9. Explain any three 7QC tools of problem solving.
10. Define Quality Circle and discuss the advantages and limitation of quality circle.
11. Discuss 5 S for housekeeping and its role in quality control.
12. Give two definitions of Quality and explain various dimensions of quality with suitable examples.
13. What is Quality Circle? Discuss advantages of quality circle.
14. Write short note on PDCA cycle.

15. Explain following quality control tools in brief;
    a) Pareto diagram.
    b) Histogram.
    c) Mean Time To Failure.
    d) Mean Time Between Failure.
16. Discuss Reliability in terms of Hazard rate and failure density.
17. Suggest the various dimensions, customers are influenced by in determining quality levels.
18. Write short notes on following.
    (a) Define quality circles.
    (b) Discuss organization structure of Quality Circles.
    (c) Discuss various costs associated with quality.
    (d) Discuss ways of reducing cost of quality.
19. Discuss Juran's triology and Define MTBF, MTTR and Random variable .
20. Discuss reliability bathtub curve in brief with its sketch.
21. What is Quality Management? Briefly discuss the objectives of quality management?
22. Define Quality Control. What are the objectives of quality control?
23. Explain PDCA cycle in detail.
24. Define the term Quality. State and explain the different dimensions of quality.
25. Discuss 5-S for house keeping with suitable example and explain its role in improving quality and productivity.
26. State only the usefulness of following quality control tool:
    (a) Pareto diagram.
    (b) Cause and effect diagram.
    (c) Histogram and Check sheet.
27. Draw and explain the Bathtub curve
28. Discuss views of different Quality Gurus.
29. Explain about 7 New Quality Improvement Tools
30. Explain Reliability in terms of Hazard rate and failure density,Conditional probability and multiplication rules
31. Enlist the various phases of quality revolution and discuss inspection, quality control and quality assurance.
32. Discuss failure characteristics of product depicting its three major life regions.
33. Discuss system reliability for the assembly in which components are connected in series.

34. Based on data given in the following table (f or product A and B) answer the following;

(a) Which product has higher reliability?

(b) Which product has greater maintainability?

Product	MTBF	MTTR
A	100	6
B	140	4

35. The process of making component X for productY involves six workstations. The cycle time is 4 minutes per item for each workstation. What will be the impact of reducing lot size from12 units to 4 units on process flow through time and work in process inventory level. A lot of 25 articles contain 3 defective. A sample of 5 is selected at random from the lot for inspection. What are the respective probabilities of 0 and 1 defective occurring in the sample of 5?

36. Define Reliability. Draw a typical Bath-Tub Curve and explain its phases.

37. A system has MTBF of 1000 hrs. and MTTR of 40 hours. Find out the availability of that system.

38. What are the objectives of quality control

39. Mother board of a laptop has four components P,Q, R and S. 400numbers of the mother board were subjected to accelerated operation tests which have been equivalent to 3000 hours of normal operation. The following data is available from the tests.

Components	No of failures
P	8
Q	10
R	55
S	12

Find the reliability of the mother board. Also calculate the MTBF for the four components.

40. What is reliability engineering ? Define Maintainability, Availability, Failure Rate and Mean,

Time To Repair . if failure rate = 0.002 unit/hr.

What will be the reliability for 900 hours?

41. Which techniques are used to determine reliability ? 6 transformers are tested for 600 hr,2 failed after 50 and 200 hr and the remaining 4 transformer perform satisfactorily until the end of the test find the failure rate.

42. "Quality as wining strategy";Justify the sentence

43. Write a short note on views of different quality Gurus

44. Explain 7 New Quality Improvement tools

45. Solve the following problems .

(a) Three lamps are connected in parallel to produce light in a hall. The reliabilities are 0.92, 0.95 and 0.96. Find the reliability of the total lamp system. If the systems are connected in series, determine the reliability of the system.

(b) An electronic system has a MTBF of 1000 hours and a MTTR of 40 hours. Determine its availability.

46. The percent of water absorption is an important characteristic of common building brick. A certain company occasionally measured this characteristic of its product but records were never kept. It was decided to analyze the process with control chart. Twenty-five samples of four bricks each yielded these results.

Sample NO.	**1**	**2**	**3**	**4**	**5**	**6**	**7**	**8**	**9**	**10**	**11**	**12**	**13**
X	15.3	12.2	7.6	8.5	11.3	10.3	11.8	11.5	10.2	9.8	7.5	7.6	7.9
R	9.1	9.9	9.7	6.7	7.5	9.5	12.3	11.2	13.5	7.8	5.5	8.7	5.2

Sample NO.	14	15	16	17	18	19	20	21	22	23	24	25
X	9.3	8.2	8.6	8.5	11.5	10.2	11.8	13.01	12.2	7.8	8.8	8.0
R	17.1	10.9	4.2	4.7	5.5	6.5	4.3	7.2	8.5	6.4	6.9	6.3

Estimate the control limits for X and R chart. If any point lies out of the control limits, estimate the revised control limits

47. Explain in detail about the Deming's fourteen points.

48. Explain briefly the product life cycle .

49. The following table indicates the time a lot of electronic parts operated until failure, The times are presented in the order in which the trials are measure included 30 measurements made with a constant failure rate of 0.0005 per hour.

Set of observation	Time until failure				
1	1	2	3	4	5
2	2737	1281	1855	1472	2638
3	2147	4522	430	3740	7714
4	961	617	6715	714	3300
5	1280	3135	505	1662	2140
6	458	3918	1846	2114	2713

Compute the respective estimates of mean life in hours that would have been approximated after 1000, 3000, 5000 and7713 hours of testing.

50. Explain Product life cycle in brief.

References

1. Mizuno Shiguro, "Company-wide total quality control", (Tokyo, Asian Productivity Organization), 1988.
2. Ishikawa Kaoru, "Guide to quality control", (Tokyo, Asian Productivity Organization), 1976.
3. E.L. Grant, "Statistical quality control, revised R.S. Leavenworth", (New York and London, McGraw- Hill, 6th éd.,), 1988.
4. J. Banks, "Principles of quality control", (London, John Wiley), 1989.
5. B.V. Gnedenko and A.Y. Khinchin, "An Elementary Introduction to the Theory of Probability", Dover Publication, New York, 1962.
6. P. Giovannetti, "Staffing methods—implications for quality", Measuring the Quality of Care, L.D. Willis & M.E. Linwood (eds), Churchill Livingstone, NY, 1984.
7. B. Lillrank and N. Kano, "Continuous Improvement: Quality Control Circles in Japanese Industries", Ann Arbor, Mich.: Univ. of Michigan, Center for Japanese Studies, 1989.
8. L.P. Sullivan, "Quality function deployment: A system to assure that customer needs drive the product design and production process", Quality Progr, Pp. 39-50, 1986.

CHAPTER 10

INVENTORY MANAGEMENT

10.0. Introduction

Inventory generally refers to the materials in stock. It is also called the idle resource of an enterprise.

Inventories represent those items which are either stocked for sale or they are in the process of manufacturing or they are in the form of materials, which are yet to be utilized. The interval between receiving the purchased parts and transforming them into final products varies from industries to industries depending upon the cycle time of manufacture. The term 'inventory' may also be defined in logical manner

Inventory is a stock of materials like – raw material , material in process , finished and semi-finished products , spare parts , storage of others devices and others stocked in order to meet an unexpected demand or for distribution in future.

Concept of inventory management :-Inventory control is a planned approach of determining what to order , when to order and how much to order and how much to stock so that costs associated with buying and storing are optimal without interrupting production and sales. Inventory deals with the problem.

- When should an order be placed ? (Order level).
- How much should be ordered ? (Order quantity).

Type of Inventory: -following are important type of inventory.

(i) **Production inventory:** - Raw material, parts and component which enters the firm's product in the production process. these may consists of two types – special items manufactured to company specifications and standard industrial items purchased " off the sell "

(ii) **MRO Inventories:** -Maintenance, Repair and Operating supply which are consumed in the production process but which do not become the parts of the products. For example – lubricating oil, soap, machine components, and machine repair parts.

(iii) **In process inventories:** - Semi-Finished product found at various stages in the production operation, also parts of inventories.

(iv) **Finished goods inventories:** - completed products, finished assembly in all respects which is ready to shipment also managed by inventory department.

10.1. Inventory Costs

According to the definition of inventory , it is associated with the various type of activity / processing , the overall processing and activity required some capital investment so therefore study of cost & its controlling will be on prime consideration for maximizing the profit aswell as proper utilization of inventory . Following three type of cost incurred with the inventory.

(i) Ordering costs or Setup costs.

(ii) Carrying costs or Holding costs.

(iii) Stock-out costs or Shortage costs.

(iv) Administrative costs.

Further Classification is also given as

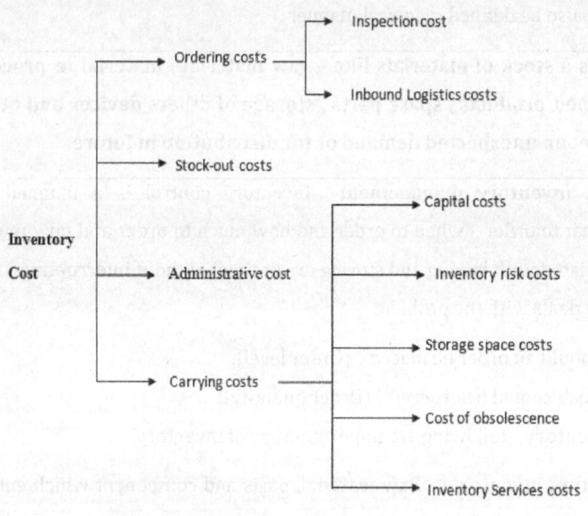

Diagram 10.1: Inventory Cost Diagram

(i) Ordering cost: -Ordering cost ((also called setup costs) are the sum of cost of procurement and inbound logistics costs form a part of Ordering Cost. Ordering Cost is dependant and varies based on two factors - The cost of ordering excess and the Cost of ordering too less. Both these factors move in opposite directions to each other. Ordering excess quantity will result in carrying cost of inventory. Whereas ordering less will result in increase of replenishment cost and ordering costs. These two above costs together are called Total Stocking Cost. The inbound logistics costs, related to transportation and reception (unloading and inspecting). Those costs are variable. How much to order is determined by the Economic Order Quantity or EOQ.

(ii) Carrying Costs: -The next component of inventory costs is carrying costs. Carrying costs are the costs that are related to keeping inventory in stock. These costs are related to the space required to hold inventory, the cost of the money needed to acquire inventory, and the risk of loss through inventory obsolescence. Most of these costs are also included in an overhead cost pool and allocated to the number of units produced in each period.

- **Capital costs:** - It is the largest component among the carrying inventory costs. It includes everything related to the investment, the interests on working capital and the opportunity cost of the money invested in the inventory.
- **Storage space costs:** - They include the cost of building and facility maintenance (lighting, air conditioning, heating, etc.), the cost of purchase, depreciation, or the lease, and the property taxes. These costs are dependents up on the kind of storage chosen, whether the warehouses are company owned or rented.
- **Inventory services costs:** - They include insurance, IT hardware and applications , physical handling with the corresponding human resources, management, etc. We can also put in this category the expenses related to inventory control and cycle counting. Finally, although they are kind of a category on their own, taxes can also be added here.
- **Inventory risk costs:** - Risks includes shrinkage, which is basically the loss of products between the purchase from the suppliers (recorded inventory) and the point of sale (actual inventory), caused by administrative errors (shipping errors, misplaced goods,), vendor fraud, pilferage and theft (including employee theft), damage in transit or during the period of storage (because of incorrect storage, water or heat damage).
- **Cost of obsolescence** :- Some inventory items may never be used or will be damaged while in storage, and so must be disposed of at a reduced price, or at no price at all. Depending on how perishable the inventory is, or the speed with which technology changes impact inventory values, this can be a substantial cost.

(iii) Stock-out costs or Shortage costs: - The third component of inventory costs is shortage costs. Shortage costs are the costs that a business incurs when they run out of stock.

(iv) Administrative costs: -The accounting department pays the wages of a cost accounting staff, which is responsible for compiling the costs of inventory and the cost of goods sold, responding to other inventory analysis requests, and defending their results to the company's internal and external auditors. The cost of cost accounting personnel is charged to expense as incurred.

10.2. Principles of Inventory Management

There are five principal of inventory management which is very important from subject point of view, these are given below.

(i) Demand forecasting.

(ii) Warehouse flow.

(iii) Inventory turns/stock rotation.

(iv) Cycle counting.

(v) Process auditing.

(i) Demand forecasting: -Demand forecasting or estimation gives businesses valuable information about the markets in which they operate and the markets they plan to pursue. Thus **Demand Forecasting** refers to the process of predicting the future demand for the firm's product. In other words, demand forecasting is comprised of a series of steps that involves the anticipation of demand for a product in future under both controllable and non-controllable factors.The business world is affected by market variation that is recognized by risk and uncertainty of both type internal or external such as technology, attrition, unrest, employee grievances, recession, inflation, modifications in the government laws, etc. Demand forecasting holds significance in the businesses where large-scale production is involved.

The following **steps in demand forecasting** are followed to facilitate a systematic estimation of future demand for product:

(i) Specifying the Objective.

(ii) Determining the Time Perspective.

(iii) Choice of method for Demand Forecasting.

(iv) Collection of Data and Data Adjustment.

(v) Estimation and Interpretation of Results.

- **Quantitative Forecasting:** – This forecasting approach is a mathematical model based on historical data. It involves using past sales data to predict future demand for goods. Data sets can go back decades or can be run for the last calendar year, however the more data available, the more accurate picture of historical demand will be attained. While it may provide a basis for forecasting, demand can be unpredictable based on variable market conditions or product seasonality. Unexpected peaks in demand can result in stock outages and quiet periods may result in costly excess stock, which can build up carrying costs resulting in diminishing profits.

- **Qualitative Forecasting:** - This method is less precise, and involves predicting demand based on less measurable factors such as market forces, economic demand and potential demand. Qualitative forecasting methods could be considered an art mastered by inventory planners over years of practice. Inventory forecasting techniques are inseparable from current stock review and reorder methods, and there are two broad models for inventory monitoring.Inventory forecasting is directly depends upon the demand forecasting that is imposed by the customer, vendors or retailers thus it is very necessary to maintain the inventory as per the future demand. Inventory Forecasting enables you to produce forecasts based on sales history, and provides the tools to measure the quality of your forecasts. It uses a set of mathematical algorithms to predict future sales based on historic demand, and enables you to easily identify those products that contribute the most to your business in terms of sales value, gross profit, cost of sales, quantity sold.

The Inventory Forecasting System Assists the Firms in the Following Important Areas

(i) It looks at past sales and matches the algorithm which would have had the best chance of success in estimating previous sales. that algorithm is then used to predict future sales .

(ii) It records Forecast errors to enable management to monitor and continually minimize those errors to acceptable corporate policy settings.

(iii) The system automatically recognizes seasonal variations and accommodates those variations.

(iv) It allows for market intelligence input to better forecast future sales.

(v) Since the system is better able to predict demand it will facilitate factory scheduling and so that manufactures are better able to optimize their factory throughput while meeting the dynamics of the marketplace.

The benefits of Inventory Forecasting: Following are some Important Benefits of Inventory Forecasting

(i) Variety of forecast algorithms, including a competition method which automatically selects the best algorithm for the collection.

(ii) analysis of product performance at stock code, warehouse, product class, supplier, planner or buyer level.

(iii) Application of seasonality factors to current forecast in requirements planning.

(iv) Review forecast accuracy to measure and improve the quality of forecast.

(v) Enhanced data selection sets which carry through the entire process into Inventory Optimization.

Depending on the Industry, there are 2 Different Models for Monitoring Inventory

Continuous Review: – Stock is reviewed on a continuous basis, and a certain amount of stock is reordered once levels have dropped below a certain level known as the reorder point. Having the right reorder point will allow for supplier variability and should be supported by a safety stock level "safety net" to ensure service level rates are maintained. Companies with fast moving products are highly encouraged to invest into tools that support continuous inventory review.

Periodic Review: – Stock levels across all storage points are monitored at set periods, and replacement stock is ordered. This usually results in higher inventory levels as stock needs to be ordered in sufficient quantity to reach the next review date. This model requires more available cash flow and working capital to maintain and the risk of overstocking and under stocking items is greater.

Businesses with slower inventory turnover can leverage this model successfully with the help of inventory optimization solutions that provide inventory alerts when actions need to be taken to avoid undesirable stocking situations.

(i) **Warehouse flow:** -The old concept of warehouses is out dated and costly. Lean manufacturing concepts, including 5"S have found a place in warehousing. **Sorting, setting order, systemic cleaning, standardizing, and sustaining,** which helps in cut down the capital investment for inventory.

(ii) **Inventory Turns/Stock Rotation :-** In certain industries, such as pharmaceuticals, foodstuffs and even in chemical warehousing, managing inventory down to lot numbers can be critical to minimizing business costs. Inventory turns is one of the key metrics used in evaluating how effective your execution is of the principles of inventory management. Defining the success level for stock rotation is critical to analyzing your demand forecasting and warehouse flow.

(iii) **Cycle Counting:** - One of the key methods of maintaining accurate inventory is cycle counting. This helps measures the success of your existing processes and maintain accountability of potential error sources. There are financial implications to cycle counting. Some industries require periodic 100% counts. These are done through perpetual inventory count maintenance or though full-building counts.

10.3. Inventory Model

Inventory models deal with idle resources like men, machines, money and materials. These models are concerned with two decisions: how much to order (purchase or produce) and when to order so as to minimize the total cost.

1. How much to order: There are two basic costs are considered these are, inventory carrying costs and the ordering or acquisition costs. As the quantity ordered is increased, the inventory carrying cost increases. Economic order quantity is calculated by balancing the two costs. Economic Order Quantity (EOQ) is that size of order which minimizes total costs of carrying and cost of ordering.

Determination of EOQ by Tabulation (Trial & Error) Method

- Select the number of possible lot sizes to purchase.
- Determine average inventory carrying cost for the lot purchased.
- Determine the total ordering cost for the orders placed.
- Determine the total cost for each lot size chosen which is the summation of inventory carrying cost and ordering cost.
- Select the ordering quantity, which minimizes the total cost.

The data calculated in a tabular column can plotted showing the nature of total cost, inventory cost and ordering cost curve against the quantity ordered

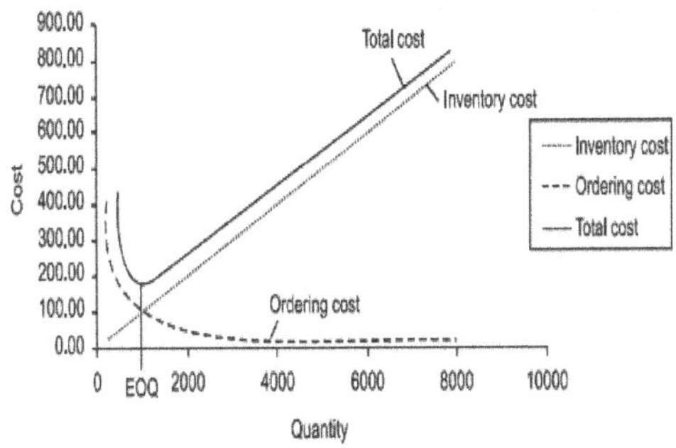

Diagram 10.2: Economic Order Quantity (EOQ)

Determination of EOQ by Analytical Method: -One of the important decisions to be taken by a firm in inventory management is how much quantity is to buy at a time, or say, for how much inventory to place order at a time. This is called 'Economic Ordering Quantity.

Assumptions: Following Assumption Should be Considered

(i) The firm knows with certainty how much items of particular inventories will be used or demanded for within a specific period of time.

(ii) The use of inventories or sales made by the firm remains constant or unchanged throughout the period.

(iii) The moment inventories reach to the zero level, the order of the replenishment of inventory is placed without delay.

(iv) The above assumptions are also called as limitations of EOQ model.

(v) Production or supply of commodity is instantaneous.

(vi) Lead-time is zero.

Inventory carrying cost is C1 = CI, where C is the unit cost and I is called inventory carrying cos The relation between time & quantity are represented on inventory-time diagram, where

- Q is represented on the vertical axis (how much quantity is required)
- T is represented on the horizontal axis (1 – year time period), the given time is divided into n- parts as per convenient.

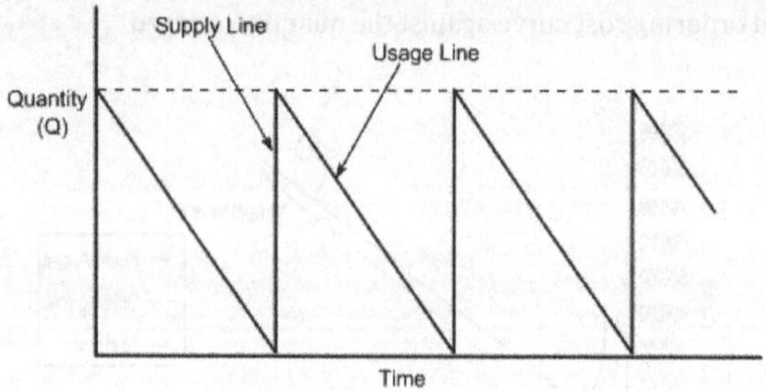

Diagram 10.3: Inventory-Time Diagram

Since total inventory cost is combination of two costs whose estimation must be considered.

1. Carrying cost.
2. Annual ordering cost or set-up cost.

Average inventory = [maximum inventory level + minimum inventory level] / 2

(Q + 0)/2 = Q/2

$$\text{Total inventory carrying cost} = \text{Average inventory} \times \text{Inventory carrying cost per unit}$$

$$\text{Total inventory carrying cost} = Q/2 \times C_c = \mathbf{Q\,[C_c/2\,]} \quad \mathbf{(1)}$$

$$\text{Total annual ordering costs} = \text{Number of orders per year} \times \text{Ordering cost per order}$$

$$\text{Total annual ordering costs} = (O_p / Q) \times O_c = \mathbf{[(O_p \times O_c) / Q]\ (2)}$$

$$\text{Total inventory cost} = \text{Total inventory carrying cost} + \text{Total annual ordering costs}$$

$$\mathbf{Q\,[C_c/2\,] + [(O_p \times O_c) / Q]}$$

$$\text{Inventory carrying costs} = \text{total annual ordering costs}$$

Then total inventory cost will be minimum

$$Q\,[C_c/2\,] = [(O_p \times O_c) / Q]$$

$$Q^2 = \{\, 2\,(O_p \times O_c) / C_c$$

$$Q = \sqrt{2\,(O\,p \quad \times\ O\,C)/(C\,C\,)}$$

It is an optimal quantity, for given model

Optimum number of orders, (N0) = $[O_c / Q\,]$

Problem:-An oil engine manufacturer purchases lubricants at the rate of Rs. 42 per piece from a vendor. The requirements of these lubricants are 1800 per year. What should be the ordering quantity per order, if the cost per placement of an order is Rs. 16 and inventory carrying charges per rupee per year is 20 paise ?

SOLUTION: Given data are:

Number of lubricants to be purchased, D = 1800 per year

Procurement cost, C3 = Rs. 16 per order

Inventory carrying cost, CI = C1 = Rs. 42 × Re. 0.20 = Rs. 8.40 per year

Then, optimal quantity (EOQ), Q_0 = $Q = \sqrt{2\,(O\,p \quad \times\ O\,C)/(C\,C\,)}$

$$= Q = \sqrt{2\,X16X1800/8.4}$$

Q = 83 lubricants

10.4. Inventory Control System

The inventory control techniques represent the operational aspect of inventory management and help realize the objectives of inventory management and control. Several techniques of inventory control are in use and it depends on the convenience of the firm to adopt any of the techniques. What should be stressed, however, is the need to cover all items of inventory and all stages, i.e. from the stage of receipt from suppliers to the stage of their use. The techniques most commonly used are the following.

(i) **ABC Analysis:** ABC analysis is a business term used to define an inventory categorization technique often used in materials management. It is also known as 'Selective Inventory Control.'ABC analysis provides a mechanism for identifying items which will have a significant impact on overall inventory cost. Whilst also providing a mechanism for identifying different categories of stock that will require different management and controls. When carrying out an ABC analysis, inventory items are valued (item cost multiplied by quantity issued/consumed in period) with the results then ranked. The results are then grouped typically into three categories.

s.no	Category	Description
01	'A'	"A class" inventory will typically contain items that Consumes about 80% of total value, or 20% of total items. "
02	'B'	"B class" inventory will have around 15% of total Value or 30% of total items.
03	'C'	"C class" inventory will account for the remaining 5%, or 50% of total items.

ABC Analysis is similar to the Pareto principle in that the "A class" group will typically account for a large proportion of the overall value but a small percentage of the overall volume of inventory.

Application of ABC – Analyses

- In day to day warehouse operations, materials are some time under issued, over issued, issued and not accounted into the system, misplaced, stolen etc. This results into inaccuracy in the inventory. Cycle counting is the process to count and reconcile the materials. Every material in the warehouse should be counted during a fixed interval (every year) for maintaining 100% accuracy, but counting & reconciling every material is not cost effective and very expensive. To count the accuracy of the inventory in a cost effective manner, it is recommended to count the materials based on inventory classification.
- If A class materials are counted within a fixed interval (could be six months or a year) then the firm needs to count only 5% to 10% of the total materials and it will cover 60% to 80% of the inventory value. It means that firm only counts 5 % to 10% of the materials and remove the inaccuracy from the inventory value from 60% to 80%.
- Similarly B class materials can also be counted on a less frequency (from once in 18 months to 24 Months) as the number of materials become higher
- C class materials at even lesser frequency (once in 27 months to 36 months) as number of material becomes more (60% to 85% of the total materials).

- An inventory controller shall be concentrating more on the A class items for reducing the inventory as he/she shall be concentrating only 5% to 10% of the total items and shall be getting the opportunity to reduce inventory on 60% to 80% of the value.
- Any reduction in lead time of A class items shall result in reduction in inventory, so procurement manager will work out with suppliers to reduce the lead time.
- On issue of materials, tight control on A class, Moderate control on B class, Loose Control on C class. So 'A' class items may be issued after getting the approvals from Senior Executives of the company. B may be moderately controlled. Very little control can be exercised while issuing C class item.

(i) VED Analysis : VED Analysis attempts to classify the items used into three broad categories, namely.

Vital , Essential , and Desirable. The analysis classifies items on the basis of their criticality for the industry or company.

Vital: - Vital category items are those items without which the production activities or any other activity of the company may be stopped or affected.

Essential: -Essential items are those items whose stock – out cost is very high for the company.

Desirable: - Desirable items are those items whose stock-out or shortage causes only a minor disruption for a short duration in the production schedule. The cost incurred is very nominal.

VED Analysis is very useful to categorize items of spare parts and components. In fact, in the inventory control of spare parts and components it is advisable, for the organization to use a combination of ABC and VED Analysis. Such control system would be found to be more effective and meaningful.

(ii) SDE analyses: This analyses is based on each items, in this analyses.

S – 'S' stand for scarce items , which are in short supply and their availabity is scarce.

D – 'D' stand for difficult items, which cannot be procured easily.

E – 'E 'stand for easily available items.

(iii) MNG ANALYSES: - In"MNG" analyses.

M – Stands for moving items.

N— stands for non- moving items, which could not be consumed at the last year.

G—stands for ghost items, which could not be existing items at the beginning and at the end of year, for such kind of items store keeper keeps bin – card showing nil balance.

10.5. Material Requirement Planning (MRP)

Material Requirements Planning (MRP) is a computer-based production planning and inventory control system. MRP is concerned with both production scheduling and inventory control. It is a material control system that attempts to keep adequate inventory levels to assure that required materials are available when needed. MRP is applicable in situations of multiple items with complex bills of materials.

The major objectives of an MRP system:

- Ensure the availability of materials, components, and products for planned production and for customer delivery.
- Maintain the lowest possible level of inventory.
- Plan manufacturing activities, delivery schedules, and purchasing activities.

MRP is especially suited to manufacturing settings where the demand of many of the components and subassemblies depend on the demands of items that face external demands. Demand for end items is independent. The three major inputs of an MRP system are the **master production schedule, the product structure records, and the inventory status records.** Without these basic inputs the MRP system cannot function.

(i) **Master production schedule:** The master production schedule expresses how much of each item is wanted and when it is wanted. The MPS is developed from forecasts and firm customer orders for end items, safety stock requirements, and internal orders. MRP takes the master schedule for end items and translates it into individual time-phased component requirements.

(ii) **The product structure records:** The product structure records, also known as bill of material records (BOM), contain information on every item or assembly required to produce end items. Information on each item, such as part number, description, quantity per assembly, next higher assembly, lead times, and quantity per end item, must be available.

(iii) **Inventory status record:** The inventory status records contain the status of all items in inventory, including on hand inventory and scheduled receipts. These records must be kept up to date, with each receipt, disbursement, or withdrawal documented to maintain record integrity. MRP will determine from the master production schedule and the product structure records the gross component requirements; the gross component requirements will be reduced by the available inventory as indicated in the inventory status records.

10.6. An Outline of the MRP Process

Starting with end items the MRP process goes through the following steps.

- Establish gross requirements.
- Determine net requirements by subtracting scheduled receipts and on hand inventory from the gross requirements.
- Time phase the net requirements.
- Determined the planned order releases.

Table 10.1: MRP Process Table

Week	1	2	3	4	5	6	7
Gross requirements							
Scheduled receipts							
Net requirements							
Time-phased net requirement							
Planned order releases							

The planned order releases aggregated over all the end items will result in the gross requirements for level one item, the gross requirements for this items are then netted and time phased to determined their own order releases.

(i) Computing Direct and Indirect Requirements:- Let Bij denote the number of units of item j required to make directly one unit of item i, and let Rij denote the total number of units of item j, direct or indirect, required to produce one unit of item i. Clearly Rii = 1, while for j 6= i we have,

$$R_{ij} = \sum_{k} B_{ik} R_{kj}$$

In matrix notation, we have R = I +BR

So (I –B) R = I

And R = (I –B) –1.

(ii) Expediting and Deferring Scheduled Receipts: -The process of determining net requirements, as outlined above, is to subtract scheduled receipts and on hand inventory from the gross requirements. Occasionally, because of anticipated changes in the MPS, we will find that the scheduled receipts are not enough to cover the gross requirements within a lead time. Consider, a Table, and assume that the lead time is three weeks.

- Notice that the schedule has a net requirement of 15 units in period 2
- An order placed for 15 units in period 1 will arrive in period 4, so it would need to be expedited to be ready by period 2.
- An easier alternative, is to issue an expedite notice to the schedule receipt of period 3, stating that we need 15 units by period 2.
- Suppose that it is only possible to have 10 units ready by period 2. Then we will have a shortfall of five units.
- When a shortage occurs, it is important to backtrack and identify the source of demand. It may be that 10 of the 15 units required in period 2 are for actual orders, while the other five are in anticipation of future demand.

In this case, we will allocate the 10 units to the actual order and avoid a stock out. On the other hand, there may be changes in the MPS that make scheduled receipts unnecessary. In that case the schedule receipts can be deferred to a later period.

Week	1	2	3	4	5	6	7
Gross requirements	10	15	25	30	45	20	30
Schedule receipts	10		25				
Net requirements		15		30	45	20	30

Expediting in MRP

(i) **Lot Sizing Rules: -**The problem of lot sizing is one of satisfying the requirements while trying to minimize holding and setup costs. A variety of lot sizing rules has been proposed. The lot-for-lot (LFL) is the simplest approach, and it calls for producing in period t the net requirements for period t. The LFL approach minimizes the holding cost by producing just-in-time. This approach is optimal if setup costs and setup times have been reduced to negligible levels, but it may be expensive if setup costs are significant. A variety of lot sizing algorithms have been developed to deal with the case where setup costs are significant. The Wagner-Within (WW) algorithm can be used to optimally select the lot sizes at one level. However, applying the Wagner-Within algorithm, or any other single level approach, to different levels does not guarantee that the overall policy is optimal. An alternative to the Wagner-Within policy is the Silver-Meal (SM) heuristic. Starting from the first period with positive requirements, the SM heuristic attempts to cover more and more periods with one setup while the average cost of doing this is decreasing. Once it is determined that adding the requirements of the next period increases the average cost, a new setup is incurred and the method is repeated until all the requirements are covered. Another approach, which is popular in practice, is the part period balancing (PPB) heuristic which attempts to select the number of periods covered by a setup by making the holding cost over the covered horizon as close as possible to the setup cost. The fixed order quantity (FOQ) heuristic is to order a predetermined quantity whenever an order is placed. Finally, the fixed order period (FOP) heuristics calls for covering the demand of a fixed number of periods with one setup. Vollman et. al. [1] recommend the use of different lot-sizing rules for different levels in the BOM, with FOQ for end items, either FOQ or LFL for intermediate levels, and FOP for the lowest levels. The idea is to avoid the propagation of the bullwhip effect to the lowest items.

Shortcomings of MRP:- In this clause following are important point must be considered .

(i) Capacity: -MRP expects the lead time to be constant regardless of how much work has been released into the production system, so it is implicitly assuming infinite capacity. This can create problems when production levels are at or near capacity. One way to address this problem is to make sure that the MPS is capacity feasible. Rough-cut capacity planning (RCCP) attempts to do this by checking the capacity of a few critical resources. RCCP makes use of the bill of resources (BOR) for each item on the MPS. The BOR specifies the number of hours required at each critical resource to build a particular end item and its components, and then aggregates the number of hours required at each critical resource over the end items in the MPS. RCCP then checks whether the available resources are enough to cover the MPS on each time bucket. Notice that RCCP does not perform time offsets, so the calculation of the number of hours required has to be done with time buckets that are large enough so that parts and their components can all be completed within a single time bucket. This usually makes RCCP an optimistic estimation of what can be done. Advanced MRP systems provide more detailed capacity analysis proposing alternative production schedules when the current plan is not feasible

(ii) Long Lead Times: - There are many pressures to increase planned lead times in an MRP system. MRP uses constant lead times when, in fact, actual lead times vary considerably. To compensate, planners typically choose pessimistic estimates. Long lead times lead to large work-in-process (WIP) inventories.

(iii) Nervousness:-MRP is typically applied in a rolling horizon basis. As customer orders firm up, and forecasts become better, a new MPS is fed to MRP which produces updated planned order releases that may be very different form the original. Even small changes in the MPS can result in large changes in planned order releases. Vollman et. al. [1], give an example where a small decrease in demand causes a formerly feasible MRP plan to become infeasible.

(iv) **MRP II :-**Manufacturing Resource Planning (MRP II) embeds additional procedures to address the shortcomings of MRP. In addition, MRP II attempts to be an integrated manufacturing system by bringing together other functional areas such as marketing and finance. The additional functions of MRP II include forecasting, demand management, rough-cut capacity planning (RCCP), and capacity requirement planning (CRP), scheduling dispatching rules, and input/output control. MRP II works within a hierarchy that divides planning into long-range planning, medium range planning, and short-term control.

(v) **ERP Systems and Bolt-Ons: -** Enterprise resource planning (ERP) systems are extensions of MRP systems that run on a single database in a client server environment. ERP systems support marketing and finance departments in addition to the production department. Significant coordination advantages arise when all functions draw and add to the same data. SAP is currently the leading provider of ERP systems. Many companies such as i2 Technologies and Manugistics have developed bolt-ons programs that run on top of ERP systems. These companies address specific problems that are not solved by ERP. For example, a better forecasting system or a finite-capacity scheduler can be added to SAP. Lately SAP has developed many of the capabilities that were formerly available only through bolt-ons.

Question for Discussion

1. What is inventory control ? Also explain economic order quantity.
2. What is master production scheduling ? Discuss its applications.
3. Explain and comment upon:
    a) ABC Analysis.
    b) VED Analysis.
4. Identify the factors ,which need to be considered in developing an ,inventory model. Define Inventory. Enumerate the reasons for holding inventory in the ligh to fits classical definition and idle resource'
5. It is given that the annual usage of a product is 60 units. The procurement cost per order is Rs.15 and the cost per piece is Rs 100. The cost of carrying inventory on obsolescence, taxes ,insurance and deterioration etc .is10%.Calculate the EOQ.
6. A Company needs 600 products per year.The ordering cost per order is Rs.20.The inventory carrying cost based on the average inventory is Rs.0.10per unit-year.The storage cost is Rs.0.025 per unit-year and the purchase priceis Rs.1 per unit. How many Orders should be placed per year and what total cost is involved in inventory.

7. A firm has a yearly demand for 52,000 units of a unit ,which it produces .The cost of setting up for production is Rs.80 and the weekly . production rate 1000 units .The carrying cost is Rs.3.50 per unit year .How many units should the firm produce on each production run? Identify the major reasons why firms carrying inventories.
8. Suresh Machine Company produces part that are shipped nation wide..It has an opportunity to produce plastic packaging cases.Which are currently purchased at Rs.0.70 each of Annual demand depends largely. On economic conditions ,but long run estimates are as shown.

Demand (units)	20000	30000	400000	500000	600000
Chance (%)	10	30	40	15	5

If the, company produces the cases itself, it must renovate an existing : work area and purchase a molding machine. Which will result in annual fixed cost of Rs. 8000 ? Variable costs are Rs. 0.50 per case.

a) should Suresh Machine make or buy the cases?
b) At what volume of production is it more profitable to produce in house rather than purchase from an outside source?

9. Discuss the secondary MRP outputs ?
10. What is meant by product life cycle? Explain with suitable examples.
11. Write a short note on ABC Inventory classification
12. Demand for tomato soup at a supermarket is always 25 cases a day and the lead-time is always 4 days. The shelves were just restocked with tomato soup, leaving an on-hand inventory of only 10 cases. No . backorders currently, but there is one open order for 200 cases. What is the inventory position?' Should a new order be placed?
13. Define Inventory Control System. Distinguish between fixed order quantity system and fixed time period system, Give examples of each.
14. Derive the formula for economic order quantity and total inventory related costs when the rate-of replenishment is finite.
15. An item is demanded at the rate of 3200 units per year. Ordering cost is Rs. 300. Inventory carrying cost is 20000 of the unit price per year. The supplier has offered a unit price of Rs. 150 per item but he is willing to reduce it to Rs: 85 per item if a purchase order of 1000 or more units is placed. Should the buyer accept this offer ?
16. "Inventory reduction can be used to highlight and solve the production system problems one by one". Explain.

References

1. T. Martand Telsang, "Inventory control and management", Production Management, New Delhi: S. Chand, 2007
2. Roach, "Origin of the Economic Order Quantity formula", transcription or transformation? Management Decision, Emerald Group Publishing Limited, Vol. 43, No. 9, Pp. 1262-268, 2005.
3. S. Russell Roberta and W. Taylor Bernard, "Operations Management: Quality and Competitiveness in a Global Environment", 5nd ed., John Wiley & Sons, 2006.
4. Tony Wild, "Best Practice in Inventory Management", 2nd ed., Elsevier Science & Technology Books, 2005.
5. J.W. Toomey, "Inventory Management: Principles, Concepts and Techniques", The Netherlands: Kluwer Academic publisher, 2000.
6. A.R. Cannon, "Inventory improvement and financial performance", International Journal of Production Economies, Vol. 115, Pp. 581-593, 2008.
7. T.S. Dudick and R. Cornell, "Inventory control for the financial executive", John Wiley & Sons Inc., 1980.
8. K. Howard, "Inventory management in practice", International Journal of Physical Distribution and Materials Management, Vol. 14, No. 2, Pp. 3-36, 1984.
9. R. Boute, M. Lambrecht, O. Lambrechts and P. Sterckx, "An analysis of inventory turnover in the Belgian manufacturing industry, wholesale and retail and the financial impact of inventory reduction", In Proceedings of the 14th 2006 Euroma Conference, Strathclyde, 2006.
10. H. Chen, F. Murray and W. Owen, "What actually happened to the inventories of American companies between 1981 and 2000?", Management Science, Vol. 51 No. 7, Pp. 1015-31, 2005.
11. J. Haan and M. Yamamoto, "Zero inventory management: facts or fiction?, Lessons fro Japan", International Journal of Production Economics, Vol. 59, Pp. 65-75, 1999.
12. D. Erlenkotter, "An early classic misplaced: Ford W. Harris s economic order quantity model of 1915", Management Science, Vol. 35 No. 7, Pp. 898-900, 1989.
13. P.T. Evers, "The effect of lead times on safety stocks", Production and Inventory Management Journal, 2nd Quarter, Pp. 6-10, 1999.
14. K. Howard, "Inventory management in practice", International Journal of Physical Distribution & Materials Management, Vol. 14 No. 2, Pp. 3-36, 1984.

15. R.J. Tersine, "Principles of Inventory and Materials Management", 4th ed., Prentice-Hall, Englewood Cliffs, NJ, 1994.
16. X. Chen, D. Simchi-Levi, "Coordinating inventory control and pricing strategies with random demand and fixed ordering cost: The finite horizon case", Oper. Res., Vol. 52, No. 6, Pp. 887–896, 2004a.
17. V. Gaur and S. Seshadri, "Hedging inventory risk through market instruments", Manufacturing Service Oper. Management, Vol. 7, No. 2, Pp. 103–120, 2005.
18. J. Song and P. Zipkin, "Inventory control in a fluctuating demand environment", Oper. Res., Vol. 41, No. 2, Pp. 351–370, 1993.
19. J. Hosseini, S. Baharaeen and X.T. Zhenf, "Design of knowledge based system forinventory control", Journal of EngineeringComputing and Application, Pp. 28-35, 1988.
20. Y. Liang, "The development of a knowledge-based inventory management system", PhD thesis, University of Salford, 1997.
21. E.A. Silver, "Operations research in inventory management", Operations Research, Vol. 29, No. 4, Pp. 628-45, 1981.

CHAPTER 11

PERT/CPM

11.0. Introduction

PERT was introduced in 1958 for the POLARIS missile program by the Program Evaluation Branch of the Special Projects office of the U.S.Navy, assisted by the Lockheed Missile Systems division and the Consultant firm of Booz-Allen & Hamilton. The calculations were maintained so that they could be carried out on the IBM Naval Ordinance Research Computer (NORC) at Dahlgren, Virginia.PERT is stands for Program Evaluation Review Technique which is a management tool used to schedule, organize, and coordinate tasks within a project.A PERT chart presents a graphic illustration of network diagram consisting of numbered nodes (either circles or rectangles) representing events, or milestones in the project linked by labeled vectors (directional lines) representing tasks in the project. The direction of the arrows on the lines indicates the sequence of tasks. In the diagram, for example, the tasks between nodes 1, 2, 4, 8, and 10 must be completed in sequence. These are called dependent or serial tasks. The tasks between nodes 1 and 2, and nodes 1 and 3 are not dependent on the completion of one to start the other and can be undertaken simultaneously. These tasks are called parallel or concurrent tasks. Tasks that must be completed in sequence but that don't require resources or completion time are considered to have event dependency. These are represented by dotted lines with arrows and are called dummy activities.

The PERT Chart is Sometimes Preferred Over the Gantt Chart

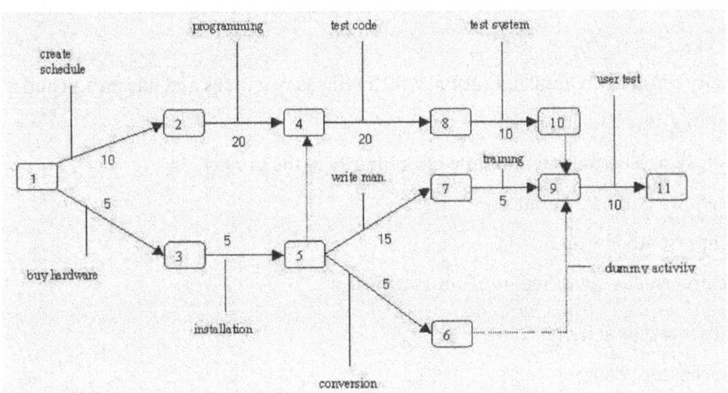

Diagram 11.1: Pert Chart

In PERT, activities are revealed as a network of precedence relationships taking use of activity-on-arrow network construction.

- Probabilistic activity times
- Multiple time estimates

USED IN: Project management - for non-recurring jobs, where the time and cost approximates tend to be somewhat doubtful. This technique utilizes probabilistic time estimates.

Basic Terms of PERT/CPM: Here are Some Important Basic Terms Applied in PERT/CPM

(i) Network :- It is a graphical representation of logical and sequentiallyconnectedactivities and events of a project. Network is also called arrow diagram PERT (Programme Evolution Review Technique) and (Critical Path Method) are the two most widely applied techniques .

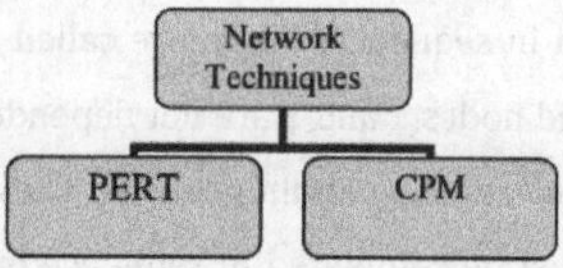

(ii) Project :- A project is defined as a combination of interrelated activities which must be executed in a certain order in for its completion.

(iii) Project Management Process :- Network analysis is the general name given to certain specific techniques which can be used for the planning, management and control of projects.

(iv) Activity :- Any individual operation, which utilizes resources and has an end and a beginning, is called activity.

- A task or a certain amount of work required in the project.
- Requires time to complete.
- Represented by an arrow.

These are usually classified into four categories:

- Predecessor activity.
- Successor activity.
- Concurrent activity.
- Dummy activity.

(v) Dummy Activity :- It Indicatesonly precedence relationshipsand does not require any time of effort.

(vi) PERT(Program Evaluation and Review Technique) :- PERT is a method to analyze the involved tasks in completing a given project, especially the time needed to complete each task, and identifying the minimum time needed to complete the total project.PERT is based on the assumption that an activity's duration follows a probability distribution instead of being a single valueThree time estimates are required to compute the parameters of an activity's duration distribution:

- **Pessimistic time (t_p):-**The time the activity would take if things did not go well .
- **Most likely time (t_m) :-**The consensus best estimate of the activity's duration.
- **Optimistic time (t_o) :-**The time the activity would take if things did go well .

$$\text{Mean (expected time)} = \frac{(t_p - 4t_m + t_o)}{6}$$

$$\text{Variance } (\sigma^2) = \left(\frac{t_p - t_o}{6}\right)^2$$

Probability computation:- Determine probability that project is completed within specified time .

$$Z = \frac{X - \mu}{\sigma}$$

Where μ = project mean time
σ = project standard mean time
x = (proposed) specified time

- Float :- Float of an activity represents the excess of available time over its duration.
- Total float (Ft) :- The amount of time by which the completion of an activity could be delay beyond the earliest expected completion time without affecting the overall project duration.
- Free Float (Ff) :- The time by which the completion of an activity can be delayed beyond the earliest finish time without affecting the earlieststart of a subsequent(succeeding) activities.

Situations in Network Diagram

(i) A must finish before either B or C can start.

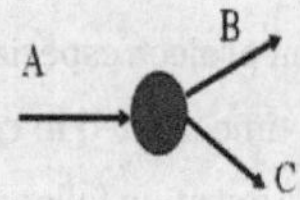

(ii) Both A and B must finish before C can star.

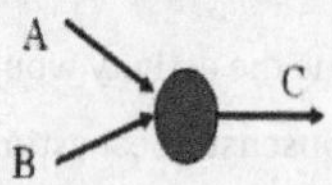

(iii) Both A and C must finish before either of B or D can start.

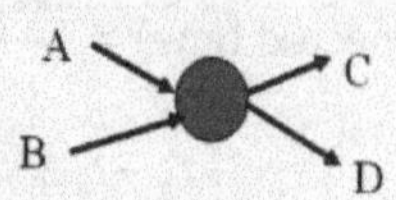

(iv) A must finish before B can start both A and C must finish before D can start.

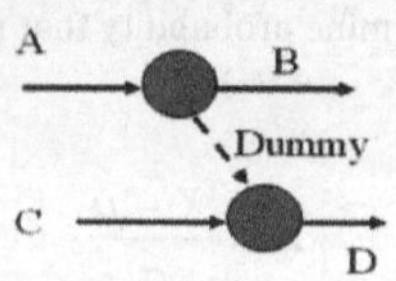

(i) Critical Path Method (CPM) :- CPM (Critical Path Method) was the invention of M.R.Walker of E.I.Du Pont de Nemours & Co. and J.E.Kelly of Remington Rand, circa 1957. The calculation was created for the UNIVAC-I computer. The first test was prepared in 1958, when CPM was applied to the development of a new chemical plant. In March 1959, the technique was applied to maintenance shut-down at the Du Pont works in Louisville, Kentucky. Unproductive time was decreased from 125 to 93 hours.CPM is stand for critical path method, which is a step-by-step project management technique for process planning that defines critical and non-critical tasks with the goal of preventing time-frame problems and process bottlenecks. The CPM is ideally suited to projects consisting of numerous activities that interact in a complex manner.In CPM activities are revealed as a network of precedence relationships taking use of activity-on-node network construction

- Deterministic activity times
- Single estimate of activity time

There are several steps that are generally applied in CPM which should be followed for proper implementation is given below.

- Define the required tasks and put them down in an ordered (sequenced) list.
- Create a flowchart or other diagram showing each task in relation to the others.
- Identify the critical and non-critical relationships (paths) among tasks.
- Determine the expected completion or execution time for each task.
- Locate or devise alternatives (backups) for the most critical paths.

11.1. Difference between PERT and CPM

Following are basic headings between **PERT AND CPM.**

Table 11.1: PERT AND CPM

S. NO.	**CPM** (critical path method)	**PERT** (Program Evaluation Review Technique)
1	CPM uses activity oriented network	PERT uses event oriented Network
2	Durations of activity may be estimated with a fair degree of accuracy	Estimate of time for activities are not so accurate and definite
3	It is used extensively in construction projects	It is used mostly in research and development projects, particularly projects of non-repetitive nature
4	Deterministic concept is used.	Probabilistic model concept is used.
5	CPM can control both time and cost when planning.	PERT is basically a tool for planning
6	In CPM, cost optimization is given prime importance. The time for the completion of the project depends upon cost optimization. The cost is not directly proportioned to time. Thus, cost is the controlling factor.	In PERT, it is assumed that cost varies directly with time. Attention is therefore given to minimize the time so that minimum cost results. Thus in PERT, time is the controlling factor.

Applications of CPM / PERT

These methods have been widely applied in industrial sector also sometimes used in government organizations. Some important are given below.

- Building of a dam or a canal system in an area.
- Building of a building or highway.
- Space flight.
- Preservation or overhaul of airplanes or oil refinery.
- Cost control of a project with the help of PERT / COST.
- Creating a prototype of a machine.
- Creation of supersonic planes.
- Construction of a new plant.

- Research and development of a new product.
- Building a ship.
- Government-sponsored projects for developing a new weapons system.
- Relocation of a major facility.
- Maintenance of a nuclear reactor.
- Installation of a management information system.

Following are Some Important Advantages of PERT/CPM

- It Helpful at various stages of project management.
- It Mathematically easy It Provide critical path and slack time.
- It Give project documentation.
- It Helpful in monitoring and controlling costs.

Both methods are fundamentally network-oriented techniques using the similar principle. PERT and CPM are essentially time-oriented methods in sense that they both move for determination of a time schedule for the project. The important difference between two approaches is that the time approximation for the dissimilar activities in CPM was implicit to be deterministic whereas in PERT these are explained probabilistically. These techniques are also known as project scheduling techniques.

11.2. PERT for Project Planning and Scheduling

PERT, the Project Evaluation and Review Technique, is a network based aid for planning and scheduling the many interrelated tasks in a large and complex project. It was developed during the design and construction of the Polaris submarine in the US in the 1950s, which was one of the most complex tasks ever attempted at the time.

Nowadays PERT techniques are routinely used in any large project such as software development, building construction, etc. Supporting software such as Microsoft Project, among others, is readily available. It may seem odd that PERT appears in a book on optimization, but it is frequently necessary to optimize time and resource constrained systems, and the basic ideas of PERT help to organize such an optimization. PERT uses a network representation to capture the precedence or parallel relationships among the tasks in the project.

The PERT formalism has these elements and rules:

- Directed arcs represent activities, each of which has a specified duration. This is the "activity on arc" formalism; there is also less-common "activity on node" formalism. Note that activities are considered to be uninterruptible once started.

- Nodes are events or points in time.
- The activities (arcs) leaving a node cannot begin until all of the activities (arcs) entering a node are completed. This is how precedence is shown. You can also think of the node as enforcing a rendezvous: no one can leave until everyone has arrived.
- There is a single starting node which has only outflow arcs, and a single ending node that has only inflow arcs.
- There are no cycles in the network. You can see the difficulty here. If an outflow activity cannot begin until all of the inflow activities have been completed, a cycle means that the system can never get started.

(i) Implementation of PERT technique

In any kind of project scheduling, there are two basic questions will be comes on the front line.

1. What will be the shortest time for completion of the project?
2. Which activities must be completed on time in order to finish in the shortest possible time? These activities constitute the critical path through the PERT diagram.

In order to get a critical path, it is very necessary to know how much time is being taken by each activity. In given diagram the, each path is denoted by a particular numbers which shows a time taken in days for completing the respective activity.

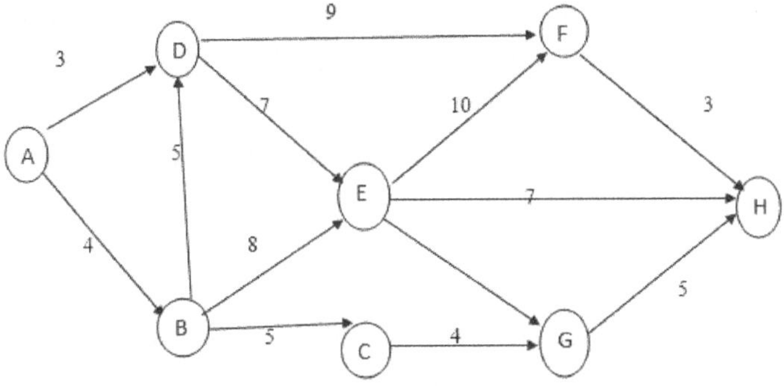

The critical path is of great interest to project managers. The activities on the critical path are the ones which absolutely must be done on time in order for the whole project to complete on time. If any of the activities on the critical path are late, then the entire project will finish late! For this reason, the critical path activities receive the greatest attention from management. The non-critical activities have some leeway to be late without affecting the overall project completion time.

The following steps must be taken for achieving critical path and for calculation, regarding the project the project.

Step 1 –Make a forward pass through the diagram, calculating the earliest time(T_E) for each event (node). TE at the starting node is zero. Thus

$$T_E = \max [(T_E \text{ of node at tail of arc}) + (\text{arc duration})]$$

Step 2-- Make a backward pass through the diagram, calculating the latest time (T_L) for each event (node).To find T_L, look at all of the activities which exit a node . The $T_L$of the ending node equals itsT_E . Thus

$$T_L = \min [(T_L \text{ of node at head of arc}) - (\text{arc duration})]$$

Step 3 --Calculate the node slack time(SN) for each node (event). This isthe amount of time by which an event could be adjusted later than its$T_E$without causing problems downstream.

$$S_N = T_L - T_E \text{ for each node}$$

Step 4-- Calculate the total arc slack time (SA) for each arc (activity). This is the amount of time by which an activity could be adjusted later than the $T_E$of the node at its tail without causing problems later. **Thus**

$$S_A = (T_L \text{ of node at arc head}) - (T_E \text{ of node at arc tail}) - (\text{arc duration})$$

Step 5--The critical path connects the nodes at which $S_N= 0$ via the arcs at which $S_A= 0$.The critical path connects the nodes and arcs which have no slack, but sometimes there is slack, then the activity does not need to be done on time.For example consider a pert diagram for that **T_E**calculated. The starting node is node **A**, and by definition the **T_E**of the starting node is 0. Thus

$$T_E(B) = T_E(A) + (\text{duration of A-B}) = 0 + 4 = 4$$

The description of T_E calculations are as follows:-

- T_E (A) = Starting node = 0
- TE(B)= TE (A) + (duration -B) = 0+4 = 4
- T_E(D) = max{T_E(A)+(duration A-D), T_E(B)+(durationB-D)}= max{0+3, 4+5} = 9
- T_E (C) = T_E (B) + (duration B-C) = 4+5 = 9
- T_E(E)= max{TE(D)+(duration D-E), T_E(B)+(duration B-E),T_E(C)+(duration C-E)}= max{9+7, 4+8, 9+6} = 16
- T_E(F)= max{T_E(D)+(duration D-F), T_E(E)+(durationE-F)}max{9+9, 16+10} = 26
- T_E (G)= max{T_E(E)+(duration E-G), T_E(C)+(durationC-G)}= max{16+7, 9+4} = 23

- $T_E(H)$= max{$T_E(F)$+(duration F-H), $T_E(E)$+(duration E-H),$T_E(G)$+(duration G-H)}= max{26+3, 16+3, 23+5} = 29
- $T_L(H)$= ending node = $T_E(H)$= 29
- $T_L(F)$= $T_L(H)$–(duration F-H)= 29–3= 26
- $T_L(G)$ = $T_L(H)$–(duration G-H)= 29–5 = 24
- $T_L(E)$= min{$T_L(F)$–(duration E-F), $T_L(H)$–(durationE-H),$T_L(G)$–(duration E-G)}= min{26–10, 29–3, 24–7} = 1
- $T_L(C)$= min$T_L(E)$–(duration -E), $T_L(G)$–(durationC-G)}= min{16–6, 24–4} = 10
- $T_L(D)$= min{$T_L(F)$–(duration D-F), $T_L(E)$–(duration D-E)}= min{26–9, 16–7} = 9
- $T_L(B)$= min{$T_L(D)$–(duration B-D), $T_L(E)$–(durationB-E),$T_L(C)$–(duration B-C)}= min{9–5, 16–8, 10–5} = 4
- $T_L(A)$= min{$T_L(D)$–(duration A-D), $T_L(B)$–(duration A-B)}= min{9–3, 4–4} = 0

In Step 3 We Find the Node Slack Time (S_N) for Each Node (Event) as Shown Below:

Node	A	B	C	D	E	F	G	H
S_N	0	0	1	0	0	0	1	0

This small PERT diagram is quite tight, only two of the nodes have non- zero slack.

In Step 4 We Find Total arc Slack Time (SA) for Each arc (activity) as Shown Below:

Arc	**AB**	AD	BC	BD	BE	CE	CG	DE	DF	EF	EG	EH	FH	GH
SA	0	6	1	0	4	1	11	0	8	0	1	10	0	1

Step 5 -- we find the critical path by linking the nodes having no slack via the arcs having no slack.

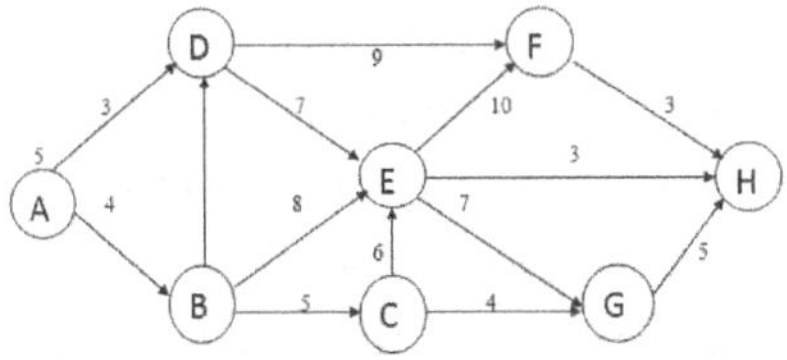

The Critical Path

Now the expected value of the total project duration is the sum of the expected activity durations along the critical path, which is found in the usual way.

11.3. Project Scheduling by Critical Path Method (CPM)

CPM is a method for planning, scheduling, and controlling such projects like Construction Management (project).

Basic Steps in CPM: - following steps must be taken by the engineers to initiate the project.

(i) Decompose project into Individual Jobs or Activities.

(ii) Determine Precedence Relationships among jobs. They come from technological constraints. For example:Suspension bridge road units cannot be lifted into place until pillars are prepared. "Painting the walls" cannot be started until "erecting walls" completed

(iii) Draw the Arrow Diagram,or AOA Diagram, or Project Network representing precedence relationships.

(iv) Estimate job durations.

(v) Use CPM to schedule activities over time to minimize project duration (DP used here).

(vi) In operation, if any job delayed, use CPM to revise schedule for remaining jobs.

Precedence Relationships

Predecessor or Ancestor	Job 1 is a Predecessor or Ancestor of job 2 if job 2 cannot be started Until job 1 completed.
Successor or Descendent	In above job 2 is a Successor or descendent of job 1.
Immediate Predecessor	If job 1 is a predecessor of job 2, and no other job is a predecessor to job 2 and successor to job 1; job 1 is an Immediate Predecessor of job 2
Immediate Successor	If job 1 is an immediate predecessor of job 2, then job 2 is an Immediate Successor of job 1.

A job may have many immediate predecessors, and many immediate successors. It can only be started after all its immediate predecessors have been completed.

Transitivity: -Job 1 predecessor of job 2, and job 2 predecessor of job 3⇒job 1 predecessor of job 3. This property called Transitivity. Given immediate predecessors of a job, we can obtain all its predecessors by transitivity. Precedence relationships define a Partial Ordering among jobs. If transitivity establishes that a job is a predecessor of itself, there is an Inconsistency in the data that should be corrected. If a distant predecessor of a job is shown as an immediate predecessor, there is Redundancy in the predecessor data, which should be corrected.

Project Networks: 2 ways to represent predecessor relationships in network form.

AON(ActivityonNode) network easytodraw, notvery useful. We discuss Activity on Arc (AOA) diagram or Arrow diagram. Here, jobs correspond to arcs. Each node corresponds to an Event, "jobs corresponding to all arcs incident into it, andAll their predecessors have been completed". To portray predecessor relationships correctly in arrow diagram, sometimes necessary to introduce Dummy arcs corresponding to dummy jobs (which always have 0 duration).

Properties Satisfied by Arrow Diagram

(i) If job 1 is a predecessor of job 2, there will be a directed path in arrow diagram from the head node of job 1 to the tail node of job 2 (i.e. predecessor relationship represented through directed paths).

(ii) If job 1 is an immediate predecessor of job 2,Either head node of job 1 is tail node of job 2Or there is a directed path from head node of job 1 to the tail node of job 2 consisting of dummy arcs only.

How to Draw Arrow Diagram ?

Put a Start node representing event of commencing project. Represent each job with no predecessor by an arc incident out of this start node. Same way, at end, represent all jobs with no successors by arcs incident into a single final node Finish Node representing event of project completion.

If a job b has a single IP a, then can be represented by an arc incident out of the head node of a. If a job b has more than one IP, let p1,...,p r be the head nodesof IPs of job b:either represent b by an arc incident out of one of the nodes186p1,...,p r, sayp1, and include dummy arcs from p2,...,printo p1.or represent b by an arc incident out of a new node Q and include dummy arcs (p1,Q),...,(pr,Q). If some jobs have identical sets of immediate successors, make the head node of arcs representing them the same. If several jobs have some common IPs, represent these commoners by arcs with the same head node, and include dummy arcs.

Coming Out of This Node as Necessary. Check

- Precedence relationships not implied by original data not introduced.
- Precedence relationship simplied by original datanot omitted.
- See if no. of nodes & no. of dummy nodes can be reduced. For example, if there is a node with a single arc (which is a dummy arc) incident at it, then the 2 nodes on this dummyarc can be merged, and that dummy arc eliminated.

Since no job can be its own predecessor, project network cannotcontain any directed cycles, so project network must be acyclicnetwork.

Acyclic Networks

- A directed network that contains no circuits (directed cycles).
- A circuit or directed cycle.
- A cycle that is not a circuit.

Theorem: A Directed network acyclic off nodes can be numbered on each arc,

Number of tail < Number of head

Such numbering of nodes called a cyclic numbering or topological ordering

How to Check If Network Acyclic, & find Acyclic Numbering of its nodes?

(i) Number all nodes with no arcs incident into them, serially beginning with 1Strike off these entire nodes & arcs incident at them.

(ii) Now continue numbering all nodes with no arcs that are not struck off, incident into them, serially. Strike off all newly numbered nodes & arcs incident at them.

(iii) Repeat Step 2 until either: all nodes numbered (in this case network acyclic, you have acyclic numbering, terminate) or some unnumbered nodes, but everyone has an arc that is not struck off, incident into it (in this case, network not acyclic, terminate).

Some Important Data Related to CPM Scheduling

This data is related to building construction.

Job	IPs Duration (days)
Site selection & purchase	30
Architectural plans	45
Revisions to AP	30
Arrange Financing	60
Construct outer frame	90
Plans for interior	25
Order furnishings (desks, etc.)	120
Order Computer equipment	120
Computer Network Design	70
Interior Construction	70
Install computer equipment	90
Move-In	05

For Constructing Single Family Home

	Job	IPs	Duration in week
0	Purchase site		16
1	Prepare blueprint	0	6
2	Excavation	1	4
3	Foundations & Basement walls	2	3
4	Frame	3	5
5	Electrical wiring	4	2
6	Insulate	4	1
7	Roofing	4	2
8	Ductwork & Plumbing	6	4
9	Drywall installation	5, 7, 8	6
10	Install siding	7	4
11	Exterior painting	10	3
12	Interior painting	9	4
13	Carpeting	12	2
14	Install interior fixtures	13	7
15	Install Exterior fixtures	11	2
16	Move-in	14, 15	1

Hydroelectric Power Station Building Project

	Job	IPs	Duration of month
1	Site Ecological survey		6.2
2	Get EPA approval	1	9.1
3	Economic feasibility study	1	7.3
4	Prel. design and cost est.	3	4.2
5	Project approval	2, 4	10.2
6	Call quotations for electrics equipments like – turbine, generators	5	4.3
7	Select suppliers for elec. Equip	6	3.1
8	Final design	5	6.5
9	Select construction contractors	5	2.7
10	Arrange materials supply	8, 9	5.2
11	Dam building	10	24.8
12	Power station building	10	18.4
13	Power lines erection	7, 8	20.3
14	Elec. equip. installation	7, 12	6.8
15	Build up reservoir water level	11	2.1
16	Commission the generators	14, 15	1.2
17	Start supplying power	13, 16	11

11.4. Scheduling

Production scheduling is defined as the order in which demand is loaded on the shop floor. It is a queuing system. The logical sequence of production is to process orders in the order of the date they were received.

It is the management and allocation of resources, events and processes to create goods and services thus scheduling requires a strong focus on the availability of the resources of a business. Resources include the raw materials used to create goods, the availability of machines and the availability of workers.

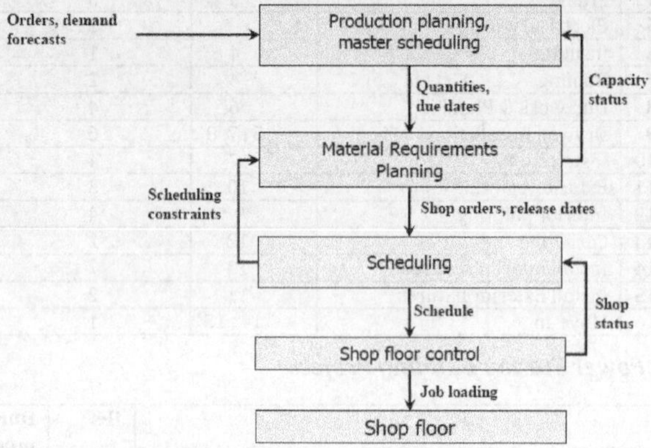

Diagram 11.2: Scheduling Diagram

There are so many definition is given by the **various management authors** regarding the Scheduling, these definitions are given below.

According to **Richard Lansburgh** and **William Spriegel.**

> **"Scheduling involves establishing the amount, of work to be done and the time each element of the work will start, or the order of work. This includes allocating the quality and rate of output of the plant or department and also the date or order of starting of each unit of work at each station along the route prescribed."**

According to **James Lundy.**

> **"Scheduling consists of the assignment of starting and completion times for the various operations to be performed."**

According to **Bethel**, Atwater, Smith and Stackman (Jr).

> **"Scheduling is that phase of production control which rates the work in the order of its priority and then provides for its release to the plant at the proper time and in the correct sequence. Thus, it is concerned with when work shall be performed on a product or part."**

According to **Leon Alford** and **Henry Beatty,**

> **"Fitting specific jobs into a general time-table so that orders may be manufactured in accordance with contracted liability, or in mass production, so that each component may arrive at and enter into assembly in the order and at the time required."**

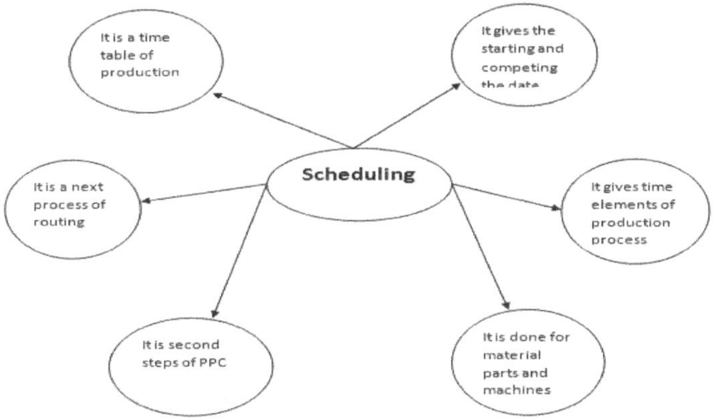

Diagram 11.3: Scheduling Diagram

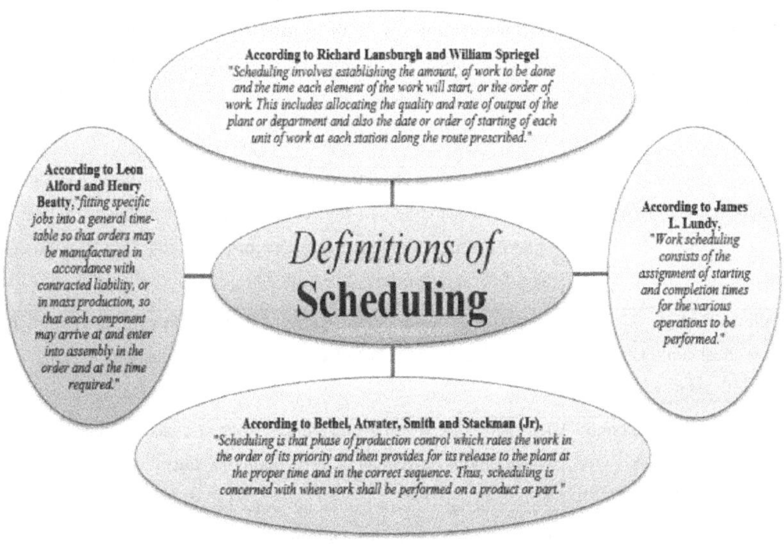

Diagram 11.4: Definition of Scheduling

The main purpose or objective of scheduling is to see that all production operations are completed in a given time. If the production is completed in an allotted time, only, then, the delivery of goods can be made on time. If this happens, it will increase the goodwill of the company.

The Functions or Procedure of Scheduling is as Follows

1. Make a list of all the production activities.
2. Fix the starting time for each activity.
3. Fix the finishing time for each activity.
4. Prepare the time-table.

11.5. Principle of Scheduling

The following are the principle of scheduling, given below with suitable description.

Table 11.2: Scheduling Principle

	Principle	Description
1	Compile your schedule just prior to the period of time it covers	Being current helps keep your schedule relevant and eliminates the need for multiple revisions.
2	Begin your schedule with the end in mind	Determine your deliverables – your required output for the period of time the schedule covers. Deliverables are always time-sensitive.
3	Schedule critical tasks first	Critical tasks are those that are both high value – important to achieving your goals – and time-sensitive. These should be added into your schedule first.
4	Recognize your controllable time	Controllable time is the time you actually have available to complete your tasks and activities. To find controllable time, estimate the time you spend dealing with unpredictable events, and deduct that amount of time from your total workday.
5	Allow time for previously unfinished critical tasks	Critical tasks always take priority. Allow time in your schedule to finish uncompleted critical tasks from your previous schedule.
6	Chunk similar tasks and activities together	It's more efficient to group similar types of tasks together, rather than waste time "task-switching" throughout the day.
7	Be flexible in your approach	Your schedule isn't set in stone. Don't be afraid to adjust and readjust your schedule until you have it just as you want it.

11.6. Factors Affecting Scheduling

There are so many factors which may affects the scheduling procedure but these factors are broadly classified into two group internal factor and external factor.

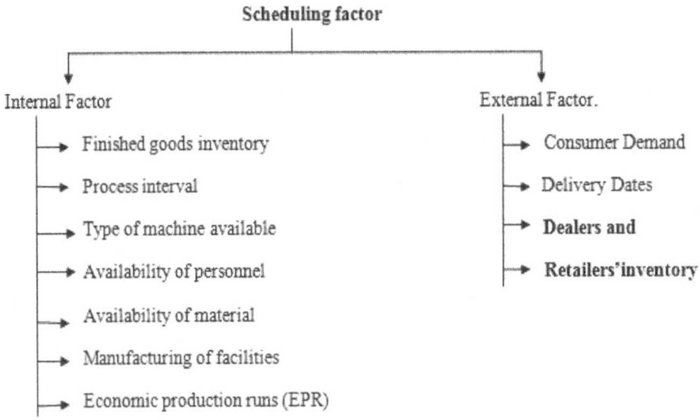

Diagram 11.5: Factors Affecting Scheduling

Internal Factors Affecting Scheduling

(i) **Finished goods inventory:** -Scheduling depends on how much stock of finished goods is kept by the company. Most companies keep, one month's supply of each product, as stock. If the company's product is fast moving or slow moving, then scheduling will have to be changed.

(ii) **Process intervals:** It depends on the process intervals of each product. Process interval is the time required to produce a product. Different products have different process intervals. For example, the process interval of a car is more than that of soap. Scheduling will be different for each process interval.

(iii) **Type of machines available:** - It also affected by the type of machines available. If the company has old and outdated machines, the schedule must keep provisions for the breakdown of machines. Modern and computerized machines make scheduling very easy.

(iv) **Availability of personnel:** - Scheduling also depends on the availability of personnel. If the company has untrained and inexperienced employees, then they will take more time to produce a product. So, the schedule must keep provisions for this. A faster schedule will be required for trained and experienced employees.

(v) **Availability of materials: -** It is also affected by the availability of materials. If a regular supply of materials is available, then the company can do normal-scheduling. However, if the supply of materials is irregular, the schedule must be made flexible. That is, when the supply is good then the schedule will be fast and vice versa.

(vi) **Manufacturing facilities: -** Scheduling depends on the manufacturing facilities available in the company. This includes space for new machines, employees, etc. It also includes the availability and supply of electricity and water, which may be required for production. If the entire required infrastructure is available, then the production schedule can be fast and vice versa.

(vii) **Economic production run (EPR) :-** It also depends on the economic production runs. Economic production runs (EPR) means the optimum lot size. That is, how many items must be produced in one lot in order to minimize the cost of production? If the company produces more or less than the optimum lot size, then the cost of production will increase. There are many formulas for calculating optimum lot size. Scheduling must be done only after calculating the optimum lot size.

External Factors Affecting Scheduling

(i) **Consumer Demand: -** Scheduling also depends on the consumer demand. Consumer demand can be found out by sales forecast. So, the production schedule is prepared according to the sales forecast. However, it has to be adjusted (changed) when the actual demand is different from the sales forecast.

(ii) **Consumer delivery dates: -** The production schedule also depends on the consumer delivery dates. The consumer is the most important person in a business. So, this factor must be given more importance than other factors. The production schedule must be made in such a way that it will guarantee timely delivery to the consumers. In case of seasonal goods, production must be spread out throughout the year; so, there will not be too much pressure in demand season.

(iii) **Dealers and retailers Inventories: -** It also depends on the stock of goods (inventories) with dealers and retailers. The production manager must find out how much stocks are held by dealers and retailers. He must also know why they are keeping this stock. Are they keeping this stock to meet current demand? If yes, then normal-scheduling can be done. However, if they are keeping stock in anticipation of future demand, the scheduling will have to be slowed down because there will be fewer orders in the future.

(viii) **Availability of materials: -** It is also affected by the availability of materials. If a regular supply of materials is available, then the company can do normal-scheduling. However, if the supply of materials is irregular, the schedule must be made flexible. That is, when the supply is good then the schedule will be fast and vice versa.

(ix) **Manufacturing facilities: -** Scheduling depends on the manufacturing facilities available in the company. This includes space for new machines, employees, etc. It also includes the availability and supply of electricity and water, which may be required for production. If the entire required infrastructure is available, then the production schedule can be fast and vice versa.

(x) **Economic production run (EPR) :-** It also depends on the economic production runs. Economic production runs (EPR) means the optimum lot size. That is, how many items must be produced in one lot in order to minimize the cost of production? If the company produces more or less than the optimum lot size, then the cost of production will increase. There are many formulas for calculating optimum lot size. Scheduling must be done only after calculating the optimum lot size.

External Factors Affecting Scheduling

(i) **Consumer Demand: -** Scheduling also depends on the consumer demand. Consumer demand can be found out by sales forecast. So, the production schedule is prepared according to the sales forecast. However, it has to be adjusted (changed) when the actual demand is different from the sales forecast.

(ii) **Consumer delivery dates: -** The production schedule also depends on the consumer delivery dates. The consumer is the most important person in a business. So, this factor must be given more importance than other factors. The production schedule must be made in such a way that it will guarantee timely delivery to the consumers. In case of seasonal goods, production must be spread out throughout the year; so, there will not be too much pressure in demand season.

(iii) **Dealers and retailers Inventories: -** It also depends on the stock of goods (inventories) with dealers and retailers. The production manager must find out how much stocks are held by dealers and retailers. He must also know why they are keeping this stock. Are they keeping this stock to meet current demand? If yes, then normal-scheduling can be done. However, if they are keeping stock in anticipation of future demand, the scheduling will have to be slowed down because there will be fewer orders in the future.

SchedulingStrategies: -Scheduling strategies vary widely among firms and range from 'no scheduling' to very sophisticated approaches. These strategies are grouped into four classes.

(i) **Detailed scheduling: -** Detailed scheduling for specific jobs that are arrived from customers is impracticable in actual manufacturing situation. Changes in orders, equipment breakdown, and unforeseen events deviate the plans.

(ii) **Cumulative scheduling: -** Cumulative scheduling of total work load is useful especially for long range planning of capacity needs. This may load the current period excessively and under load future periods. It has some means to control the jobs.

(iii) **Cumulative detailed: -** Cumulative detailed combination is both feasible and practical approach. If master schedule has fixed and flexible portions.

(iv) **Priority decision rules:-** Priority decision rules are scheduling guides that are used independently and in conjunction with one of the above strategies, i.e., first come first serve. These are useful in reducing Work-In-Process (WIP) inventory.

Types of Scheduling: - Types of scheduling can be categorized as forward scheduling and backward scheduling.

- Forward scheduling.
- Backward scheduling.

(i) Forward scheduling is commonly used in job shops where customers place their orders on "needed as soon as possible" basis. Forward scheduling determines start and finish times of next priority job by assigning it the earliest available time slot and from that time, determines when the job will be finished in that work centre. Since the job and its components start as early as possible, they will typically be completed before they are due at the subsequent work centers in the routing. The forward method generates in the process inventory that are needed at subsequent work centers and higher inventory cost. Forward scheduling is simple to use and it gets jobs done in shorter lead times, compared to backward scheduling.

(ii) Backward scheduling is often used in assembly type industries and commit in advance to specific delivery dates. Backward scheduling determines the start and finish times for waiting jobs by assigning them to the latest available time slot that will enable each job to be completed just when it is due, but done before. By assigning jobs as late as possible, backward scheduling minimizes inventories since a job is not completed until it must go directly to the next work centre on its routing. Forward and backward scheduling methods are shown in Fig.

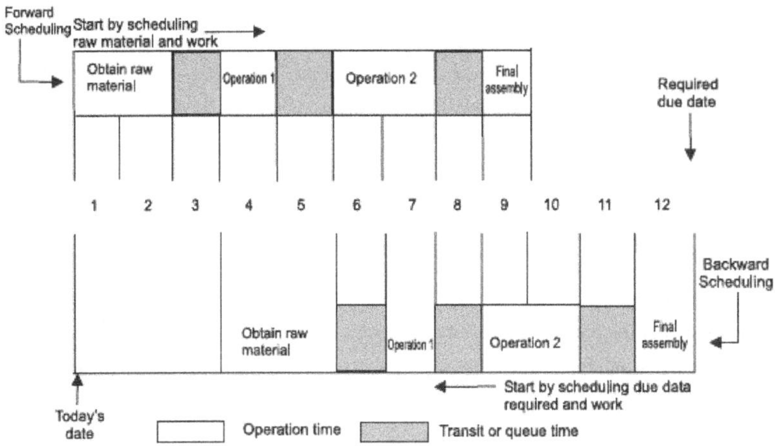

Diagram 11.6: Factors Affecting Scheduling

Forward and backward scheduling process.

11.7. The General Scheduling Problem

Production scheduling can be defined as the allocation of available production resources over time to best satisfy some set of criteria. Scheduling problems arise whenever a common set of resources (labour, material and equipment) must be used to make a variety of different products during the same period of time. The objective of scheduling is to efficiently allocate resources over time to manufacture goods such that the production constraints are satisfied and the production costs are minimized. This involves a set of tasks to be performed, and the criteria may involve both tradeoffs between early and late completion of a task, and between holding inventory for the task and frequent production changeovers. Operations require machines and material resources and must be performed according to a feasible technological sequence. Schedules are influenced by diverse factors such as job priorities, due-date requirements, release dates, cost restrictions, production levels, lot-size restrictions, machine availabilities, machine capabilities, operation precedence, resource requirements, and resource availabilities. To illustrate a production schedule, consider a one product, three-machine job-shop scheduling problem, shown in Figure. The simplest scheduling occurs when we have unlimited capacity resources for the given application while trying to minimize its latency.

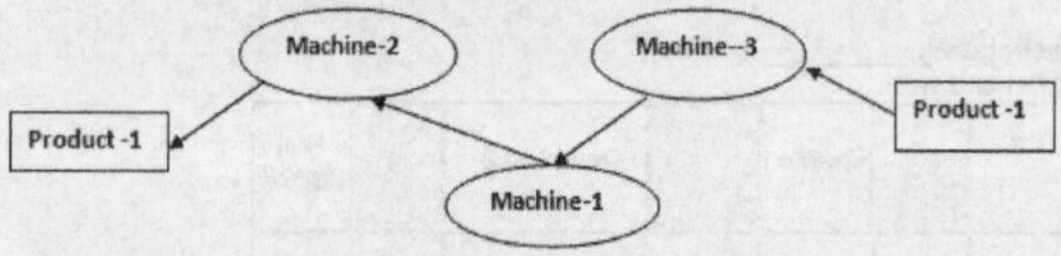

Diagram 11.7: One-Product Multi - Stage Production System

For this problem, simply solve it is to schedule an operation as soon as all of its predecessors tasks have completed, which gives it the name As Soon As Possible. It is closely related with finding the longest path between an operation and the demand due date. The first observation is that the maximum capacity of this system is equal to the lesser of the maximum capacities of the resources. Now consider Figure, In which the same three machine system, processing three different products under distinct production sequences. In this case, the maximum productive capacity of the system depends on the mix and sequence of production, the rate of resource sharing and the time profile of demand. Therefore, the production capacity of any multiproduct, multistage system is dependent on the mix of the demand and on the production schedule. This problem grows in complexity when considering various time periods for manufacturing resource planning. Note that the required ordering of operations within each job (the sequence technology) is preserved and the ordering of operations on the machine was selected to minimize the total production time. To illustrate the problem of resource allocation, Table 1 provides routes data and the processing time of a cell production, composedof three machines. The Prod1 first operation occurs on the machine M3, and requires one unit of processing time. The second operation is on the machine M1, requiring seven units of time. Finally, the third operation requires 6 units of time and occurs on machine M2. The production sequence for products 2 and 3 follow the same logic.

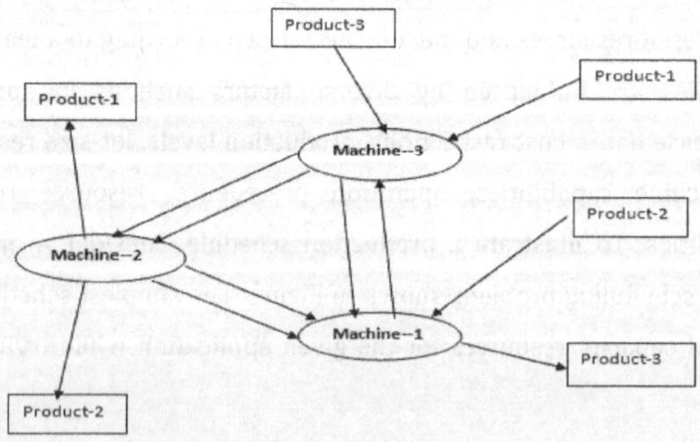

Diagram 11.8: Multi-Stage Multi-Period Multi-Product Production System

Table 11.3: Three Product Production Sequence

Operation Sequence	Product -1	Product -2	Product -3
1	1/ machine-3	6 / machine-1	3 / machine-3
2	7 / machine-1	5 / machine-3	6 / machine-2
3	6 / machine-2	3 / machine-2	10 / machine-3

One of the Gantt charts for production allocation of the three products is shown in Figure.11.9 It is assumed that all items are available at the beginning of the process and those operations are not shared.

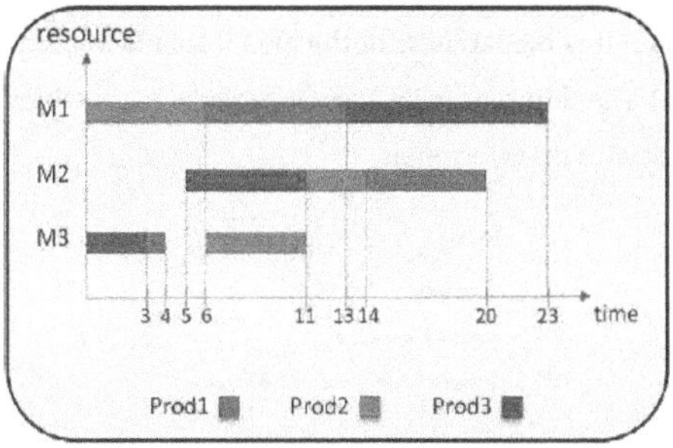

Diagram 11.9: Gantt Chart for Three Products Production Allocation

However, when considering a real problem it is not enough to simply have a heuristic to allocate the production of the items to their machines as shown above. It is essential to optimize the sequencing of production in order to properly harness the available resources, avoiding as much as possible the use of overtime, hiring services and loss of orders due to lack of unavailable capacity.

11.8. Job Shop Scheduling

Job-shop Scheduling A job-shop does not have the same restriction on workflow as a flow-shop. In a job-shop, jobs can be processed on machines in any order. The usual job shop, from a research standpoint, is one in which there are m machines and n jobs to be processed. Each job requires m operations, one on each machine, in a specific order, but the order can be different for each job3. Real job-shops are more complicated. Jobs may not require all in machines; and yet they may have to visit some machines more than once. Clearly, workflow is not unidirectional in a job-shop. Any given machine may observe new jobs arriving from outside the shop (as new inputs), and from other machines within the shop (as WIP), the same machine may be the last machine for a particular job, or it may be an intermediate processing step. Thus, the workflow can be illustrated as in Figure.

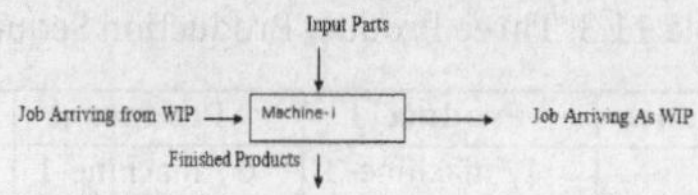

Diagram 11.10: One Machine in Job Shop

(i) **Flow time (Fj) :-** It is the amount of time job j spends in system. Make span. It is the total amount of time for all jobs to finish processing.

(ii) **Lateness (Lj) :-** It is the amount of time by which the completion time differs from the due date. Positive lateness of a job means job is completed after the due date.

(iii) **Tardiness (Tj) :-** It is the lateness of the job j if it fails to meet its due date, or zero. i.e Tj= max {0, Lj} The different measures of performances which are used in scheduling are listed below with their formula.

Mean flow time: $\bar{F} = \frac{1}{n}\sum_{j=1}^{n} F_j$

Mean tardiness: $\bar{T} = \frac{1}{n}\sum_{j=1}^{n} T_j$

Maximum flow time: $F_{max} = \max_{1 \le j \le n} \{F_j\}$

Maximum tardiness: $T_{max} = \max_{1 \le j \le n} \{T_j\}$

Number of tardy job: $N_T = \sum_{j=1}^{n} f(T_j)$

where $f(T_j) = 1$, If $T_j > 0$ and $f(T_j) = 0$, otherwise.

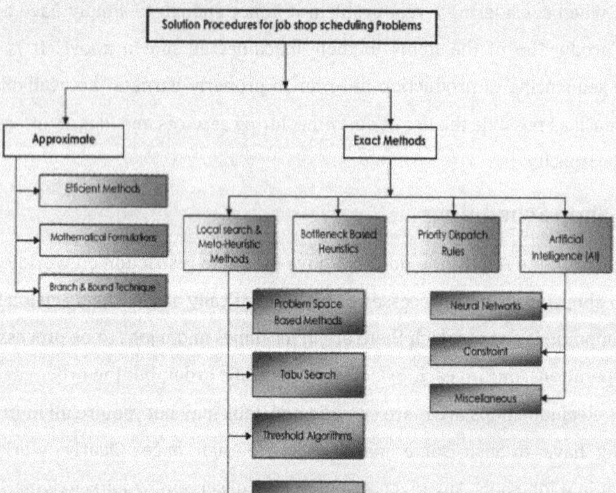

Diagram 11.11: Solution Structure of Job Shop Scheduling Problem

(i) **Flow-shop scheduling:** -A flow-shop is a shop design in which machines are arranged in series Jobs begin processing on an initial machine, proceed through several intermediary machines, and conclude on a final machine3. The series arrangement implies a linear structure to the shop, as illustrated in Figure. This figure shows a pure flow-shop in which jobs must be processed on each machine in exactly the same order. A general flow-shop is somewhat different, in that a job may skip a particular machine. For instance, although every job must proceed from left to right in Figure 1, some jobs may go from machine 1 to, say, machine 3 and then machine 4.

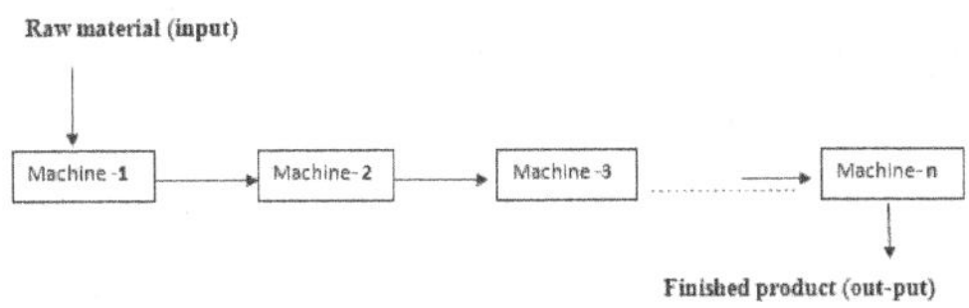

Diagram 11.12: Flow-Shop Scheduling

(ii) **Job Shop Sequencing:** -Just as many schedules are feasible for a specific group of jobs at a particular set of workstations, numerous methods can be used to generate schedules. They range from straightforward manual methods, such as manipulating Gantt charts, to sophisticated computer models for developing optimal schedules. One way to generate schedules in job shops is by using priority sequencing rules, which allows the schedule for a workstation to evolve over a period of time. The decision about which job to process next is made with simple priority rules whenever the workstation becomes available for further processing. One advantage of this method is that last-minute information on operating conditions can be incorporated into the schedule as it evolves. We already covered two important sequencing rules in Chapter 14, "Operations Planning and Scheduling." The first-come,first-served (FCFS)rule gives the job arriving at the workstation first the highest priority. The earliest due date (EDD) rule gives the job with the earliest due date based on assigned due dates the highest priority. Such rules can be applied by a worker or,

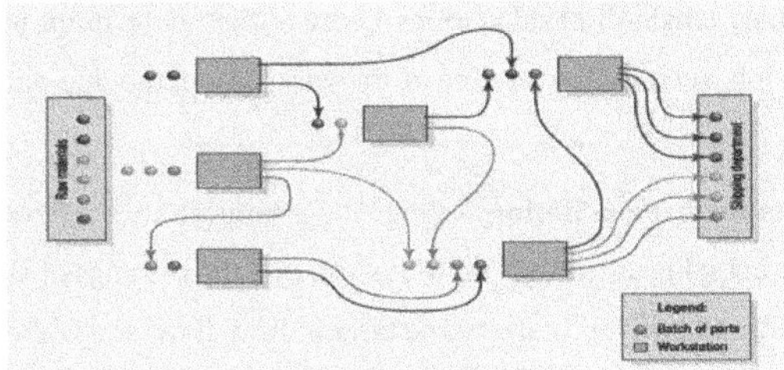

Diagram 11.13: Job Shop Sequencing

Incorporated into a computerized scheduling system that generates a dispatch list of jobs and priorities for each workstation. Additional priority sequencing rules follow:

- **Critical Ratio :-**The critical ratio (CR) is calculated by dividing the time remaining until a job's due date by the total shop time remaining for the job, which is defined as the setup, processing, move, and expected waiting times of all remaining operations, including the operation being scheduled. The formula is

$$CR = \frac{\textbf{Due date} - \textbf{Today's date}}{\textbf{Total shop time remaining}}$$

- **Shortest Processing Time:** The job requiring the shortest processing time (SPT) at the workstation is processed next.
- **Slack per Remaining Operations:** Slack is the difference between the time remaining until a job's due date and the total shop time remaining, including that of the operation being scheduled. A job's priority is determined by dividing the slack by the number of operations that remain, including the one being scheduled, to arrive at the slack per remaining operations (S/RO).

$$S/RO = \frac{(\text{Due date} - \text{Today's date}) - \text{Total shop time remaining}}{\text{Number of operations remaining}}$$

11.9. Sequencing Jobs for One Workstation

Any priority sequencing rule can be used to schedule any number of workstations. For the purpose of illustrating the rules, however, we focus on scheduling several jobs at a single workstation. We divide the rules into two categories:

- single-dimension rules.
- multiple-dimension rules. Single-Dimension Rules Some priority sequencing rules (e.g., FCFS, EDD, and SPT) base a job's priority assignment only on information about the jobs waiting for processing at the individual workstation.

We call these rules single-dimension rules because they determine priority based on a single aspect of the job, such as arrival time at theworkstation, the due date,orthe processing time.

(i) Scheduling Jobs for a Two-Station: - Flow Shop Suppose that a flow shops has several jobs ready for processing at two workstations and that the routings of all jobs are identical. In the scheduling of two or more workstations in a flow shop, the makespan varies according to the sequence chosen. Determining a production sequence for a group of jobs to minimize the makespan has two advantages:

- The group of jobs is completed in minimum time.
- The utilization of the two-station flowshop is maximized.

Utilizingthefirstworkstation continuously until it processes the last job minimizes the idle time on the second workstation. Johnson's rule is a procedure that minimizes makesspan when scheduling a group of jobs on two workstations. S. M. Johnson showed that the sequence of jobs at the two stations should be identical and that the priority assigned to a job should, therefore, be the same at both. The procedure is based on the assumption of a known set of jobs, each with a known processing time and available to begin processing on the first workstation. The procedure should be taken.

Step 1- Scan the processing times at each workstation and find the shortest processing time among the jobs not yet scheduled. If two or more jobs are tied, choose one job arbitrarily.

Step 2- If the shortest processing time is on workstation 1, schedule the corresponding job as early as possible. If the shortest processing time is on workstation 2, schedule the corresponding job as late as possible.

Step 3- Eliminate the last job scheduled from further consideration. Repeat steps 1 and 2 until all jobs have been scheduled.

11.10. Scheduling Jobs for Multiple Workstations

Priority sequencing rules can be used to schedule more than one operation. Each operation is treated independently. When a workstation becomes idle, the priority rule is applied to the jobs waiting for that operation, and the job with the highest priority is selected. When that operation is finished, the job is moved to the next operation in its routing, where it waits until it again has the highest priority. At any workstation, the jobs in the waiting line change over a period of time, so the choice of a priority rule can make quite a difference in the processing sequence. Schedules can be evaluated with the performance measures already discussed. Identifying the best priority rule to use at a particular operation in a process is a complex problem because the output from one operation becomes the input to another. The priority rule at a workstation determines the sequence of work the workstation will perform, which in turn determines the arrival of work at the next workstation downstream. Computer simulation models are effective tools to determine which priority rules work best in a given situation. Once the current process is modeled, the analyst can make changes to the priority rules at various operations and measure the impact on performance measures, such as past due, flow time, and utilization.

11.11. Line Balancing

line Balancing means balancing the production line, or any assembly line. The main objective of line balancing is to distribute the task evenly over the work station so that idle time of man of machine can be minimized. Lime balancing aims at grouping the facilities or workers in an efficient pattern in order to obtain an optimum or most efficient balance of the capacities and flows of the production or assembly processes. Assembly Line Balancing (ALB) is the term commonly used to refer to the decision process of assigning tasks to workstations in a serial production system.

The task consists of elemental operations required to convert raw material in to finished goods. Line Balancing is a classic Operations Research optimization technique which has significant industrial importance in lean system. The concept of mass production essentially involves the Line balancing in assembly of identical or interchangeable parts or components into the final product in various stages at different workstations. With the improvement in knowledge, the refinement in the application of line balancing procedure is also a must. Task allocation of each worker was achieved by assembly line balancing to increase an assembly efficiency and productivity.

The basic function of line balancing is to minimization of the number of workstations and maximization of the production rate is the most common goals.

In general Line balancing operated under two condition when,

- **Precedence Constraint:** Products cannot progress to other station if it doesn't complete necessary task at that station. It should not across other station because certain part needs to be performed before other activities.
- **Cycle time Restriction:** Cycle time is maximum time for products spend in every workstation. Different workstation has different cycle time.

(i) **Line Balancing: -** Line Balancing is leveling the workload across all processes in a cell or value stream to remove bottlenecks and excess capacity. A constraint slows the process down and results if waiting for downstream operations and excess capacity results in waiting and absorption of fixed cost.

(ii) **Single-Model Assembly Line: -**In early times assembly lines were used in high level production of a single product. But now the products will attract customers without any difference and allows the profitable utilization of Assembly Lines. An advanced technology of production which enables the automated setup of operations and it is negotiated time and money. Once the product is assembled in the same line and it won't variant the setup or significant setup and it's time that is used, this assembly system is called as Single Model Line.

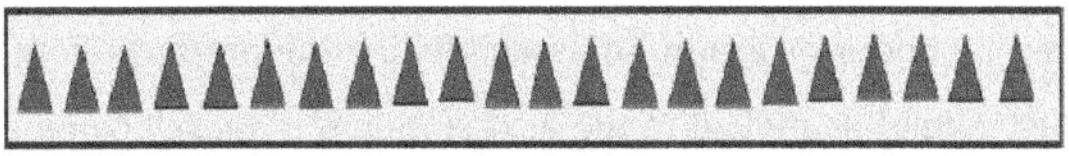

Diagram 11.14: Single-Model Assembly Line

(iii) Mixed Model Assembly Line: -In this model the setup time between the models would be decreased sufficiently and enough to be ignored. So this internal mixed model determines the assembled on the same line. And the type of assembly line in which workers work in different models of a product in the same assembly line is called Mixed Assembly Line.

Diagram 11.15: Mixed Model Assembly Line

(iv) Multi Model Assembly Line: - In this model the uniformity of the assembled products and the production system is not that much sufficient to accept the enabling of the product and the production levels. To reduce the time and money this assembly is arranged in batches, and this allows the short term lot-sizing issues which made in groups of the models to batches and the result will be on the assembly levels.

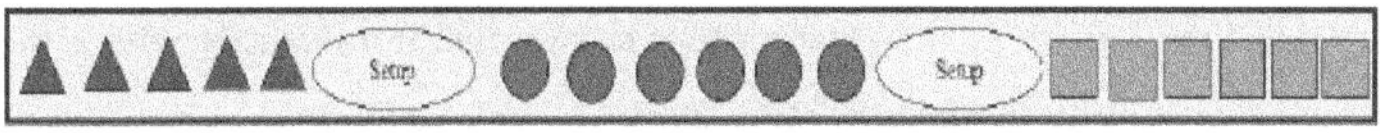

Diagram 11.16: Multi Model Assembly Line

Terms used in line balancing: - Following are some important terms that are usually applied in line balancing techniques.

- **Cycle Time:** It is a Maximum amount of time allowed at each station. This can be found by dividing required units to production time available per day. This is the time expressed in minutes between two simultaneous products coming of the end of production. The calculation of cycle time takes into consideration of the entire production quantities. If multiple lines are producing the same product, then the composite cycle time is less than the actual lapse time of any individual line.
- **Lead Time:** Summation of production times along the assembly line

- **Bottleneck:** Delay in transmission that slow down the production rate. This can be overcome by balancing the line.
- **Task Precedence:** It is the sequence by which tasks are carried out. It can be represented by nodes or graph. In assembly line the products have to obey this rule. The product cannot be moved to the next station if it doesn't complete at the previous station
- **Idle time:** A period when system is not in used but is available.
- **Productivity:** Defined as ratio of output over input. Productivity depends on several factors such as workers skills, jobs method and machine used.
- **Takt times:** The time needed by competent worker or unattended machine to perform a task. This is usually expressed in minutes. Takt time is the swiftness of production that aligns production with client demand. It shows how fast the need to manufacture product in order to fill the customer orders. Producing faster than takt time results in overproduction which is a type of waste whereas producing slower than takt time results in bottlenecks where the customer orders may not be fulfilled in time.

Benefits of Takt Times

- Achieve a steady and continuous flow of production.
- Eliminate the waste of overproduction by producing actual customer demand.
- Improves accuracy of planning.
- Encourage the development of standardize work instructions, promoting quality and efficiency.
- Set real time targets for production that shows operators exactly where their work output should be at any given point of time.
- Establish what-if scenario for customer demand based on flexible manning.

Work station: A physical area where a worker with tools / one or more machines or unattended machines such as robot perform specific task in a production line.

Downtime: Downtime explained as the time that is non value added.

- **Defects:** Defect is direct costs of a company.
- **Overproduction:** One of the severe wastes discourages a smooth flow of goods and services, which may lead to unnecessary lead and storage time. It will lead to the defects which cannot be detected earlier and then the products may deteriorate. It will also lead to excess work-in-progress stocks.

- **Waiting:** It occur when the goods are not moving or being worked on. It affects both goods and workforce where the waiting time should be used for some value added activities such as training and maintenance.
- **Transportation:** Any movement in factory can be considered as waste. Double handling and excessive movements are likely to cause damage and deterioration with the distance of communication between processes proportional to time taken. It takes to feedback reports of poor quality and needs corrective action.
- **Unnecessary inventory:** There is a problem with extra inventory. Inventory will increase the lead time, preventing quick identification of problems and increasing space. Significant storage costs are wasted which absolutely lower the competitiveness of the organization of value stream
- **Unnecessary motion:** Involve the ergonomics of production where operation might have to stretch, bend and pick up when these actions actually could be avoided. It not only tires the workers but also leads to poor productivity.
- **Inappropriate processing:** Over-complexity of a process discourages ownership and encourages the employees to over produce to recover the large investment in the complex machines. It encourages poor quality and takes corrective action.

(i) Determining Cycle Time: - Cycle time is longest time allowed at each station. This can be expressed by this formula:

$$\text{Cycle time} = \frac{\text{Available time}}{\text{Desired output}}$$

(ii) Assigning tasks to workstation: - The tasks distributions should be taken after completing a time cycle. It's good to allocate tasks to workstation in the order of longest task times.

$$\text{Number of work Stations} = \frac{\Sigma \text{ Task Time}}{\text{Desired Actual Time}}$$

(iii) Calculating an Efficiency Line: - This is done to find effectiveness of the line. The formula is given by

$$\text{Line Efficiency} = \frac{\text{Sum of task times}}{\text{Number of workstation X Desired cycle time}}$$

Question for Discussion

1. Scheduling prepares a time table indicating time and rate of production?Explain critically.
2. What is the use of line balancing? Give example
3. What is line of balancing? discuss in detail assembly line balancing taking suitable example
4. Discuss – briefly about Assembly Line Balancing .
5. Discuss the method of job sequencing to minimize the machine idle time.
6. Explain loading and scheduling with its examples
7. What is the difference between scheduling and sequencing?
8. What are the types of schedules in a job shop scheduling?
9. Discuss the concept of line balancing? Classify line balancing with suitable sketch
10. Explain about the Jobs Scheduling for a Two-Station & for Multiple Workstations .
11. Discuss about the external and internal factors which affects the machine scheduling.
12. What is scheduling? Explain the various principle of scheduling .
13. compare and contrast the project evaluation and review technique (PERT) with the critical path method (CPM).
14. Using the information in Table 1, assuming that the project team will work a standard working week (5 working days in 1 week) and that all tasks will start as soon as possible:

Task	Description	Duration (Working Days)	Predecessor/s
A	Requirement Analysis	5	
B	Systems Design	15	A
C	Programming	25	B
D	telecoms	15	B
E	Hardware Installation	30	B
F	Integration	10	C , D
G	System Testing	10	E , F
H	Training/Support	5	G
I	Handover and Go-Live	5	H

(i) Determine the critical path of the project.

(ii) Calculate the planned duration of the project in weeks Identify any non-critical tasks and the float (free slack) on each.

References

1. Antonio Prensa, "The Critical Path Method" in Management, 2002.
2. G.W.F. Hegel, "A Brief History of Scheduling Back to the Future", Introduction to Philosophy of History, Pp. 1770-1831.
3. Patrick Weaver, "Why Critical Path Scheduling (CPM) is Wildly Optimistic",.

4. J.J. Moder and C.R. Philips, "Project management with cpm and pert",.
5. Edward Wilson Davis, "An introduction to PERT/CPM Bruce", N. Baker, Rene L. Eris3 Project Management.
6. Dr.B.C. Punmia and K.K. Khandelwal, "Project Planning and Control with PERT & CPM",.
7. Kerzner and Harold, "Project Management: A Systems Approach to Planning", Scheduling, and Controlling (8th Ed. ed.). Wiley, 2003.
8. D.G. Malcolm, J.H. Roseboom, C.E. Clark and W. Fazar, "Application of a Technique for Research and Development Program Evaluation Operations Research", Vol. 7, No. 5, 2004.
9. J.K. Sharma, "Project Management: PERT and CPM", Operations Research: Theory and Applications, Rajiv Beri (Ed.), New-Delhi: Macmillan Publishers India Ltd., Pp. 425-482, 2009.
10. K.G. Lockyer, "An Interlocution to Critical Path Analysis", London: Pitman, 1970.
11. Project planning and control with PERT & CPM Dr. B.C.Punammia
12. S.C. Sharma, "Operation Research PERT/CPM & Cost Analysis", Discovery Publishing House, Pp. 126-208, 2006.
13. R. Sivarethinamohan Operations Research, Tata Mc Graw Hill, 2008.

4. H. [illegible], Project management with gantt and pert.
5. Edward Wilson Davis, "An Introduction to CPM/PERT", New York: [illegible] Project Management.
6. Dr. B.C. Punmia and K.K. Khandelwal, Project Planning and Control with PERT & CPM.
7. Kerzner and Harold, Project Management: A Systems Approach to Planning, Scheduling, and Controlling (9th Ed.ed.) Wiley, 2006.
8. D.G. Malcolm, J.H. Roseboom, C.E. Clark and W. Fazar, "Application of a Technique for Research and Development Program Evaluation, Operations Research, 7 vol. No. 5, 1959.
9. J.K. Sharma, Project Management: PERT and CPM, Operations Research: Theory and Applications, [illegible] New Delhi: Macmillan Publishers India Ltd., [illegible] 2009.
10. K.G. Lockyer, An Introduction to Critical Path Analysis, London: Pitman, 1976.
11. Project planning and scheduling with PERT & CPM by [illegible].
12. S.D. Sharma, Operations Research, PERT, CPM & Cost Analysis, Discovery Publishing House, [illegible] 2006.
13. [illegible], Operations Research, Tata McGraw Hill, 2008.

CHAPTER12

PRODUCTION PLANNING AND CONTROL

12.0. Introduction

Production plans and control can be viewed as nervous system of the production operation. This function aims at efficient utilization of material resources, people and facilities in any undertaking through planning, coordinating and controlling the production activities that transform the raw material into finished products or components as a most optimal manner. All the activities in manufacturing or production cycle must be planned, coordinated, organized and controlling to its objectives. Production planning and control as a department plays a vital role in manufacturing organizations. It is clear from name that it is something about planning. Planning is defined as setting goals. Production planning and control provides different kinds of information to different departments. It provides information about available manufacturing resources to marketing department. Marketing department receives orders according to that information. Similarly, it coordinates with other departments and provides relevant information. In general it is obvious that, Production planning and control composed of three basic parts, production / production system, planning and control. The brief description of all these three terms is given below.

Production: Production is defined as "the step-by-step conversion of raw material into finished product by applying pre-defined process to achieved required level of quality. Thus production is a value addition process. At each stage of processing, there will be value addition.

According to Edwood Buffa production is defined as

A process by which goods and services are created.

In any kind of production, the inputs will be the available resources, and then transformation takes place thereby it convert in to finished product thus it is a desired output according to the design specification. The schematic diagram of production is given below.

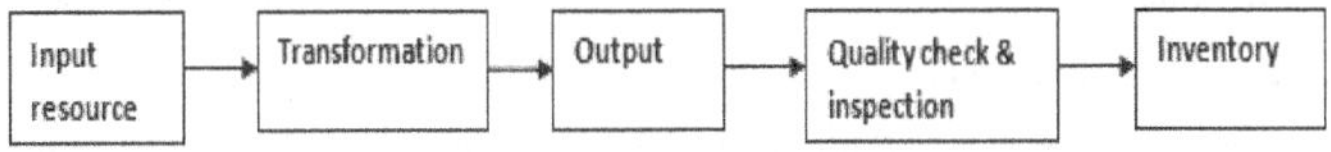

Diagram 12.1: Production System

- Input resources are men, machine , material , capital , time , space

- Transformation process consists product design, production planning and control.
- Output is finished product
- Send to quality control department for quality checking
- Then finally storage to inventory.

Planning : - it is a second stage of the operation, in which we will consider following steps.

1. Determining what products are to be produced and the quantities to be produced in defined time periods.
2. Dividing the complete product in to its component assemblies and individual parts. This process is called the planning breakdown.
3. Evolving the optimum methods for economic and timely production of the assemblies and their individual parts.
4. Assigning the task to be performed to identify groups such as divisions of company, department of a division, and work centers of a department.
5. Scheduling the sequence, priorities and time periods for each task for each group.
6. Arranging to provide the resources in time for execution of the tasks .7
7. Setting targets and standards of performance for each task in terms of time / cost/quality.
8. Authorizing the execution of work

Control :- In this section we will monitoring the progress of execution of the task in each group against the specified targets or standard of the performance , analyzing the variances between actual achievement and the set targets and standard , evolving and take corrective action , in case of short falls in achievements when compared with the set targets or standards .

Production system: **Production system is defined as the method, procedure or arrangement which includes all functions required to accumulate the input, process the input and create a products or output according to the management policies.**

Properties of production system: Following are some important properties of production system.

1. Production is an organized activity by the manufacturing firms, so therefore production system has an objective.
2. The basic concept of production system is to transforms the various inputs to useful outputs.
3. It does not operate in isolation from the other organization system.
4. Feedbackoftheactivities controls and improve the performance of production system.

12.1. Classification of Production System

Production systems usually classified as:

(i) Job Shop Production System.

(ii) Batch Production System.

(iii) Mass and Continuous Production systems.

The diagrammatic arrangement of production system is given below.

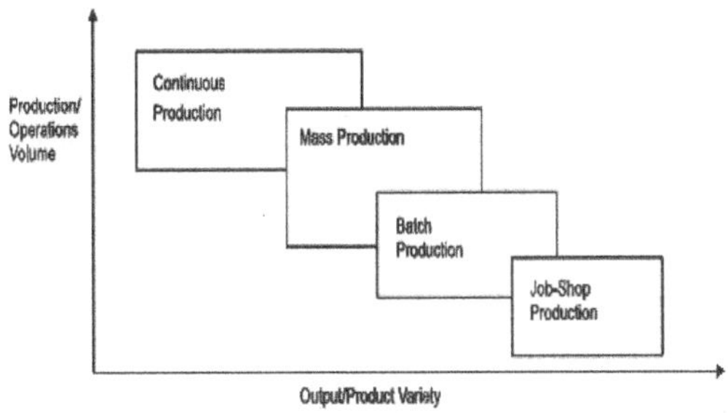

Diagram 12.2: Diagrammatic Arrangement of Production System

(i) Job shop production: -Job shop production is a manufacturing process in which product is ordered / demanded by the customer, customers gives a design specification, cost as well as time for that product. The quantity of products may be single or few. The distinguishing feature of this is low volume and high variety of products. A job shop comprises of general purpose machines arranged into different departments.

Characteristic of job shop production system: following are some important characteristics of job shop production.

1. High variety of products and low volume.
2. Use of general purpose machines and facilities.
3. Highly skilled operators who can take up each job as a challenge because of uniqueness.
4. Large inventory of materials, tools, parts.
5. Detailed planning is essential for sequencing the requirements of each product, capacities for each work centre and order priorities.

Advantage of job shop production: following advantages are counted in this cadre.

1. High flexibility of production.
2. High expansion flexibility, machines can be easily substituted or added.

3. Low obsolescence of machines.
4. Multiple different batches can go on parallel.

Disadvantages of Job Shop Production

1. High skilled man power required, who can work on different kind of machines
2. Hard scheduling of product, since production flow changes with every product.
3. Low capacity utilization relative to flow manufacturing or mass production system.

(ii) Batch production: According to American Production and Inventory Control Society (APICS).

A form of manufacturing in which the job passes through the functional departments in lots or batches and each lot may have a different routing.

This method is generally similar to job production except the quantity of production. Instead of making one single product as in case of job production, a batch or group of products are produced at one time. It should be remembered here that one batch of products may not resemble with the nextbatch. Under batch system of production the work is divided into operations and one operation is done at a time. After completing the work on one operation it is passed on to the second operation and so on till the product is completed.

The Batch Production Method Possesses the following Characteristics

1. The work is of repetitive nature.
2. There is a functional layout of various manufacturing processes.
3. One operation is carried out on whole batch and then is passed on to the next operation and so on.
4. Same type of machines is arranged at one place.
5. It is generally chosen where trade is seasonal or there is a need to produce great variety of goods.

Advantage of Batch Production System

1. Better utilization of plant and machinery.
2. Promotes functional specialization.
3. Cost per unit is lower as compared to job order production.
4. Lower investment in plant and machinery.
5. Flexibility to accommodate and process number of products.
6. Job satisfaction exists for operators.

Limitation of of batch production system: Followingare the limitations of batch production:

1. Material handling is complex because of irregular and longer flows.
2. Production planning and control is complex.
3. Work in process inventory is higher compared to continuous production.
4. Higher set up costs due to frequent changes in set up.

(iii) Mass production system: Manufacture of discrete parts or assemblies using a continuous process are called mass production. This production system is justified by very large volume of production. Standardized products are produced under this method by using standardized materials and equipment thus Process layout is best suited method for mass production units. Flow production / Mass production is the manufacture of a product by a series of operations, each article going on to a succeeding operation as soon as possible. The manufacturing process is broken into separate operations. The product completed at one operation is automatically passed on to the next till its completion. There is no time gap between the work done at one process and the starting at the next. The flow of production is continuous and progressive

Characteristics of Mass Production System

1. The units flow from one operation point to another throughout the whole process.
2. There will be one type of machine for each process.
3. The products, tools, materials and methods are standardized.
4. Production is done in anticipation of demand. Production volume is usually high.
5. Machine set ups remain unchanged for a considerable long period.
6. Any fault in flow of production is immediately corrected otherwise it will stop the whole production process.
7. Production planning and control is easy.
8. Material handling can be completely automatic.

Advantages of Mass Production System

1. The product is standardized and any deviation in quality etc. is detected at the spot.
2. There will be accuracy in product design and quality.
3. It will help in reducing direct labour cost.
4. There will be no need of work-in-progress because products will automatically pass on from operation to operation.
5. There may not be any need of keeping work-in-progress, hence storage cost is reduced.

Limitations of Mass Production

1. Breakdown of one machine will stop an entire production line.
2. Line layout needs major change with the changes in the product design.
3. High investment in production facilities.
4. The cycle time is determined by the slowest operation.

(iv) Continuous production system: -In Continuous production system, goods are produced constantly as per demand forecasting. Goods are produced on the large scale for stocking and selling. They are not produced on customers 'orders. Here the inputs and outputs are standardized along with the production process and sequence.

Features of Continuous Production System

- 1. Flow of Production is Continous and not intermittent.
- 2. Products are Standardized.
- 3. Products are produced as per Quality Standards.
- 4. Products are produced in Anticipation of Demand.
- 5. Standardized routing sheets and schedules are prepared.

Diagram 12.3: Feature of Continuous Production System

Characteristics of Continuous System

1. Standard products are manufactured, which have large demand throughout the year.
2. Standardized inputs and standardized sequence of operations, machine tools and equipment are used.
3. Division of labour is made more efficient.
4. Minimum and constant material handling.
5. Minimum flow of work at any point of time.
6. Small work in progress is involved.
7. Use of productivity techniques is feasible.
8. Minimum cost of production per unit is possible.
9. Rigid quality control is exercised..
10. More maintenance is required.

Advantages: Following are the advantages of continuous production:

1. Standardization of product and process sequence.
2. Higher rate of production with reduced cycle time.
3. Higher capacity utilization due to line balancing.
4. Manpower is not required for material handling as it is completely automatic.
5. Person with limited skills can be used on the production line.
6. Unit cost is lower due to high volume of production.

Limitations: Following are the Limitations of Continuous Production System.

1. Flexibility to accommodate and process number of products does not exist.
2. Very high investment for setting flow lines.
3. Product differentiation is limited.

Type of continuous system: There are three types of continuous production.

(i) Mass production system

(ii) process production system

(iii) Assembly production system.

(i) Mass production: In this type of continuous production, only one type of product or a maximum of two or three types are manufactured in large quantities, as much emphasis is not given to orders of the consumers. Standardization of product, process, materials, machine and uninterrupted flow of materials are the main characteristics of this system.Mass production system is employed in several industries where the production is carried on without any interruption. Electronics, electrical, automobiles, bicycles and container industries are a few examples of mass production industries.

(ii) Process Production: This system is used for the manufacture of those items whose demand is continuous and high. Here, single raw material can be transformed into different kinds of products at different stages of the production process e.g., processing of crude oil in refinery — we get kerosene, gasoline, etc., at different stages of production.

(iii)Assembly Production: Assembly line is a type of flow production which was developed in the automobile industry in the U.S.A. A manufacturing unit prefers to develop and employ the assembly line because it helps to improve the efficiency of production. The use of flow production methods results in cost reduction. Assembly line is particularly useful when a limited variety of similar products is to be produced on a mass scale or in fairly large batches on a regular or continuous basis. In assembling process, two or more components are combined to manufacture a finished product. Manufactured parts are joined into sub-assemblies or final assemblies. Such process is employed in assembling automobiles, radio sets, television sets, bicycles, watches, cameras.

12.2. Factor Affecting Production Planning and Control

(i) **Use of Computers:** Modern factories are using office automation equipment like PC, punch cards etc. It helps accurate computation of required of men and machine.

(ii) **Seasonal Variations:** Demand of certain products is affected by seasons, for instance umbrellas and raincoats during the monsoons and outputs. Production planning and control must take such changes into consideration while planning and control activities of inputs and outputs.

(iii) **Test Marketing:** In an aggressive marketing strategy new products are to be test marketed in order to know the trends. This is a short- cycle operation, intermittent in nature and often upsets regular production.

(iv) **After Sales Service:** This has become an important parameter for success. In after sales services, many items are returned for repair. These are unscheduled work and also overload the production line.

(v) **Losses due to Unpredictable:** Factors Losses occur due to accidents, fire and theft of production inputs, mainly materials and components. These are unpredictable. Shortage of input due to such factors upset the planned production schedule in time and quantity.

(vi) **Losses due to Predictable Factors:** There are losses of inputs, due to natural d engineering phenomena like production losses and changes in consumption of materials and occurrence of defectives.

(vii) **Production of Order:** There are occasions when last minute prioritization of existing orders due to external pressure takes place. These changes in priority are often decided by sufficiently high level of management.

(viii) **Design Changes:** Design changes are issued by R & D and the engineering department. This will necessarily force production planning control change the input materials and

(ix) **Process. Rejection and Replacement:** There are occasions when sub-assemblies or finished goods are rejected during stage or final inspection. Production planning and control must cater for contingent plans to take care of rework without affecting scheduled quality.

Objective of Production Planning and Control

(i) Nature of Inputs: To manufacture a product, different types of inputs are used. The quality of the product depends upon the nature of the inputs are used. Hence the planning is done to determine the nature of various types of inputs which is a complicated process.

(ii) Quantity of Inputs: To achieve a level of production, determination of quantity of the inputs and their composition is very important. A product can be prepared only when there is an estimate of the required composition of inputs.

(iii) Proper Coordination: It ensures the proper coordination among the workforce, machines and equipment. This leads to avoidance of wastages and smooth flow of production.

(iv) Better Control: Production planning is the method of control. For a better control, planning is a precondition. Only then, one can compare the performance and calculate the deviations which lead control of the production.

(v) Ensure Uninterrupted Production: The planning of materials ensures the regular supply of raw materials and other components. The regular flow of materials and supplies are helpful in the uninterrupted production.

(vi) Capacity Utilization: There is a need to use the available resources effectively. It is helpful in bringing down various costs of production.

(vii) Timely Delivered: If there is good production planning and control, there will be timely production and the finished product will be rushed to the market in time. This also ensures the better relationship with the customers.

12.3. Manufacturing Systems

Manufacturing systems consist of human workers, automation, and various material handling technologies, configured in ways that create specific manufacturing system typologies. More specifically, a manufacturing system is a collection of integrated equipment and human resources, whose function is to perform one or more processing and/or assembly operations on a starting raw material, part, or set of parts.

(i) Flexible manufacturing system: A flexible manufacturing system (FMS) is a highly automated group technology machine cell, consisting of a group of processing workstations that are interconnected by an automated material handling and storage system, and controlled by a distributed computer system. Flexible manufacturing systems (FMSs) are the most automated and technologically sophisticated of the machine cell types used to implement cellularManufacturing. An FMS usually has multiple automated stations and is capable of variable routings among stations, while its flexibility allows it to operate as a mixed model system.It consists of consists of four or more processing stations connected mechanically by a common parts handling system and electronically by a distributed computer system. FMS is larger than the flexible manufacturing cell, not only in the number of workstation may contain, but also in the number of supporting stations in the system, such as part/pallet washing stations, co-ordinate measuring machines, storage stations and so on. Computer control is also more sophisticated; it includes functions not found in the flexible manufacturing cell such as diagnostics and tool monitoring.

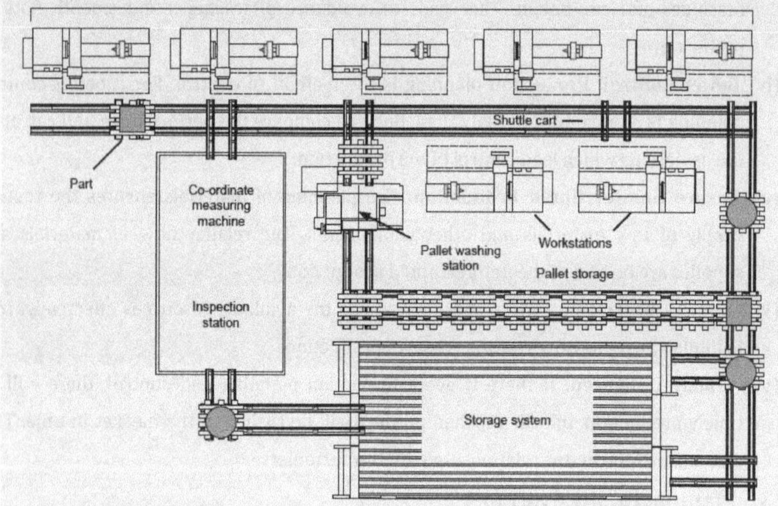

Diagram 12.4: A Seven-Station Flexible Manufacturing System

Flexible Manufacturing System Components: The basic components of an FMS are: workstations, material handling and storage systems, computer control system, and the personnel that manage and operate the system.

- **Work-stations:** Following are the basic purpose of work- station.
- **Load/Unload:** Physical interface between the FMS and the rest of the factory, it is where raw parts enter the system, and completely-processed parts exit the system. Loading and unloading can be performed manually by personnel, or it can be automated as part of the material handling system. Should be designed to permit the safe movement of parts, and may be supported by various mechanical devices (e.g.

cranes, forklifts).The station includes a data entry unit and monitor for communication between the operator and computer system, regarding parts to enter the system, and parts to exit the system. In some FMSs, various pallet fixtures to accommodate different pallet sizes may have to be put in place at load/unload stations.

- **Machining : The** most common FMS application occurs at machining stations. These are usually CNC machine tools with appropriate automatic tool changing and tool storage featuresto facilitate quick physical changeover, as necessary.Machining centers can be ordered with automatic pallet changers that can be readily interfaced with the FMS part handling system. Machining centers used for non-rotational parts; for rotational parts turning centersare used. Milling centers may also be used where there are requirements for multi-tooth rotational cutters.
- **Assembly:** The assembly operation usually consists of a number of workstations with industrial robots that sequentially assemble components to the base part to create the overall assembly. They can be programmed to perform tasks with variations in sequence and motion pattern to accommodate the different product styles assembled in the system.
- **Supporting :** stations may include inspection stations where various inspection tasks may be carried out. Co-ordinate measuring machines, special inspection probes, and machine vision may all be used here. Other supporting stations may include pallet and part washing stations for particularly dirty or oily FMSs, and temporary storage stations for both parts and pallets
- **Other:** Other possible stations may be found in specific industries, such as—for example—sheet metal fabrication, which has stations for press-working operations, such as punching, shearing, and certain bending and forming processes. Forging is another labour intensive operation which may be broken into specific station categories, such as heating furnace, forging press, and trimming station
- **Material Handling and Storage System:** The basic Functions of the material handling and storage system in FMSs are the allowance of random, independent movement of work-parts between stations; the handling of a variety of work-part configurations; the provision of temporary storage; the provision of convenient load and unload stations; and the creation of compatibility with computer control. There is a primary and secondary material handling system used in most FMSs. The primary handling system establishes the FMS lay-out and is responsible for moving parts between stations in the system.

- The secondary handling system consists of transfer devices, automatic pallet changers, and other mechanisms to transfer parts from the primary material handling system to the work head of the processing station, or to a supporting station. The secondary handling system is responsible also for the accurate positioning of the part at the workstation, so that the machining process may be performed upon the part in the correct manner. Other purposes of the secondary handling system include: (1) re-orientation of the part if necessary to present the surface that is to be processed; and (2) to act as buffer storage as the workstation.

12.4. Flexible Manufacturing System Layout

There are five type of FMS layout, the brief description is given below.

(i) In-line layout: The machines and handling system are arranged in a straight line, the parts progress from one workstation to the next in a well-defined sequence with work always moving in one direction and with no back-flow. Similar operation to a transfer line except the system holds a greater variety of parts. Routing flexibility can be increased by installing a linear transfer system with bi-directional flow, here a secondary handling system is provided at each workstation to separate most of the parts from the primary line. Material handling equipment used: in-line transfer system; conveyor system; or rail-guided vehicle system.

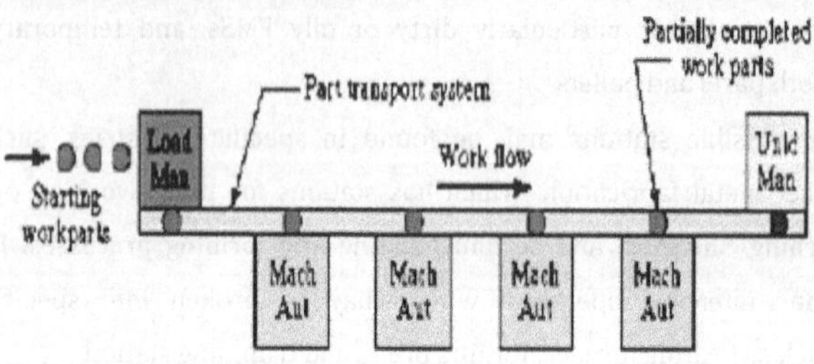

Diagram 12.5: In-Line Layout

(ii) Loop layout: Workstations are organized in a loop that is served by a looped parts handling system. The parts are usually flow in one direction around the loop with the capability to stop and be transferred to any station. Each station has secondary handling equipment so that part can be brought-to and transferred-from the station work head to the material handling loop. Load/unload stations are usually located at one end of the loop.

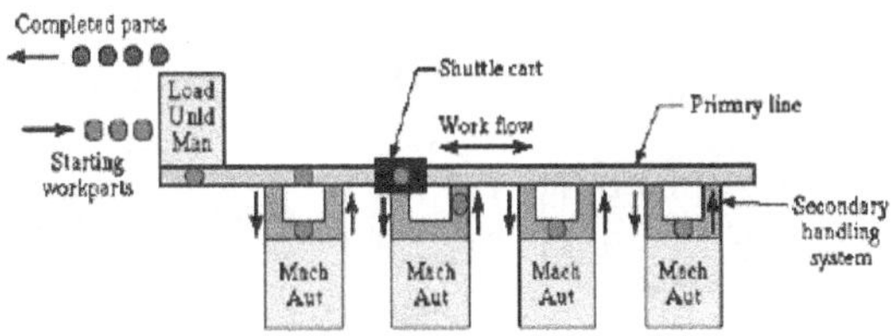

Diagram 12.6: Loop Layout

(iii) Ladder layout: This consists of a loop with rungs upon which workstations are located. The rungs increase thenumber of possible ways of getting from one machine to the next, and obviates the need for a secondary material handling system. It reduces average travel distance and minimizes congestion in the handling system, thereby reducing transport time between stations.

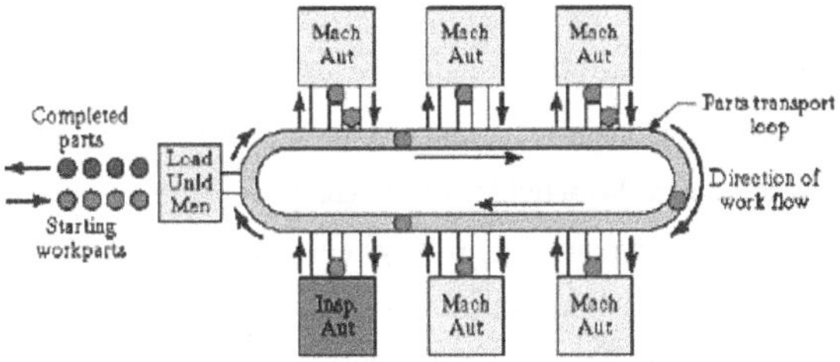

Diagram 12.7: Ladder Layout

(iv) Open field: Consists of multiple loops and ladders, and may include sidings also. This layout is generally used to process a large family of parts, although the number of different machine types may be limited, and parts are usually routed to different workstations—depending on which one becomes available first.

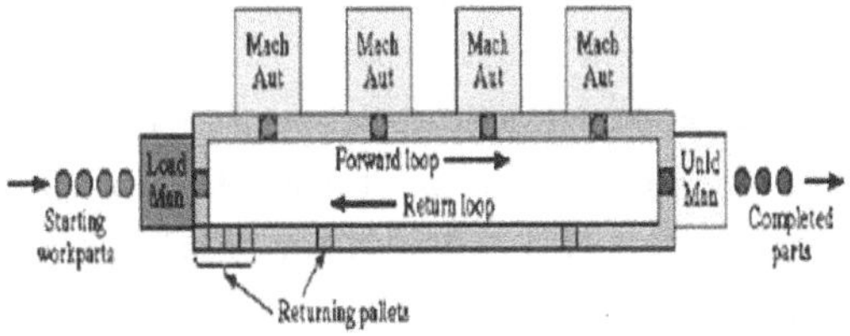

Diagram 12.8: Open Field Layout

Robot-centered: This layout uses one or more robots as the material handling system.

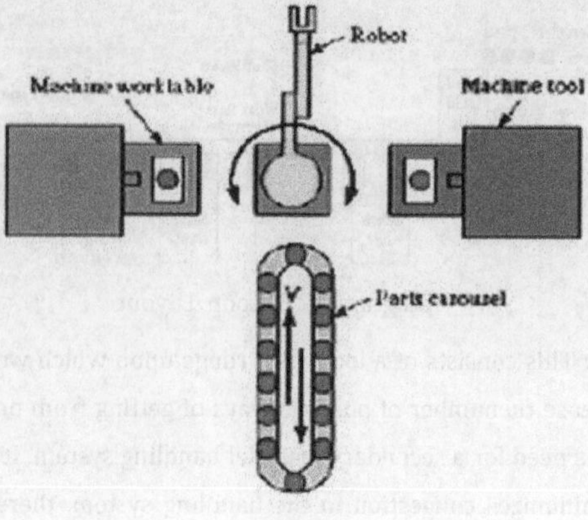

Diagram 12.9: Robot Centered Layout

Computer Control System: To operate, the FMS uses a distributed computer system that is interfaced with all workstations in the system, as well as with the material handling system and other hardware components. It consists of a central computer and a series of microcomputer co-ordinates the activities of the components to achieve smooth operational control of the system. The computer control system in an FMS has the following functions: workstation control; distribution of control instructions to workstations; production control; traffic control; shuttle control; work piece monitoring; tool control; performance monitoring and reporting; and diagnostics.

- **Workstation control**—Fully automated FMSs use some form of workstation control at each station, often in the form of CNC control
- **Distribution of control instructions to workstations**—A central computer is required to handle the processing occurring at disparate workstations; this involves the dissemination of part programmers to individual workstations, based upon an overall schedule held by the central computer
- **Production control**—Management of the mix and rate at which various parts are launched into the system is important; alongside data input of a number of essential metrics, such as: daily desired production rates, number of rawworkparts available, work-in-progress etc.
- **Traffic control**—Management of the primary handling system is essential so that parts arrive at the right location at the right time and in the right condition
- **Shuttle control**—Management of the secondary handling system is also important, to ensure the correct delivery of the workpart to the station's workhead

- **Work piece monitoring—**The computer must monitor the status of each cart or pallet in the primary and secondary handling systems, toensure that we know thelocation of every element in the system
- **Tool control—**This is concerned with managing tool location (keeping track of the different tools used at different workstations, which can be a determinant on where a part can be processed), and tool life Performance monitoring and reporting—the computer must collected data on the various operations on-going in the FMS and present performance findings based on this
- **Diagnostics—**The computer must be able to diagnose, to a high degree of accuracy, where a problem may be occurring in the FMS

(i) Single machine cell : It contains one machine (often a CNC machining centre) connected to a parts storage system, which can load and unload parts to and from the storage system .It is designed to operate in batch mode, flexible mode, or a combination of the two. When in batch mode, the system processes parts of a single style in specific lot sizes before physical and programme changeover to the next batch specifications; in flexible mode the system satisfies three of the four tests for flexibility—the exception being error recovery, since, if the CNC machine centre breaks down, the system stops.

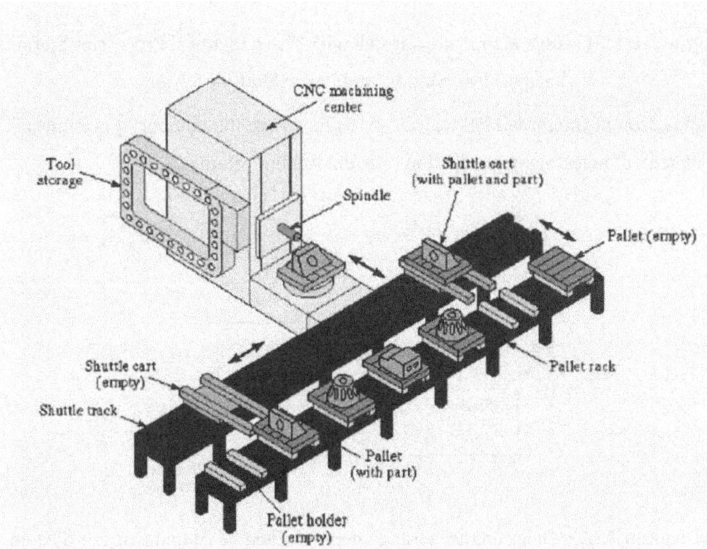

Diagram 12.10: Single Machine Cell with one CNC Machining Centre and Parts Storage Unit

(ii) Flexible manufacturing cell: it contains two or three processing workstations (often CNC machining or turning centers), plus a parts handling system. This set-up can operate in flexible mode and batch mode, as necessary, and can readily adapt to evolving production schedule and increased production volumes. Since there is more than one machine, error recovery is possible by re-routing the failed machine's intended parts for processing to the other two machines in the system; and new part designs can be introduced with relative ease into the set-up. The flexible manufacturing cell satisfies all four flexibility tests.

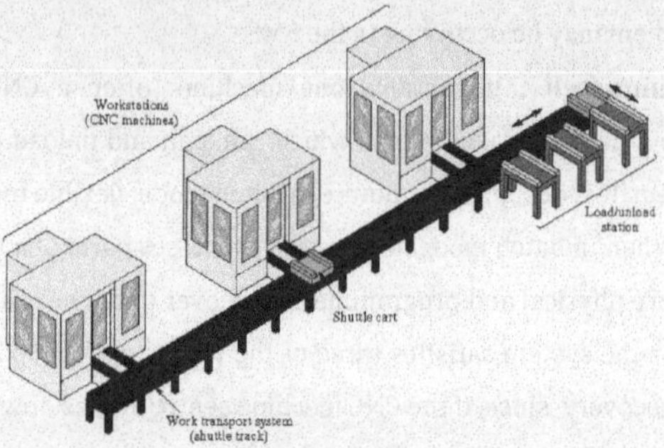

Diagram 12.11: Flexible Manufacturing Cell with Three Identical Processing Stations, A Load/Unload Station, and Parts Handling System

A comparison of the three FMS types is given in, where the number of machines is plotted against metrics of investment, production rate and Annual volume.

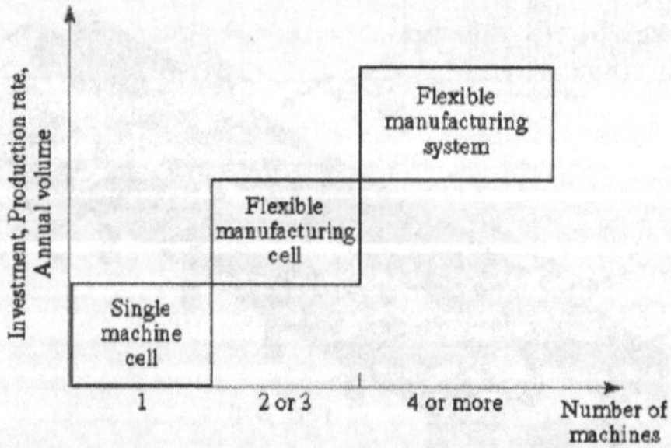

Diagram 12.12: Diagramatic Arrangement of Flexible Manufacturing System

A comparison of the two FMS types is indicated in Figure given below, where the production rate and annual volume is plotted against metrics of flexibility and part variety

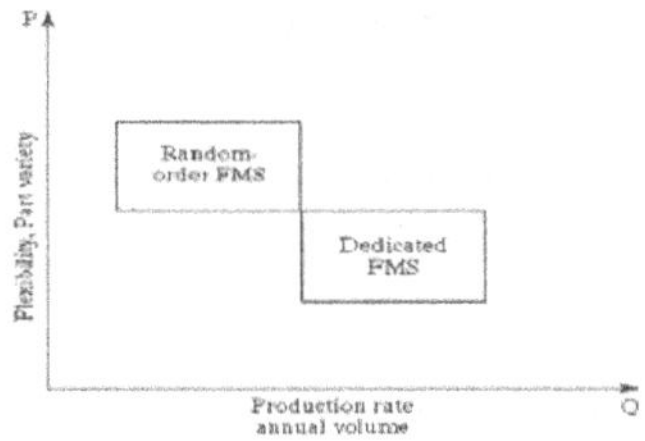

Diagram 12.13: Comparison of Dedicated and Random - Order FMS

12.5. Flexibility Concept

The terms flexibility means to produce reasonably priced customized products of high quality that can be quickly delivered to customers. There are three levels of manufacturing flexibility which is described below.

(i) Basic flexibilities.

- **Machine flexibility** - The ease with which a machine can process various operations.
- **Material handling flexibility** - A measure of the ease with which different part types can be transported and properly positioned at the various machine tools in a system.
- **Operation flexibility** - A measure of the ease with which alternative operation sequences can be used for processing a part type.

(ii) System flexibilities.

- **Volume flexibility** - A measure of a system's capability to be operated profitably at different volumes of the existing part types.
- **Expansion flexibility** - The ability to build a system and expand it incrementally.
- **Routing flexibility** - A measure of the alternative paths that a part can effectively follow through a system for a given process plan.
- **Process flexibility** - A measure of the volume of the set of part types that a system can produce without incurring any setup.
- **Product flexibility** - The volume of the set of part types that can be manufactured in a system with minor setup.

(iii) Aggregate flexibilities.

- **Program flexibility** - The ability of a system to run for reasonably long periods without external intervention.
- **Production flexibility** - The volume of the set of part types that a system can produce without major investment in capital equipment.

- **Market flexibility** - The ability of a system to efficiently adapt to changing market conditions.

(iv) Advantages and disadvantages of FMSs implementation.

- Faster, lower- cost changes from one part to another which will improve capital utilization.
- Lower direct labor cost, due to the reduction in number of workers.
- Reduced inventory, due to the planning and programming precision.
- Consistent and better quality, due to the automated control.
- Lower cost/unit of output, due to the greater productivity using the same number of workers.
- Savings from the indirect labor, from reduced errors, rework, repairs and rejects.

Disadvantages

- Limited ability to adapt to changes in product or product mix (ex. machines are of limited capacity and the tooling necessary for products, even of the same family, is not always feasible in a given FMS).
- Substantial pre-planning activity.
- Expensive, costing millions of dollars.
- Technological problems of exact component positioning and precise timing necessary to process a component.
- Sophisticated manufacturing systems.

12.6. Lean Manufacturing or Lean Production

The term lean manufacturing or lean production is comes out from the book " **The Machine that changed the world** "which has been published **whose author are Jim Womack, Daniel Jones and Daniel Roos' book and examination is performed on manufacturing activities.** The basic concept of lean manufacturing is to eliminate the "waste" systematically, as the name implies, lean is focused at cutting "fat" from production activities. It has also been successfully applied to administrative and engineering activities as well.

The 3 M's of Lean: - Lean manufacturing is a Japanese method focused on 3M's. These Ms are:

- **'Muda** the Japanese word for **waste.**
- **Mura,** the Japanese word for **inconsistency.**
- **Muri,** the Japanese word for **unreasonableness.**

Waste is broadly defined as anything that adds cost to the product without adding value to it, waste are classified as.

- Waste from overproduction.
- Waste of waiting.
- Transportation waste.
- Processing waste.
- Inventory waste.
- Waste of motion.
- Waste from product defects.

(i) **Waste of over-producing: -** Producing components that are neither intended for stock nor planned for sale immediately.

(ii) **Waste of waiting: -** Refers to the idle time between operations.

(iii) Waste **of transport: -** Moving material more than necessary.

(iv) **Waste of processing: -** Doing more to the product than necessary and the customer is willing to pay.

(v) **Waste of inventory: -** Excess of stock from raw materials to finished goods.

(vi) Waste **of motion: -** Any motion that is not necessary to the completion of an operation.

(vii) **Waste of defects and spoilage: -** Defective parts that is produced and need to be re-worked.

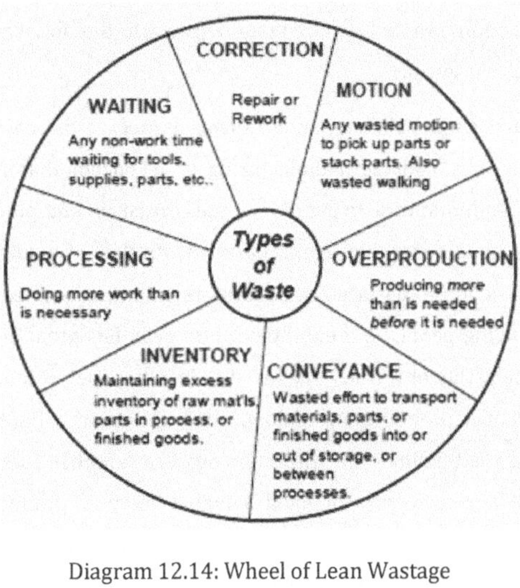

Diagram 12.14: Wheel of Lean Wastage

The second "M" is stand for **Mura**, or **inconsistency**. Inconsistency is a problem that increases the variability of manufacturing. Mura is evidenced in all manufacturing activities ranging from processing to material handling to engineering to management.

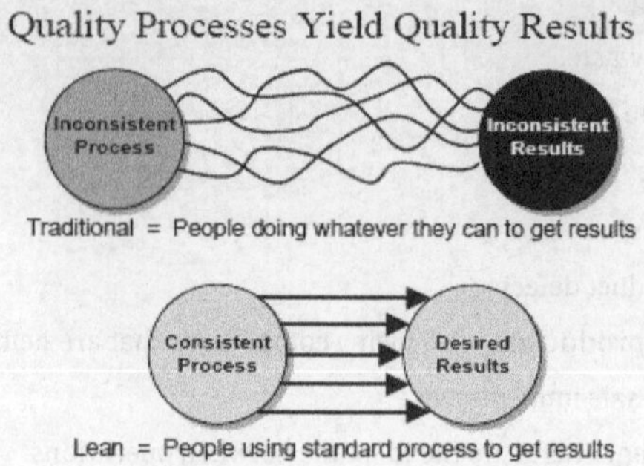

Diagram 12.15: Consistency & Inconsistency in Manufacturing

Inconsistency is a problem in manufacturing : The final **"M"** is for **Muri** or unreasonableness. **Muri** applies to a variety of manufacturing and management activities.

12.7. Lean Manufacturing Tools and Techniques

Once companies pinpoint the major sources of waste, tools such as continuous improvement, just-in-time production, production smoothing, and others will guide companies through corrective actions so as to eliminate waste. In the following sections a brief description of such tools is given.

(i) Cellular Manufacturing: Cellular manufacturing is one of the cornerstones when one wants to become lean. Cellular manufacturing is a concept that increases the mix of products with the minimum waste possible. A cell consists of equipment and workstations that are arranged in an order that maintains a smooth flow of materials and components through the process. It also has assigned operators who are qualified and trained to work at that cell. Arranging people and equipment into cells has great advantage in terms of achieving lean goals. One of the advantages of cells is the one product moves through the process one unit at a time without sudden interruption, at a pace determined by the customer's need. Extending the product mix is another advantage of cellular manufacturing. When customers demand a high variety of products as well as faster delivery rates, it is important to have flexibility inthe process to accommodate their needs. This flexibility can be achieved through grouping similar products into families that can be processed on the same equipment in the same sequence. This will also shorten the time required for changeover between products, which will encourage production in smaller lots. Other benefits associated with cellular manufacturing include.

- Inventory (especially WIP) reduction.
- Reduced transport and material handling.
- Better space utilization.
- Lead time reduction.
- Identification of causes of Defects and machine problems.
- Improved productivity.
- Enhanced teamwork and communication.
- Enhanced flexibility and visibility.

(ii) Continuous Improvement : Continuous improvement is another fundamental principle of lean manufacturing. Kaizen, which is the Japanese word for a continuous endeavor for perfection, has become popular in the west as a paramount concept behind good management. Kaizen is a systematic approach to gradual, orderly, continuous improvement. In manufacturing settings improvements can take place in many forms such as reduction of inventory, and reduction of defective parts. One of the most effective tools of continuous improvement is 5S, which is the basis for an effective lean company. 5S is a first, modular step toward serious waste reduction. 5S consists of the Japanese words Seiri (Sort), Seiton (Straighten), Seiso (Sweep and Clean), Seiketsu(Systemize), and Shitsuke (Standardize). The underlying concept behind 5S is to look for waste and then to try to eliminate it. Waste could be in the form of scrap, defects, excess raw material, unneeded items, old broken tools, and obsolete jigs and fixtures (Monden, 1998). The first S, Seiri, deals with moving those items that are not currently being used on a continuous basis (e.g., items that will not be used for the next month or so) away from those that are. Moving those items and tossing away needless items will make material flow smoothly, and workers move and work easily (Feld, 2000). Seiton has to do with having the right items in the right area. Items that do not belong to a given area must not be in that area. For a given workplace area tools must be marked andarranged as belonging in that area. This willsmake it easier to move those items that are notlabeled from that area. Arranging items in the right place will make tools, jigs, fixtures, and resources noticeable, detectable, and easy to use (Feld, 2000). Seiso deals with cleaning and sweeping the work place methodically. The workplace should look neat and clean and ready to use for the next shift. The work place should be maintained on a regular basis (e.g., daily). All tools and items should being the right place and nothing should be missing. A well-maintained workplace creates a healthy environment to work with (Feld, 200Seiketsu is maintaining a high standard of housekeeping and workplace arrangement. A regular audit should be run and scores should be assigned for areas of responsibilities. If ever area has people assigned to it then everyone has responsibility to maintain a high standard of housekeeping and cleaning (Feld, 2000). Shitsuke is management's accountability to train people to follow housekeeping rules. Management should implement the housekeeping rules in a practiced fashion so that their people can buy into it. Management should walk the shop floor, explain what they want from people, reward those who follow and instruct those who do not (Feld, 2000). Taken together, 5S means good housekeeping and better workplace organization. Kaizen tools such as 5S are not only a means to increase profitability of a firm but also allow companies to reveal potential strengths and capabilities that were hidden before (Hirai, 2001). Sweeny (2003), and Cox (2002) have reported good results implementing 5S. Further, benefits of implementing 5S will be described later.

(iii) Just-In-Time:-Just-In-Time is Closely associated with lean manufacturing is the principle of just-in-time, since it is a management idea that attempts to eliminate sources of manufacturing waste by producing the right part in the right place at the right time. This addresses waste such as work-in-process material, defects, and poor scheduling of parts delivered (Nahmias, 1997). Inventory and material flow systems are typically classified as either push (traditional) or pull (just-in-time) systems. Customer demand is the driving force behind both systems. However, the major difference is in how each system handles customer demand. Just-in-time is a tool that enables the internal process of a company to adapt to sudden changes in the demand pattern by producing the right product at the right time, and in the right quantities (Monden, 1998). Moreover, just-in-time is a critical tool to manage the external activities of a company such as purchasing and distribution. It can be thought of as consisting of three elements: JIT production, JIT distribution, and JIT purchasing. More details are given for each in the following sections.

(iv) Just-In-Time Production:- Lean manufacturing is about eliminating waste wherever it is. One of the most important steps in the implementation of lean manufacturing is JIT. Monden (1998) and Levy (1997) both agree that JIT production is the backbone of lean manufacturing. Just-in-time production is about not having more raw materials, work in process or products than what are required for smooth operation. JIT utilizes what is known as a "pull system". Customer demand, which is the generator of the order, sends the first signal to production. As a result, the product gets pulled out of the assembly process.

The final assembly line goes to the preceding process and pulls or withdraws the necessary parts in the necessary quantity at the necessary time (Monden, 1998). The process goes on as each process pulls the needed parts from the preceding process further up stream. The whole process is coordinated through the use of a kanban system. Shipments under JIT are in small, frequent lots. A kanban is used to manage these shipments. Kanban is an information system that is used to control the number of parts to be produced in every process (Monden, 1998). The most common types of kanbans are the withdrawal kanban, which specify the quantity that the succeeding process should pull from the preceding process, and the production kanban, which specifies the quantity to be produced by the preceding process (Monden, 1998).

(v) **Just-In-Time Distribution: -** JIT effectiveness depends heavily on having a strategic alliance between buyers and suppliers. By having a third-party logistics distributor, companies can focus on their core competencies and areas of expertise leaving the logistics capability to logistics companies (Simchi-Levi, D. et al., 2000; Quinn and Hilmer, 1994). Third-party logistics (3PL) refers to the use of an outside company to perform all or part of the firm's materials 17management and product distribution functions (Simchi-Levi, D. et al., 2000). 3PL can support just-in-time distribution (JITD) by providing on time delivery to customers or distributors, technological flexibility such as EDI and flexibility in geographical locations. (Simchi-Levi, D. et al., 2000; Raia, 1992). JITD requires the exchange of frequent, small lots of items between suppliers and customers, and must have an effective transportation management system because the transportation of inbound and outbound material can have a great effect on production when

There is no buffer inventory (Spencer, Daugherty and Rogers, 1994). Under JITD having a full truckload sometimes is difficult due to the frequent delivery of smaller lots, which in turn will increase the transportation cost. However, to get over the problem Monden (1998) states that instead of having one part loading, using a mixed loading strategy makes it possible to have full truckloads and increase the number of deliveries. Another important factor that is essential to JITD is EDI. In order to have effective product deliveries between suppliers and their distributors or customers, an EDI system must be in place. In the traditional product delivery system suppliers. Always have to keep finished goods inventory or have to alter their production schedules to respond to demand surges. Under EDI suppliers can look at all the shipment and inventory data and adjust their production schedule accordingly (Simchi-Levi, D. et al., 2000). To stay competitive under JITD, it is very important to share information in the whole supply chain because suppliers can adjust their production schedules and narrow their delivery windows as more product data become available to them. Other benefits of EDI include cost reduction, cycle time reduction, stock out reduction, and inventory reduction.

(vi) Just-In-Time Purchasing: - Ansari and Mondarress (1986)and Gunasekaran (1999) define just-in-time purchasing (JITP) as the purchase of goods such that their delivery immediately precedes their demand, or as they are required for use. The idea of JITP runs counter to the traditional purchasing practices where materials are brought well in advance before their use. Under JITP activities such as supplier selection, product development and production lot sizing become very critical. Customer-supplier relationships are a very important part of JITP. Under JITP it is necessary to have a small number of qualified suppliers. Having quality-certified suppliers shifts the inspection function of quality and piece-by-piece count of parts to the supplier's site where the supplier must make sure that parts are defect free before they are transported to the manufacturer's plant. Another important factor of JITP is product development. Buyers must have a "Black Box" relationship with the suppliers where suppliers participate heavily in design and development. The benefits of sharing new product development and design innovation include a decrease in purchased material cost, increase in purchased material quality, a decrease in development time and cost and in manufacturing cost, and an increase in final product technology levels (Simchi-Levi, D. et al., 2000). EDI is very important under JITP. The ultimate goal of JITP is to guarantee that production is as close as possible to a continuous process from the raw material reception until the distribution of the finished goods (Gunasekaran, 1999). EDI can support JITP by reducing the transaction processing time and meeting the specialized needs of buyers by helping them to synchronize their material movement with their suppliers. Although under JITP the carrying cost of materials is increased due to frequent small lots, this cost is offset by a decrease in the cost of processing a purchase order and by to decreased inventory holding cost

Some of the Benefits of JIT are

- Eliminating unnecessary work-in-process, which results in reduction of inventory costs?
- Since units are produced only when they are needed, quality problem can be detected early.
- Since inventory is reduced, the waste of storage space will be reduced.
- Preventing excess production can uncover hidden problems.

(vii) Production Smoothing: - In a lean manufacturing system it is important to move to a higher degree of process control in order to strive to reduce waste. Another tool to accomplish this is production smoothing. Heijunka, the Japanese word for production smoothing, is where the manufacturers try to keep the production level as constant as possible from day to day (Womack et al., 1990). Heijunka is a concept adapted from the Toyota production system, where in order to decrease production cost it was necessary to build no more cars and parts than the number that could be sold. To accomplish this, the production schedule should be smooth so as to effectively produce the right quantity of parts and efficiently utilize manpower. If the production level is not constant this leads to waste (such as work-in-process inventory) at the workplace.

(viii) Standardization of Work: - A very important principle of waste elimination is the standardization of worker actions. Standardized work basically ensures that each job is organized and is carried out in the most effective manner. No matter who is doing the job the same level of quality should be achieved. At Toyota every worker follows the same processing steps all the time. This includes the time needed to finish a job, the order of steps to follow for each job, and the parts on hand. By doing this one ensures that line balancing is achieved, unwarranted work-in-process inventory is minimized and non-value added activities are reduced. A tool that is used to standardize work is what is called "takt" time. Takt (German for rhythm or beat) time refers to how often a part should be produced in a product family based on the actual customer demand. The target is to produce at a pace not higher than the takt time (Mid-America Manufacturing Technology Center press release, 2000). Takt time is calculated based on the following formula (Feld, 2000):

Takt time = it is defined as the ration between **available work time per day** to **customer demand per day**

12.8. Total Productive Maintenance

Machine breakdown is one of the most important issues that concern the people on the shop floor. The reliability of the equipment on the shop floor is very important since if one machine breaks down the entire production line could go down. An important tool that is necessary to account for sudden machine breakdowns is total productive maintenance. In almost any lean environment setting a total productive maintenance program is very important. There are three main components of a total productive maintenance program: preventive maintenance, corrective maintenance, and maintenance prevention. Preventive maintenance has to do with regular planned maintenance on all equipment rather than random check ups. Workers have to carry out regular equipment maintenance to detect any anomalies as they occur.

By doing so sudden machines breakdown can be prevented, which leads to improvement in the throughput of each machine (Feld, 2000). Corrective maintenance deals with decisions such as whether to fix or buy new equipment. If a machine is always down and its components are always breaking down then it is better to replace those parts with newer ones. As a result the machine will last longer and its uptime will be higher. Maintenance prevention has to do with buying the right machine.

If a machine is hard to maintain (e.g., hard to lubricate or bolts are hard to tighten) then workers will be reluctant to maintain the machine on a regular basis, which will result in a huge amount of lost money invested in that machine. Researchers including Nicholls (1994), Taylor (1996), Suehiro (1992), Ljungberg (1998), Nakajima (1989) and others have reported good results implementing TPM. Further, benefits of implementing TPM will be described later.

(i) Other Waste Reduction Techniques: -Some of the other waste reductions tools include zero defects, setup reduction, and line balancing. The goal of zero defects is to ensure that products are fault-free all the way, through continuous improvement of the manufacturing process (Karlsson et al., 1996). Human beings almost invariably will make errors. When errors are made and are not caught then defective parts will appear at the end of the process. However, if the errors can be prevented before they happen then defective parts can be avoided. One of the tools that the zero-defect principle uses is poka-yoke. Poka-yoke, which was developed by Shingo, is an autonomous defect control system that is put on a machine that inspects all parts to make sure that there are zero defects. The goal of poka-yoke is to observe the defective parts at the source, detect the cause of the defect, and to avoid moving the defective part to the next workstation (Feld, 2000). Ohno at Toyota developed SMED in 1950. Ohno's idea was to develop asystemthat could exchange dies in a more speedy way. By the late 1950's Ohno was able to reduce the time that was required to change dies from a day to three minutes (Womack et al, 1990). The basic idea of SMED is to reduce the set up time on a machine. There are two types of setups: internal and external. Internal setup activities are those that can be carried out only while the machine is stopped while external setup activities are those that can be done while the machine is running. The idea is to move as many activities as possible from internal to external (Feld, 2000). After all activities are identified then the next step is to try to simplify these activities (e.g., standardize setup, use fewer bolts). By reducing the setup time many benefits can be realized. First, die-change specialists are notneeded. Inventory can be reduced by producing small batches and more variety of product mix can be run. Line balancing is considered great weapon against waste, especially the wasted time of workers. The idea is to make every workstation produce the right volume of work that is sent to upstream workstations without any stoppage (MidAmerica Manufacturing Technology Center press release, 2000). This will guarantee that each workstation is working in a synchronized manner, neither faster nor slower than other workstations

12.9. The 5 S's of Lean

Much of Lean manufacturing is applying "**common sense**" to manufacturing environments. In implementing Lean, 5 S's are frequently used to assist in the organization of manufacturing. The 5 S's are from Japanese and are:

- Seiri (sort, necessary items).
- Seiton (set-in-order, efficient placement).
- Seison (sweep, cleanliness).
- Seiketsu (standardize, cont. improvement).
- Shitsuke (sustain, discipline).

Laying out a Lean Production Facility : -Anothercritical aspect of Lean is the organization of the production facility. Since one of the keys to lean is waste elimination, the layout of any system should be arranged in such manners that waste of motion (material handling and material transport) and elimination of inventory is part of the object for the layout. You may recall that there are two traditional forms of layout in manufacturing: process and product. In a process layout (or job shop as it is informally called), machines are organized and clustered by type, where typically all mills are in one department, all lathes in another, etc. In a product layout (or flow shop), machines are located so that sequential operations are performed at adjacent machines.

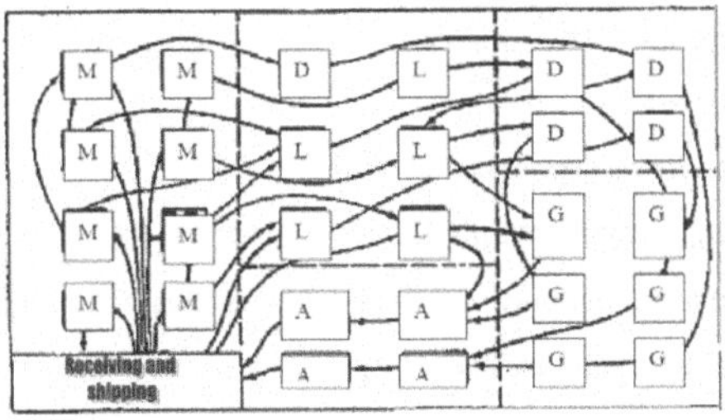

A Typical Process or Job Shop Layout

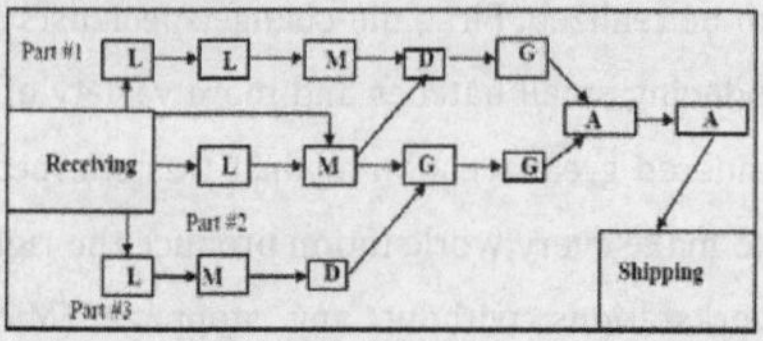

A Typical Product or Flow Shop Layout

Process layout is typically employed for a large variety of products that are made in very small batches (ones or twos).

The Advantages of Process Layout Are

- The flexibility of the system to product almost any part that fits within the volume metric boundaries of the machines.
- An in depth understanding of a specific process can be obtained.
- Some tooling and fixtures can be shared.

The Disadvantages of Process Layout Are

- The spaghetti flow is difficult to manage and control,
- There is usually a lot of inventory in front of each machine.
- Set up is usually expensive.
- Material handling times are large.
- It is difficult to automate these types of system.

Product layout systems are used effectively for the economic production of high volume goods.

The Advantages of These Systems Are

- Large batches can be produced inexpensively.
- Material handling is minimal.
- In-process materials are minimized.
- It is easy to control these systems.
- Automation is more achievable and justifiable.

The Disadvantages of these Systems Are

- They are inflexible, in that only one or very few products can be produced on them.
- Set up time for these systems is very large.
- Duplicate tooling is required to replace worn tooling so that maintenance can be minimized.

12.10. The Five Steps of Lean Implementation

The process used to implement lean manufacturing is a straightforward one. However it is critical that lean is implemented in a logical manner. The steps associated in implementing lean follow:

Step 1: Specify Value

Define value from the perspective of the final customer. Express value in terms of a specific product, which meets the customer's needs at a specific price and at a specific time.

Step 2: Map

Identify the value stream, the set of all specific actions required to bring a specific product through the three critical management tasks of any business: the problem-solving task, the information management task, and the physical transformation task. Create a map of the Current State and the Future State of the value stream. Identify and categorize waste in the Current State, and eliminate it.

Step 3: Flow

Make the remaining steps in the value stream flow. Eliminate functional barriers and develop a product-focused organization that dramatically improves lead-time.

Step 4: Pull

Let the customer pull products as needed, eliminating the need for a sales forecast.

Step 5: Perfection

There is no end to the process of reducing effort, time, space, cost, and mistakes. Return to the first step and begin the next lean transformation, offering a product that is ever more nearly what the customer wants.

12.11. Principle of Lean Manufacturing

All kind of production process having certain rules and regulation which helps to implementing and follow the procedure, the complete set of rules and regulation is termed as basic principle of that particular process. Thuslean manufacturing also working on certain guiding principle which is described below.

(i) Principle related to Continuous Flow of material.

(ii) Principle related to Lean Machines/Simplicity

(iii) Principle related to Workplace Organization.

(iv) Principle related to Parts Presentation.

(v) Principle related to Re-configurability.

(vi) Principle related to Product Quality.

(vii) Principle related to Maintainability.

(viii) Principle related to Ease of access.

(ix) Principle related to Ergonomics.

The Detailed Description of all the Elements Are Given Below

(i) Principle related to Continuous Flow of material:- The continuous flow of material , from work station to workstation is very necessary for uninterrupted production , to fulfill the desired purpose heavy parts may be transported on belt conveyors, manual push or gravity conveyors are ideal for moving parts between workstations. Their minimum complexity makes them easy to service and minimizes down time. In addition, they are easy to connect end-to-end, which makes it easy to move workstations within a work cell.In this procedure lean workcell is used which is of U-shaped , which has a interior working area of cell from work station to work station thereby it gained minimum movement requirement to move the workpiece or assembly from one workstation to the next . The schematic diagram is given below. The advantage of such kind of cell is to Elimination of non-value added movements, work in process, and inventory

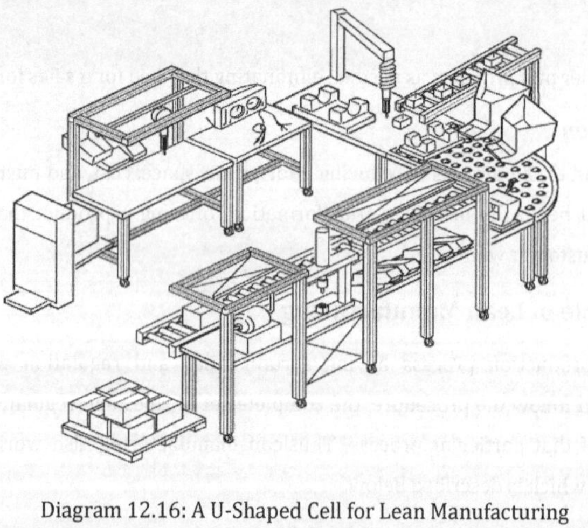

Diagram 12.16: A U-Shaped Cell for Lean Manufacturing

(ii) Principle related to Lean Machines/ Simplicity: -TheContinuous-flow of material is on prime consideration in each and every production process, in lean manufacturing has great importance including this aspect machine simplicity also same. Hence it is important that each workstation or machine be designed to fit within a minimal working area which reduces flat space at the workstation or machine.This is done to avoid the possibility of storing parts or subassemblies at the machine. Storing parts increases work in process and results in "batch" processing, which subsequently defeats the purpose of lean manufacturing. In addition, smaller, minimal size workstations and machines eliminate unnecessary steps taken by the worker between subprocesses.Finally, significant floor space may be saved by properly sizing workstations and machines. The advantage of such kind of machinesimplicitly gives us One-at-a-time manufacture, quick production changeover, reduced WIP, easily modified, customizable production.

(iii) Principle related to Workplace Organization: -A smooth, uninterrupted flow of completed work-pieces is the desired result of a properly designed lean work cell. Nothing can slow or stop this flow faster than the loss or misplacement of tools. Thus, all tools used at a workstation should have their own holder. There should be exactly as many holders as there are tools so that the absence of a tool is quickly noticed. Using a modular tool holder system with a specific holder for each tool is ideal. If holders can easily be added to or taken away from a workstation, this simply adds to the flexibility of the workstation and increases its usefulness in a lean manufacturing process. To minimize downtime, backup tools should also be available at any automated work-stations .we have to put here another things regarding the current position of work piece that should be displayed on the notice board by the supervisor , the position may be of assembly processes, work instructions, repair procedures, or even production targets. The advantage of workplaceorganizationresults Minimize downtime, reduce wasted motion, uninterrupted work-piece flow, improve quality .

(iv) Principle related to Parts Presentation: - During the work shift, additional parts will be required for the work-cell. Those parts must be provided for the operator to achieve the minimum number of work interruptions. All the parts should be supplied to each workstation from outside the work-cell. The use of gravity feed conveyors or bins fit the simplified design of the lean work-cell. Parts bins should load from behind so that the worker may continue production without interruption. Gravity carries the parts to the worker's reach area. Bins should also be reconfigurable, although bins are ideal for small parts, many assemblies require larger parts. These may be delivered in bins or boxes. Again the parts should be delivered to the work-cell without entering the work space. In case where parts are very heavy, lift assist devices are recommended. Heavy parts or boxes of parts can be loaded onto a case lifter and raised to the proper work height with electric, pneumatic, or hydraulic power. The advantage of parts presentation is easy re-configuration; reduce wasted motion, uninterrupted production, and quickchangeover.

(v) **Principle related to Re-configurability:-**A properly designed lean work-cell must be easy to reconfigure. In fact, the ability to change the process and go from good part to good part as quickly as possible is a must. The faster the changeover, the less production time is lost. Quick-change fixtures are one way this can be accomplished. With a good quick-change fixture, changeover can be accomplished in a matter of seconds. A number of different fixtures may be stored at the workstation and swapped **as** the situation requires. The advantage of Re-configurability is Minimize downtime, quick changeover, and uninterrupted work-piece flow

(vi) **Principle related to Product Quality:-**A flawed process or malfunctioning machine may be the source of the quality problems. As each part is produced, visual inspection by the worker can verify that it is correctly assembled. If verification is required through gages, they should be mounted to the machine or workstation and be easily replaced. Quick release of fixtures using star knobs or locking levers is a necessity. A malfunctioning machine may also be easily replaced, particularly if quick disconnects for all pneumatic or electric lines are provided for when the lean cell is designed. In addition, no pneumatic or electrical connections should be present between machines in the lean cell .Ease of reconfiguration and changeover eliminate any reluctance on the part of the worker or management to try to "make do" with machines or processes that are "almost" right. This change in attitude can contribute greatly to true quality production.

(vii) **Principle related to Maintain-ability:** -Ease of service is another requirement of a lean cell. Long down times cannot be tolerated in a pull-through system. When customer demand exists, the product must be produced. A modular structural framing system provides the ultimate in maintainability .The structural framing system also provides a source for common components for all machine bases, guards, and workstations. With standardized components, a minimum number **of** tools is required to maintain a structure. With a structural framing system, three or four simple hand tools are sufficient to build or repair any structure .Another benefit is that common components eliminate the need for a large inventory of spare parts. It is possible to standardize on one size of aluminum profile, one or two connectors, and a few accessories to create virtually any structure. The advantage of Maintain-ability is Minimum down time, easy-to-source replacement parts, and quick service

(viii) **Principle related to ease of access:** -Using an aluminum framing system as the foundation of a lean cell, all necessary work components can be mounted in easily accessible locations because each surface is a potential mounting surface. Parts bins, tools, shelves, and fixtures can all be positioned in the optimum location for efficient work. The T-slot on the framing system's surface also allows quick repositioning of pneumatic or hydraulic components if clearance space is critical. The benefit of such is to Minimum down time, easy serviceability, and optimum ergonomic design.

(ix) **Principle related to Ergonomics:** -Finally, the worker must be protected from ergonomic problems. Any properly designed lean work-cell must, by definition, be ergonomically designed. Maintaining the work at the ergonomically correct height throughout the work-cell is always important. Although it is frequently not taken into consideration, designing for the average worker height is also a necessity. The benefit of Ergonomics is fewer work-related injuries, lower employee turnover, and better working environment.

12.12. Agile Manufacturing System

Agile manufacturing is a term applied to an organization that has created the processes, tools, and training to enable it to respond quickly to customer needs and market changes while still controlling costs and quality. Agile manufacturing is seen as the next step after Lean manufacturing in the evolution of production methodology. In the beginning of globalization of the market in 21st century the agile manufacturing paradigm was formulated in response to the constantly changing `new economy' and as a basis for returning to global competitiveness. Agile manufacturing (AM) has emerged after lean production the word agile is used means the ability fast thinking with a clever method and in fact it is concept of maximum flexibility and an agile organization should be able to respond to possible changes that may occur in the organization.

A lot definitions provided abut agility, but in general we can say agility is responding to available challenges in business environment that this challenge is via change and uncertainty to the business environment. A key feature of agility organization is flexibility. Agile manufacturing is an approach to manufacturing which is focused on meeting the needs of customers while maintaining high standards of quality and controlling the overall costs involved in the production of a particular product. This approach is geared towards companies working in a highly competitive environment, where small variations in performance and product delivery can make a huge difference in the long term to a company's survival and reputation among consumers. In fact the main origin of "agility" as a concept in business and work is rooted in flexible manufacturing systems. At first was thought the path toward flexibility in production and manufacturing is move from automation towards create ability for rapid changes such as reducing the startup time and thus faster reaction to changes in product mix or volume. Soon after this theory about flexibility in the production was extended greater fields of business).a organization for responding to changes occurs in its business requires ability that called agility and is including speed, flexibility, competence and accountability.

Some Important Features of Agile Manufacturing

(i) Agile Manufacturing should primarily be a business concept. Its aim is quite simple - to put our enterprises way out in front of our primary competitors. In Agile Manufacturing our aim is to develop agile properties. We will then use this agility for competitive advantage, by being able to rapidly respond to changes occurring in the market environment and through our ability to use and exploit a fundamental resource -knowledge.One fundamental idea in the exploitation of this resource is the idea of using technologies to lever the skills and knowledge of our people. We need to bring our people together, in dynamic teams formed around clearly identified market opportunities, so that it becomes possible to lever one another's knowledge. Through these processes we should seek to achieve the transformation of knowledge and ideas into new products and services, as well as improvements to our existing products and services.

(ii) The concept of Agile Manufacturing is also built around the synthesis of a number of enterprises that each have some core skills or competencies which they bring to a joint venturing operation, which is based on using each partner's facilities and resources. For this reason, these joint venture enterprises are called virtual corporations, because they do not own significant capital resources of their own. This, it is believed, will help them to be agile, as they can be formed and changed very rapidly. Central to the ability to form these joint ventures is the deployment of advanced information technologies and the development of highly nimble organizational structures to support highly skilled, knowledgeable and empowered people.

(iii) Agile Manufacturing enterprises are expected to be capable of rapidly responding to changes in customer demand. They should be able to take advantage of the windows of opportunities that, from time to time, appear in the market place. With Agile Manufacturing we should also develop new ways of interacting with our customers and suppliers. Our customers will not only be able to gain access to our products and services, but will also be able to easily access and exploit our competencies, so enabling them to use these competencies to achieve the things that they are seeking.

(iv) Agility theory: - In order to plan for agile business processes a thorough understanding of what can be termed `agility theory' is necessary [10]. The manufacturing environment has undergone several transitions, from the craft industry, to mass production, and now the newest archetype, agility.

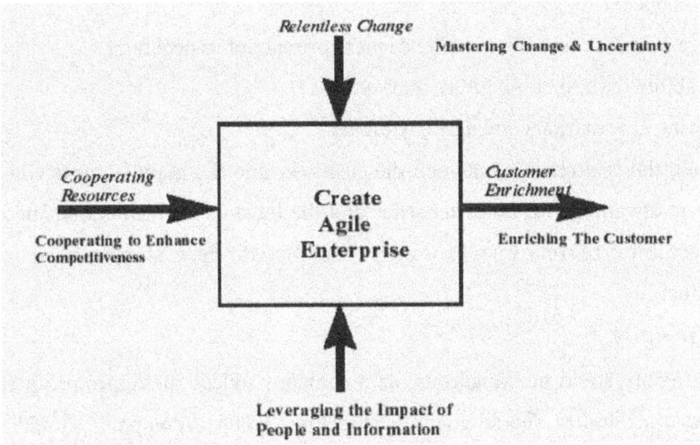

Diagram 12.17: Agility Process

To facilitate the understanding of the new paradigm of agility, the Agility medium has introduced four dimensions of agility to convey the holistic approach involved. The dimensions may be viewed systemically as shown in Figure. The four basic elements include Cooperating to Enhance Competitiveness, Enriching the Customer, Mastering Change and Uncertainty, and Leveraging the Impact of People and Information.

Agile Manufacturing Philosophy

With fast changes occurring in the demand of the client, the manufacture must adapt itself so that it would not diminish. Survival will depend on the capability to keep up with the continuous and unexpected change. This could be described on how agile an environment can be. Being agile will allow the company to seek for new opportunities as well as to ignite new innovations. In agile production if a product life ends prematurely the system is quickly redesigned and retooled for the new/different product. Instead of shutting the systems or postponing the process a continuously evolution of the work floor is done to meet the latest demands. Even though these changes are radical the quality of the product is not affected. Hence agile manufacturing will deal with things that cannot be controlled. This means that a agile competitor to be successful, he must not only understand the current market and customers well but also understand the potential for future buyers and markets.

Agile Manufacturing Deals

(i) The fast and unpredictable change in the market demand.

(ii) Low volumes (due to unpredictable change), yet a high quality and a specific product.

(iii) Products with a short life-cycle, therefore a short development and production lead time is needed.

(iv) Meeting each client's demand hence mass production is declining.

(v) The ability to predict the future market.

(vi) Creating new markets and new customers.

As a result, the relationship between the customer and the manufacturer will strengthen, resulting in an advantage for both the seller and the buyer. This means that the target of an agile manufacturer is to come up with a specific solution for the customers need and not just a generic product.

When to Use Agile

Adopting agility could be dangerous for a company unless the appropriate studies have been carried out. Studies should identify guidelines when a company should adopt agile manufacturing. These should be carried out by the right people, being developers and customers, as well as experienced consultants in the agile manufacturing field. Agility should be implemented:

- When rapid changes in the customers' requirements produces pressure upon the company's production and needs to respond fast.
- When the customer requires the requested product within a fixed time frame.
- When innovative and modern technologies give rise to emerging requirements.

After each project, the company needs to evaluate if it has reached the desired directives such as its objectives, goals, and deadlines.

How to become Agile

Agility is all about satisfying the customer's latest needs. Thus, the first step to become agile is to research the customers' expectations; mainly the time they are willing to wait for their product to be delivered to them. This will continuously change the company's position which makes it more competitive in its sector. Keeping in mind that agile manufactures work within specified deadlines, communication responsiveness is a key aspect when coming to agile manufacturing? While the customers are stating their requests, the management of the company can immediately distribute the work amongst its suppliers, designers and labourers. This is commonly referred to as strategic agility. Also feedback from each department is essential so that a data base of information could be created pointing out problems which need to be tackled preventing them from getting themselves into bigger problems. Being capable to solve such problems requires the company to have flexible human resources and organizations which are ready to adapt to any changes needed.

Enablers for Agile Manufacturing

(i) **Virtual Enterprise :** This is an organization that has a virtual organizational arrangement meaning it is an organization made up of contributors that add value to a product. As customers' needs change VEs also change. Contributors that do not add enough value to the new product leave and terminate the virtual enterprise while other companies join to add valued to the new product. A virtual enterprise is valued since entering new markets is easier since resources can be easily changed.

(ii) **Rapid Prototyping :** This refers to the generation of a product so that design and functionality can be tested. In virtual prototyping, the prototype is stored in a virtual environment like a computer rather than being physically created. This allows for observation, analysis and manipulation while it reduces costs and the time of product development.

(iii) **Electronic Commerce :** Electronic Data Interchange helps companies to interchange business related information between computers within same companies, different companies, **government** organizations as well as banks. Using the internet, companies can go global and serve customers by improving the response time through online systems.

(iv) **Rapid Partnership :** In large company's work must be distributed. This is done by making use of teams that work together with the appropriate partnership. Partnership involves the usage of partner company's potentials to come up with innovative ideas and concepts that lead to new products. Apart from the rapid response, good marketing skills are essential to make sure that the company remains competitive.

(v) **Physically Distributed Teams :** This unites partner enterprises located all over the world; thus, having a distributed manufacturing environment. Every enterprise contributes to take advantage of a specific business opportunity and works against the market threat. Some reasons for change are the increasing flexibility of global transport, improvement in manufacturing processes and economic conditions for example the shifting of production to countries with cheap labour costs.

(vi) **Concurrent Engineering :** CE ensures that new products are designed and all factors are concerned. This includes marketing, quality and serviceability. Customers tend to go for products which give them better assurance, service and testing processes. Several tools and techniques are available to improve the design and manufacturing of a product. This includes functionality analysis, solid modeling and finite element analysis.

(vii) Integrated Production Information System : A fully integrated strategy system is required to set up a standard and avoid communication problems. This enables users from different firms with different capabilities to understand the shared data and information. Software representation includes data sources, data bases, packages and testing equipment.

Advantages and Disadvantages of Agile

Agile manufacturing refers to the ability of a company to change its production according to the unforeseen changes in the customers' demands. Since there is always a rapid change in the production line this leads to advantages as well to disadvantages.

Advantages of Agile Manufacturing

- The company is given a competitive advantage since it is continuously changing its approach to satisfy its customers.
- Innovative designs based on the customers' demands are provided. This in turn gives a wider variety for the customer to choose from.
- Responds quickly to emerging crisis.
- Even though the production could change rapidly, mass production could still be reached while flexibility is still possible.

Disadvantages of Agile Manufacturing

- Sudden increase in demand (more than expected due to the liking of the new product) will cause shortage. Meanwhile if the demand of a product with a high production rate decreases drastically this could result in many products that could not be sold.
- To become an agile manufacturer the company will need to invest in trained and highly skilled labourers who are competent to be agile.
- The change in machinery to produce the new products requires prohibitive costs.
- Continuous need to keep the machinery and workers up-to-date to modern technologies, and to keep the company competitive due to the short product life cycles.
- The complexity of the new machinery could lead to a higher increase in breakdown of the same machinery, which will lead to an increase in production down time.
- The maintenance to keep the machinery in good working condition increases the costs due to higher costs of the parts.

- Intensive planning and management of such systems is required, since a shift is being introduced from mass production ideology to agile manufacturing.

Question for Discussion

1. What is production control ? Discuss computer applications in production control .
2. What is production planning? How it is related with capacity planning
3. Define Production Planning and Control (PPC), identify the objectives of PPC and also
4. discuss the importance of PPC for an industrial organization.
5. Discuss the main characteristics of the JIT system and the advantages it offers ?
6. Differentiate between job process and batch process.
7. What is a production system? Discuss the main characteristics of batch production system.
8. Compare and contrast between Project and Mass production systems.
9. What do you understand by lean manufacturing ? explain about lean wheel .
10. Discuss the main objectives and functions of PPC ?
11. What is batch production system? What are the advantages and drawbacks of batch production systems ?
12. What is Waste? Explain seven types of wastes in production systems as identified by Toyota production systems.
13. Describe briefly various *types* of manufacturing systems .
14. What is capacity planning? Explain the steps in capacity planning. Explain some of the strategies to modify the capacity in short run.
15. Explain the steps of JIT implementation in an Organization.
16. What is TPM? Discuss its objectives and benefits of implementing TPM in an industry.
17. Explain Methods of Total Productive Maintenance.
18. What is JIT production system? Discuss merits and demerits of JIT production system.
19. Briefly explain the concepts of Lean and Agile Manufacturing. Discuss the advantages offered by these systems to the industries.
20. Which are the main functions of production planning and control department
21. What is process planning ? Discuss the steps involved in process planning
22. What is lean manufacturing ? explain about various principles associated to lean.
23. Discuss the significance of term , the 5 S's of Lean manufacturing.
24. Write short notes on:
    a) Agile manufacturing.
    b) Just-In-Time Production.
    c) Cellular Manufacturing.

d) Flexible Manufacturing System Layout.

References

1. Warner and Malcolm, "New technology and manufacturing management", strategic choices for flexible production systems, New York, John Wiley, 1990.
2. Schonberger and J. Richard, "World class manufacturing, the lessons of simplicity applied", NewYork, Free Press, 1986.
3. Monden and Yasukiro, "Applying Just in Time: The American Japanese experience", Atlanta, 504 Georgia, Institute of Industrial Engineers, 1986.
4. Mayer and E. Raymond, "Production and operations management", New York and London, McGraw-Hill, 4th éd., 1982.
5. H.B. Maynard, "Production: An international appraisal of contemporary manufacturing systems and the changing roloe of the worker", New York and London, McGraw-Hill, 1975.
6. Industrial engineering handbook, New York and London, McGraw-Hill, 3rd éd., 1971.
7. F.P.M. Biemans, "Manufacturing planning and control: A reference model", (Amsterdam, Elsevier, 1990.
8. S. Agurén, "The Volvo Kalmar plant: The impact of new design on work organization", Stockholm, Rationalization Council-Swedish Employers' Confederation-Swedish Trade Union Confederation, 1976.
9. B.S. Elwood, Sarin and K. Rakesh, "Modern production operations management", New York, John Wiley, 8thed., 1987.
10. F. Butera and J.E. Thurman, "Automation and work design", Amsterdam, North-Holland, 1984.
11. G.B. Carson, "Production handbook", New York, Ronald Press, 3rd éd., 1972.
12. J. Carpentier and P. Cazamian, "Night work : Its effects on the health and welfare of the Worker", Geneva, ILO, 1977.
13. A.C. Laufer, 1984, Production and Operations Management, 3rd edn, South-Western Publishing Co., Cincinnati.
14. F.E. Meyers and J.R. Stewart, "Time and Motion Study: For Lean Manufacturing", 3rd ed. Upper Saddle River, NJ: Prentice Hall,2001.
15. YW. McEwan, "Flexible working arrangements in continuous shift production", Pers Rev., Pp. 12-19, 1978.

CHAPTER 13

PRODUCT DESIGN AND DEVELOPMENT

13.0. Product Design and Development

Product design deals with conversion of ideas into reality. Every business organization have to design, develop and introduce new products as a survival and growth strategy. Developing the new products and launching them in the market is the biggest challenge faced by the organizations. The entire process of need identification to physical manufactures of product involves three functions: **marketing, product development, manufacturing.** Product development translates the needs of customers given by marketing into technical specifications and designing the various features into the product to these specifications. Manufacturing has the responsibility of selecting the processes by which the product can be manufactured.

Product design and development provides link between marketing, customer needs and expectations and the activities required to manufacture the product. Organization success is dependent on customer satisfaction and delight. Customer satisfaction is achieved through development of product and service, which have all attributes required by the customer. A success product or services do not only have attractive package design but should be also able to provide robust performance. Thus, product design must be practical enough for production and powerful enough to provide a competitive advantage.

13.1. Common Stage &Common Features of Good Product Design

Common stage of product design: - there are three common stage of product design.

(i) First stage: His stage involves brainstorming, bringing ideas and analysis of customer and market feedback.

(ii) Second Stage: Idea is converted into a feasible solution to satisfy the customer expectation, using available resource and technology.

(iii)Third Stage: This is the last stage in which the product is introduced in the market.

The essence of product design is to satisfy customer and maximizes the value for the customer at minimum cost.

The product or service should also be able to meet primary needs and desire of the customer. This may not require development of new product, but enhancement to existing product or service.

Common Features of Good Product Design

(i) **Utility:** The product design should make product utility as per expectation of customers and provide steady performance through the product life.

(ii) **Aesthetics:** Product aesthetics is important in success of the product. The product aesthetics is dependent on market and end customer.

(iii) **Producible:** Product design should enable effective production of product through available production methods.

(iv) **Profitability:** Product design should make economic sense as to deliver value to customer and sustainability to the organization.

(v) **Differentiable:** A good product design should enable product to be differentiate among its competition. This can be achieved by attractive packaging and also by providing additional service on the product.

13.2. Product Function

Products play a key role in the attempts to reduce the total environmental impact of human activities. This is because every human activity - and consequently its impacts on the environment – can be related to specific needs and the fulfilment of these needs by material or non-material products. Different products, causing different environmental impacts, may fulfil the same need. Thus, environmental impacts may be reduced by substituting between products. The purpose of life cycle assessments is to assess the possible environmental impacts of such product substitutions, i.e. the choice of one product instead of another (or the choice of a specific product instead of refraining from this product). All products are produced to perform one or more functions, providing one or more services, fulfilling one or more customer requirements.

For example, the function of an office chair can be expressed in terms of "seating support for one computer workstation for one year" with some additional minimum requirements with respect to durability, strength, stability, safety and comfort, including adjustments of seat and backrest, arm rests etc. For a specific life cycle study, the different product functions and ways of fulfilling them are not all equal relevance. Therefore, as part of defining the scope of the study it is crucial to determine the studied functions, the relevant alternative solutions to fulfill these functions, and to ensure the equivalence of the compared alternatives.A functionalunits a quantified description of the performance of the product systems, for use as a reference unit. The performance described by the functional unit.

Properties of the product may be related to:

- Functionality -related to the main function of the product.
- Technical quality-such as stability, durability, ease of maintenance.
- Additional services-rendered during use and disposal.
- Aesthetics-such as appearance and design.
- Image(of the product or the producer).
- Costs-related to purchase, use and disposal.
- Specific environmental properties.

13.3. New Product Design Process

New product design is very complex and critical task, it consists of following steps to achieve the purpose.

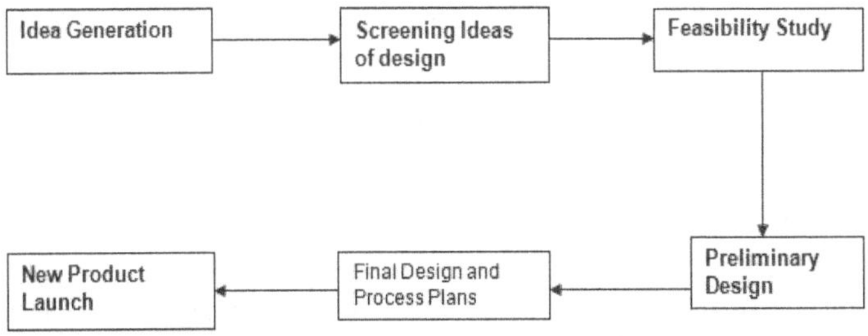

Diagram 13.1: New Product Design Process

(i) Idea Generation: Ideas for new products can come from a variety of sources both within and outside the firm. Internal sources include employees, research and development, market research sales force and reverse engineering. The external sources include customers, legislation, environment, technology and strategic position of the organization. Competitors are also the source of ideas for new products or services. Perceptual maps, bench marking and reverse engineering can help companies learn from their competitors. The design process depends upon the customers and their needs.

Description of Internal Sources Are

- **Basic research:** many companies, have several scientists who are assigned the task of developing new product ideas and related technology.
- **Manufacturing:** people who manufacture products often have ideas about modifications and improvements, as well as completely new concepts.

- **Salesperson:** company salespeople and representatives can be a most helpful source of ideas, since they not only know the customer best, but they also know the competition and the relative strengths and weaknesses of existing products.
- **Top management:** the good top executive knows the company's needs and resources, and is a keen observer of technological trends and 0; competitive activity.

Description of Secondary Sources Are

- **Secondary sources of information:** -there are published lists of new products, available licenses, and ideas for new product ventures.
- **Competitors :** good inferences about competitive product development can be made on the basis of indirect evidence gained from salespeople and from other external sources, including suppliers, resellers, and customers.
- **Customers:** frequently customers generate new product ideas, or at least relay information regarding their problems that new and improved products would help to solve.
- **Resellers :** a number of firms use "councils" or committees made up of representative resellers to assist in solving various problems, including product development.
- **Foreign markets:** -many companies look toward foreign markets, especially Western Europe, because they have been so active in product development.

(ii) Screening Ideas: The second step in the product development process is screening. It is a critical part of the development activity. Product ideas that do not meet the organization's objectives should be rejected. :f a poor product idea is allowed to pass the screening state, it wastes effort and money in subsequent stages until it :s later abandoned. Even more serious is the possibility of screening out a worthwhile idea.

(iii) Feasibility Study; Feasibility study consists of a market analysis, an economic analysis, and technical and strategic analysis. Marketing takes the ideas that are generated and the customer needs that are identified from the first stage of the design process and develops alternative product concepts. Market analysis is based on customer analysis and market survey. Technical and strategic analysis is concerned with technical viability of the product with respect to technology, process of manufacture, availability of materials etc.

(iv) Preliminary Design: The process of preliminary design involves building a prototype, testing the prototype, revising the design, retesting and so on until a viable design is determined. Design incorporates both form and function. Form design refers to the physical appearance of a product, its shape, size, color, styling etc. Aesthetics aspects such as image, market appeal, special identification, finish etc. will also form a part of the form design . Production design is concerned with how the product will be made. Design, which are difficult to make result in poor quality products. During the design stage itself the manufacturing aspects should be considered. The production design or design for production includes simplification, standardization and modularity.Design simplification attempts to reduce the number of parts, subassemblies and options into a product. Standardization refers to use of commonly available and interchangeable parts and subassemblies. Modular design consists of combining standardized building blocks or modules in a variety of ways to create a unique finished product. Modular design is common in electronics and automobile industry.

(v) **Final Design and Process Plans:** In the preliminary design stage, prototypes are built and tested after several iterations. The final design consists of detailed drawings and specifications for the new product. The accompanying process plans are workable instructions for manufacture including necessary equipment's and tooling, component sources job descriptions, work instructions and Programmers for computer-assisted machines.

(vi) **New Product Launch:** Launching a new product or service involves ramp up production, In ramp up, production starts at a relatively low level of volume as the organization develops confidence in its abilities to execute production consistently and marketing's abilities to sell the product, the volume increases. Launching the new product or service involves co-coordinating the supply chain and rolling out marketing plans. Marketing and production will work in a co-coordinated way during this phase.

13.4. Process Design and Analysis

The objective of organization is to provide service and product, which satisfy customer and create value for them. A product and service designed is based on the customer feedback and requirement of the market. Process design is where the product is broken down into parts, which further can be helpful in the actual manufacturing process.

(i) **Process Planning :** Process development for process design can be summarized through following steps:

- **Process Requirement:** The very 1st step is to collect and gather information to give structure with the end objective. That is to make process requirement document highlighting various stages, risk and stakeholders for production. This will include assessment of available technology, raw material requirement, factory/plant layout and demand forecast.

- **Team Building:** Once the process requirements are finalized, for each objective, a team is finalized based on skill level and experience. Function of the team is to get familiarize with the whole process.
- **Planning and Implementation:** Process planning team will develop module; policies and procedure require for production, which are after required approval internal as well as external is implemented.
- **Audit:** A regular audit is carried out to ensure that process thus implemented is in line and delivering value to customers.
- **End of Life:** Over a course of time there may be enhancement of the product or product may get discontinued in these circumstances, process thus develop is discontinued.

(ii) Production Process: Based on the nature of product and service production or conversion process can be divided into two broad categories, continuous production (assembly line, oil refinery) and intermittent production (job work, service).Production process for both manufacturing industry and service industry can be classified into broad categories based on standardization of product or service. It can range from single project assignment like a building or bridge (manufacturing) to interior design (service) and mass production project like a car (manufacturing) to a fast-food joint (Services).

(iii) Process Design: A successful process design has to take into account the appropriateness of the process to overall organization objective. Process design requires a broad view of the whole organization and should not have a myopic outlook. And the process should deliver customer constant involvement of the management at various stages.In order to achieve a good process design, effective process strategy is required, which deals with singular line items required to manufacture the end product. Effective process strategy deals with raw material procurement, customer participation, technology investment, etc . Over a period of time process design has undergone change and new concepts like Flexible Manufacturing Systems have been developed, which delivers efficient and effective production design and analysis.

13.5. Product Design Concept Development

Good product concept development is crucial. During this stage, the needs of the target market are identified, competitive products are reviewed, product specifications are defined, a product concept is selected, an economic analysis is done, and the development project is outlined. This stage provides the foundation for the development effort,

(i) **Identify Customer Needs:** Through interviews with potential purchasers, focus groups, and by observing similar products in use, researchers identify customer needs. The list of needs will include hidden needs, needs that customers may not be aware of or problems they simply accept without question, as well as explicit needs, or needs that will most likely be reported by potential purchasers. Researchers develop the necessary information on which to base the performance, size, weight, service life, and other specifications of the product. Customer needs and product specifications are organized into a hierarchical list with a comparative rating value given to each need and specification.

(ii) **Establish Target Specifications:** Based on customers' needs and reviews of competitive products, the team establishes the target specifications of the prospective new product. While the process of identifying customer needs is entirely a function of marketing, designers and engineers become involved in establishing target specifications. Target specifications are essentially a wish-list tempered by known technical constraints. Later, after designers have generated preliminary products concepts, the target specifications are refined to account for technical, manufacturing and economic realities.

(iii) **Analyze Competitive Products:** An analysis of competitive products is part of the process of establishing target specifications. Other products may exhibit successful design attributes that should be emulated or improved upon in the new product. And by understanding the shortfalls of competitive products, a list of improvements can be developed that will make the new product clearly superior to those of others. In a broader sense, analyzing competitive products can help orient designers and provide a starting point for design efforts. Rather than beginning from scratch and re-inventing the wheel with each new project, traditionally, the evolution of design builds on the successes and failures of prior work.

(iv) **Generate Product Concepts:** Designers and engineers develop a number of product concepts to illustrate what types of products are both technically feasible and would best meets the requirements of the target specifications. Engineers develop preliminary concepts for the architecture of the product, and industrial designers develop renderings to show styling and layout alternatives. After narrowing the selection, non-functional appearance models are built of candidate designs.

(v) **Select a Product Concept:** Through the process of evaluation and tradeoffs between attributes, a final concept is selected. The selection process may be confined to the team and key executives within the company, or customers may be polled for their input.

Candidate appearance models are often used for additional market research; to obtain feedback from certain key customers, or as a centerpiece of focus groups.

(vi) **Refine Product Specifications:** In this stage, product specifications are refined on the basis of input from the foregoing activities. Final specifications are the result of tradeoffs made between technical feasibility, expected service life, projected selling price, and the financial limitations of the development project. With a new luggage product, for example, consumers may want a product that is lightweight, inexpensive, attractive, and with the ability to expand to carry varying amounts of luggage. Unfortunately, the mechanism needed for the expandable feature will increase the selling price, add weight to the product, and introduce a mechanism that has the potential for failure. Consequently, the team must choose between a heavier, more costly product, or one that does not have the expandable feature. When product attributes are in conflict, or when the technical challenge or higher selling price of a particular feature outweighs its benefits, the specification may be dropped or modified in favor of other benefits.

(vii) **Perform Economic Analysis:** Throughout the foregoing activities, important economic implications regarding development expenses, manufacturing costs, and selling price have been estimated. A thorough economic analysis of the product and the required development effort is necessary in order to define the remainder of the development project. An economic model of the product and a review of anticipated development expenses in relation to expected benefits is now developed.

(viii) **System Level Design**: System-level design, or the task of designing the architecture of the product, is the subject of this stage. In prior stages, the team was focused on the core product idea, and the prospective design was largely based on overviews rather than in-depth design and engineering. Once the development plan is approved, marketing may begin to develop ideas for additional product options and add-ons, or perhaps an extended product family. Designers and engineers develop the product architecture in detail, and manufacturing determines which components should be made and which should be purchased, and identifies the necessary suppliers. The product architecture defines the product in *chunks,* or the primary functional systems and subsystems, and how these systems are arranged to work as a unit. For example, an automobile is comprised of a body and a chassis with an engine, a transmission, final drive, frame, suspension and braking system. The architecture of an automobile design determines the platform layout, whether the vehicle is front-wheel-drive or rear-wheel-drive, the size and location of the engine, transmission and final drive, the overall design of suspension system, and the layout and type of other necessary subsystems such as brakes, wheels, and steering. The architecture may determine the layout of the exhaust system, but it would not provide the detailed engineering needed to determine the diameter and thickness of the exhaust pipe, the detailed design of mufflers, nor the engineering of motor mounts and exhaust hangers needed to isolate vibrations from the passenger compartment. The architecture of the product, how it is divided into chunks and how the chunks are integrated into the total product, impacts a number of important attributes such as standardization of components, modularity, options for change later on, ease of manufacture, and how the development project is divided into manageable tasks and expenses. If a family of products or upgrades and add-ons are planned, the architecture of the product would determine the commonality of components and the ease with which upgrades and add-ons can be installed. A system or subsystem borrowed from another product within the company's line will economize on development, tooling and manufacturing costs. With outsourced components, the supplier may contribute much of the associated design and engineering.

(ix) **Detail Design**: Detail design, or design-for-manufacture, is the stage wherein the necessary engineering is done for every component of the product. During this phase, each part is identified and engineered. Tolerances, materials, and finishes are defined, and the design is documented with drawings or computer files. Increasingly, manufacturers and developers are turning to three-dimensional solid modeling using programs such as Pro-Engineer. Three-dimensional computer models form the core of today's rapid prototyping and rapid manufacturing technologies. Once the database has been developed, prototype components can be rapidly built on computerized machines such as CNC mills, fused deposition modeling devices, or stereo lithography systems.

(x) **Testing and Refinement**: During the testing and refinement stage, a number of prototypes are built and tested. Even though they are not made from production components, prototypes emulate production products as closely as possible. These *alpha* prototypes are necessary to determine whether the performance of the product matches the specifications, and to uncover design shortfalls and gain in-the-field experience with the product in use. Later, *beta* prototypes are built from the first production components received from suppliers.

(xi) **Production Ramp-up :** During production ramp-up, the work force is trained as the first products are being assembled. The comparatively slow product build provides time to work out any remaining problems with supplier components, fabrication, and assembly procedures. The staff and supervisory team is organized, beginning with a core team, and line workers are trained by assembling production units.

(xii) Technology-Push Products: The generic development process is used with technology-push products, but with slight modification. With technology-push products, the company acquires or develops a new technology and then looks for appropriate markets in which to apply the technology. Consequently, an extra phase is added at the beginning during which the new technology is matched to an appropriate market opportunity. When the match has been made, the generic development process is carried out as described.

(xiii) Models and Prototypes: The terms prototype and model are often used interchangeably to mean any full-scale pre-production representation of a design, whether functional or not. I prefer to use the term model to describe a non-functional representation and the term prototype to describe a functional item. An appearance model is a full-scale, non-functional representation that looks, as closely as possible, identical to the prospective new product. Modeling and prototyping serve a variety of purposes throughout the development effort. Early on, engineering prototypes may be built of systems and subsystems to bench-test performance and debug the system before proceeding with the design. Appearance models prove out styling and ergonomics. A full-scale mockup of an automobile interior, for example, provides a real-world test of ease of ingress, seating position, access to controls, visibility and appearance. Models and prototypes are necessary because of the limitations of theoretical work and artificial mediums. A product can be designed and put into simulated use on computer, but one doesn't really know how it will work until the item is built and tested in its intended environment. Prototyping and modeling efforts begin virtually at the inception of the project and continue into production ramp-up.

13.6. Economics of a New Design

The following steps must be followed by the design engineer in order to estimate the economic design regarding the product.

- Define the problem and the objective.
- Identify feasible alternatives for accomplishing the objective, taking into account any constraints.
- Determine whether an economic analysis is necessary, and if so, the level of effort which is warranted.

Diagram 13.2: Economic Design Process

- Select a method or methods of economic analysis.
- Select a technique that accounts for uncertainty and/or risk if the data to be used with the economic method are uncertain.
- Compile data and make assumptions called for by the economic analysis method(s) and risk analysis technique.
- Compute a measure of economic performance.
- Compare the economic consequences of alternatives and make a decision, taking into account any non-quantified effects and the risk attitude of the decision maker.

13.7. Product Cost Analyses

A functional cost analysis is a method that can be applied to examine the component costs of a product in relation to the value as perceived by the customer. The outcome of the analysis is to improve the value of the product while maintaining costs and/or reduce the costs of the product without reducing value

(i) Functional Cost Analysis is Done in Following Manners

- Identifying product needs and requirements.
- Conceptual through to detail design.
- Pre-production.
- Post-design product improvement.

(ii) Other Tools that is Useful in Functional Cost Analysis

- Competitive Analysis.
- Requirements Management.
- Kano Model Analysis.
- Brainstorming.
- Prioritization Matrices.
- Evaluation Matrices.
- Quality Function Deployment.
- Configuration Management.

The functional cost analysis is a value engineering method that aims to increase the difference between the cost and the value of a product. The cost is the amount that is incurred in the production and delivery of the product. This expense can include the price of parts, labour, overhead (e.g., building, power), packaging, shipping, and advertising, among others. What the product is worth in the eyes of the customer is considered the value. When completing a functional cost analysis, remembering this definition of value is extremely important. The design team may not perceive a certain product feature to be valuable, however if it is important to the customer, then that feature must be regarded as valuable.Value engineering can be applied to new designs, extensive redesigns or in an evaluation of a competitor's product. Ideally, the application of value engineering should begin at the conceptual design stage and continue through the design process as part of the designto cost effort.

(iii) Who Should Contribute to the Functional Cost Analysis?

The design team, representatives from marketing, purchasing, costing and production should be invited to participate – in short, anyone who can provide detailed cost information or provide insight into the value customers place on individual functions. If the design team has access to the customer, including them in the exercise can be valuable; however if the analysis will be conducted on a competitive product, be cautious about inviting the customer to the table.

(iv) Steps Applied in Cost Analysis

Following steps must be closely understood by the design engineer while designing the new products.

Step 1: Create a List of Components

The first step in a functional cost analysis is to list the component parts of the product. If a tangible product exists (e.g., existing product, competitor's product, prototype), it may be beneficial to dismantle the product to support this exercise. If a tangible product does not exist, exploded diagrams can be used. Parts lists and engineering drawings can also be used but may be of lesser value because they do not offer the same ease of visualization.

Step 2: Determine the Cost of Each Component

The determination of product costs is often oversimplified and therefore inaccurate, even for relatively simple products. This explains why many companies employ specialists in cost estimating. It is not sufficient to merely determine the cost of the individual components and estimate the labour involved. A number of other factors must be considered and included. Some of these factors are:

- Cost of parts including the purchase price as well as any associated shipping costs.
- Cost of consumables that are used in the fabrication of the product (e.g., glue, cleaners, finishes).
- Labour costs for assembly as well as for any modification or finishing to parts Equipment,
- tooling costs Handling.
- inventory control Packaging and shipping costs,

Equipment or tooling costs are items that may require some calculation since there is not a one-to-one ratio between the unit of equipment and unit of product. Therefore, it may be necessary to calculate equipment costs based on projected volumes. Depreciation of equipment may also need to be taken into consideration. If the equipment is used for more than one product line or component, the equipment costs are further subdivided.Very detailed cost calculations may also include indirect or overhead costs. These involve facility costs (e.g., heat, hydro, taxes), non-direct labour (e.g., finance, human resources, marketing), or any other cost that cannot be directly tied to a product. There are a number of different methods by which companies apply overhead costs to products. One method is to add an overhead percentage to direct labour costs while another is to divide the overhead costs amongst all products. In a functional cost analysis, it is easiest to assume that all components equally share the overhead costs and therefore leave these figures out of the calculations

Cost Calculations: Once the costs have been determined for each component or sub-assembly in relation to the function(s) they support, the total cost of the component or sub-assembly is determined. The total product cost is calculated by summing the totals for all components. This total is then used to calculate the percentage of the total cost that each part represents.

Step 3: Determine the Value of Each Component

This step is one of the most difficult in the functional value analysis because much of it is subjective and it requires an estimate of the value as perceived by the customer. When determining the value of the components and subassemblies, it is important to consider the value of aspects other than those that contribute to function or performance. One method to determine value in terms of currency is to compare the costs and characteristics of similar products.

Step 4: Consider the Value of Functions

Before looking at ways of reducing the cost of the product or increasing its value, the value of the current functions should be examined. If a certain function is not perceived to be valuable to the customer and the function is determined not to be necessary for reasons such as performance or reliability, then perhaps that function should be simplified or eliminated.

Step 5: Generate Alternatives that Lead to Reduced Costs and/or Increased Value

Once the existing product has been assessed, the next step is to find design or component alternatives that serve to reduce costs without risking value or increase value without adding cost. The most ideal situation is to find alternatives that reduce costs while adding value.

Table 13.1: Cost Reducing Procedure

Process	Description
Reduction	• Reduce the number of components • Combine components • Reduce specifications (e.g., material strength requirements) • Reduce or change packaging
Elimination	• Eliminate redundant or unnecessary components
Simplification	• Simpler design • Simpler shapes • Simpler assembly process or sequence • Relaxed tolerances or dimensions
Modification	• Less expensive materials • More efficient manufacturing • Less expensive finishes • Use alternative transportation methods
Standardization	• Use standardized or bought-out rather than specialized components • Standardized dimensions • Duplicated components

Step 6: Evaluate Alternatives

Not all alternatives suggested for decreasing the costs or increase value will be feasible or compatible therefore a careful evaluation should be made before any implementations are made. Evaluation matrices or quality function deployment may be helpful in considering the alternatives. If the functional cost analysis results in changes to the product design, be sure to manage these changes carefully and update all necessary documentation including specifications.

13.8. The Product Life Cycle

A company has to be good at both developing new products and managing them in the face of changing tastes, technologies, and competition.

Evidence suggests that every product goes through a lifecycle with predictable sales and profits, as illustrated in Figure.

As such, the manager must find new products to replace those that are in the declining stage of the product lifecycle and learn how to manage products optimally as they move from one stage to the next. The five stages of the PLC and their components can be defined as follows:

(i) **Product development:-** the period during which new product ideas are generated, operational zed, and tested prior to commercialization.

(ii) **Introduction:-** the period during which a new product is introduced. Initial distribution is obtained and promotion is obtained.

(iii) **Growth:-**the period during which the product is accepted by consumers and the trade. Initial distribution is expanded, promotion is increased, repeat orders from initial buyers are obtained, and word-of-mouth advertising leads to more and more new users.

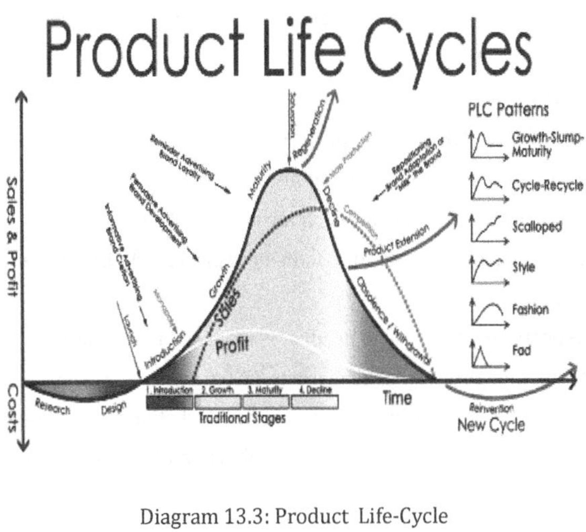

Diagram 13.3: Product Life-Cycle

(iv) Maturity:- the period during which competition becomes serious. Towards the end of this period, competitors' products cut deeply into the company's market position.

(v) Decline:- the product becomes obsolete and its competitive disadvantage result in decline in sales and, eventually, deletion.

Different Aspects of the Products

For proper recognition of any products, it involve so many aspect among them some aspect are given below.

(i) Functional aspects: -functional aspects are relatedto performed, the intended function by product for that it has been created . In a functional sense, the key question is, **"Does the product do what the consumer wants it to do?" Does it get clothes clean? Does it quench your thirst? Does it save you money?** Some of these questions can be answered only through product research, but consumer research provides more answers. Product features include such factors as form, color, weight, odor, material, and tactile qualities. The product must also be aesthetically pleasing. When the entire product is put together, it must create an appealing, visually attractive and distinctive need satisfier.

(ii) Aesthetic aspects: -Aesthetics of product is a valuable element because many consumers purchase not only a product but also value and experience.The concept of visual product aesthetics usually plays a basic role for ideas about the sensorial character of the productsFurthermore, visual aesthetics is a one of the most important factors affecting consumer perception in many ways, Centrality of visual product aesthetics also indicates the durable interest that creates the aesthetic advantages provided from the product. Centrality of visual product aesthetics includes four related components. Degree of Aesthetic aspects.

- The value offered by the product appearance.
- The acumen, ability of recognition, classify and product design evaluation.
- The degree of response to visual design forms of products.
- Definiteness in visual aesthetics that influences product choice and sense of satisfaction in purchasing.

(iii) Marketing aspect: - whenever new product is launched in the markets, it passes through the so many stages for survival in the markets. Called marketing aspect of the product. The elements of the marketing aspect are given below. Marketing function people are frontline representative of the company and provide insights to real product needs of customers.

Table 13.2: Marketing Aspect

s.no	Marketing Aspect	Elements -1	Elements -2
1	**Advertising**	Direct Marketing	Loyalty Marketing
2	**Branding**	Event Planning	Market Research
3	**Copyrighting**	Graphic Design	Marketing Communications
4	**CRM**	Internet Marketing	Media Relations
5	**Merchandising**	New-Product Development	Pricing
6	**Product Management**	Promotion	Public Relations
7	**Sales Management**	Sales Support	Search Optimization
8	**Media Optimization**	Strategic Planning	Supply-Chain Management

(iv) Production aspect: -Any Successful organizations have well defined and efficient line function and support function. Production aspect comes under the category of line function which directly affects customer experience and there by future of organization itself .Aim of production function is to add value to product or service which will create a strong and long lasting customer relationship or association. And this can be achieved by healthy and productive association between Marketing and Production people. An effective planning and control on production parameters to achieve or create product & value for customers is recognize as production aspect .

13.9. Profit and Competitiveness

(i) Profits: - The aim of any business is to make profit. If the cost of production and advertising are too high then the profit margins continue to shrink. Keeping these expenses on the lower side is possible. A good way to do this is by using cheaper raw materials to make high quality products, and advertising through better targeting media. This way, high quality and effective advertising, become expenses that a manufacturer is willing to pay in exchange for future rewards , When a firm has the opportunity to make a profit, this provides an incentive for them to go ahead and enter the market. For example, if a company is thinking about producing bubblegum, and they learn that they can make money doing so, they will use that money incentive to enter the competitive market and begin to produce the bubblegum .

(ii) Competitive Market: - A competitive market is one where there are numerous producers that compete with one another in hopes to provide goods and services us, as consumers, want and need. In other words, not one single producer can dictate the market. Also, like producers, not one consumer can dictate the market either. This concept is also true where price and quantity of goods are concerned. One producer and one consumer can't decide the price of goods or decide the quantity that will be produced. A great example of competitive market is farming. There are thousands of farmers and not one of them can influence the market or the price based on how much they grow. All the farmer can do is grow the crop and accept whatever the current price is for that product. They do not get to determine the price they want to sell the crop for.A market with a large number of buyers and a large number of sellers, such that no single buyer or seller is able to influence the price or any other aspect of the market -- no one has any market control. A competitive market achieves efficiency in the use of our scarce resources if there are no market failures present.

Competitiveness: - it is defined as how effectively an organization *meets the needs* of customers *relative to others* that offer similar goods or services, Organizations compete through some combination of their marketing and operations functions, consists following question

- What do customers want?
- How these customers can needs best be satisfied?

Activities under this heading are oriented towards the removal of supply-side constraints, the promotion of the manufacture of products with high export potential and the provision of assistance related to:

Developing Productive Capacities

- Developing a conducive policy environment for investment and private sector development.
- Identifying key export areas facing supply-side constraints and value chain analysis.
- Upgrading industrial structures and mechanisms for value addition.
- Advising on product design, technology, upgrading and quality control.
- Establishing technology support institutions to improve technology acquisition.
- Improving business efficiency and performance, especially quality management.
- Introducing energy-saving, cleaner technologies, minimizing waste and utilizing by-product.
- Enhancing capacity to meet international standards and client quality and safety requirements.

Factors Affecting Product Design

Following factors are very important, that's affect the product design.

(i) **Correct Team Selection:** This is very essential to get the correct team in place which has expert designers who are not only aware and comfortable with technology but also understanding of customer expectation.

(ii) **Customer Involvement:** Involvement of customer in product design and testing can provide insight into the direction of the project

(iii) **Prototyping and testing:** Product design is high risk concept as it involves commitment of capital and man-power; therefore, it is imperative that extensive prototyping and testing are done with customer and market.

(iv) **Raw Material:** It is essential that raw material to be used in the production meets the quality standards of the end product. Furthermore, procurement system needs to be in place to ensure continuous, cost effective supply.

(v) **Production method and process layout:** Feasibility of production method and process layout determines future success of the product.

(vi) **External Factors:** Environmental and government regulations plays an important part in product design. And these norms are updated from time to time, so product design should have the flexibility to adapt.

13.10. Break-Even Analysis

Break-Even Analysis is perform for the cost based pricing analyses of the products. The break-even price is the price that will produce enough revenue to cover all costs at a given level of production. Total cost can be divided into fixed and variable (total cost = fixed cost + variable cost). Recall that fixed cost does not change as the level of production goes up or down. for example rent paid for the building. No cost is fixed in the long run, but in the short run, many expenses cannot idealistically be changed. Variable cost does change as production is increased ordecreased. For example

- The cost of raw material to make the product will vary with production.
- A second shortcoming of break-even analysis is it assumes that variable costs are constant. However, wages will increase with overtime and shipping discounts will be obtained.
- Third, break-even assumes that all costs can be neatly categorized as fixed or variable. Where advertising expenses are entered, break-even analysis will have a significant impact on the resulting break-even price and volume.

(i) Variable Costs: Variable cost changes according to the quantity of a good or service being produced. It includes inputs like labor and raw materials. Variable costs are also the sum of marginal costs over all of the units produced, For exampleIn the case of a clothing manufacturer, the variable costs would be the cost of the direct material (cloth) and the direct labor. The amount of materials and labor that is needed for each shirt increases in direct proportion to the number of shirts produced. The cost "varies" according to production.

(ii) Fixed Costs: Fixed cost is incurred independent of the quality of goods or services produced. They include inputs (capital) that cannot be adjusted in the short term, such as buildings and machinery. Fixed costs tend to be time related costs, including salaries or monthly rental fees. An example of a fixed cost would be the cost of renting a warehouse for a specific lease period.

(iii) Economic Cost: Economic cost is the sum of all the variable and fixed costs plus opportunity costs.

(iv) Total Cost: The total cost is the total economic cost of production. It consists of variable costs and fixed costs. Total cost is the totalopportunity cost of each factor of production as part of its fixed or variable costs.

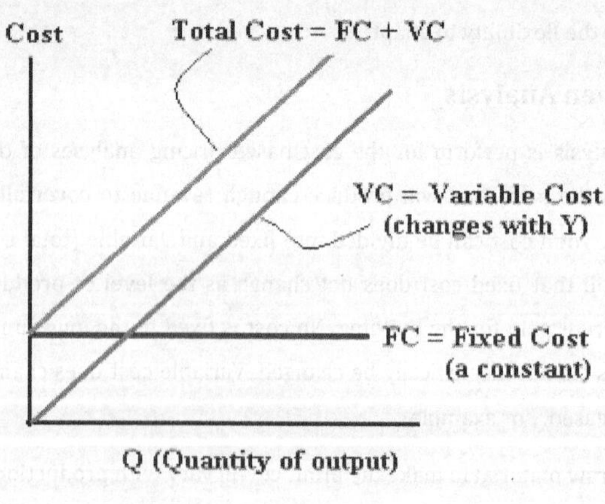

Diagram 13.4: Relationship between Variable and Fixed Cost

In Break-Even Analysis, Let be Q_{BE} denote the break-even output level.

According to definition

TR (at QBE) = TC (at QBE)

or **TR** (at QBE) = **TFC + TVC** (at QBE) ....equ – 1

See A break-even diagram with constant price and constant unit cost.

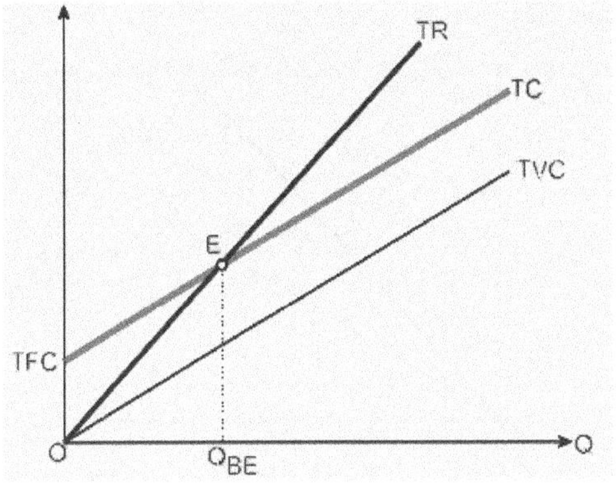

Diagram 13.5: A Break-Even Diagram with Constant Price and Constant Unit Cost

The **break-even condition** (equ--1) holds true for *any* cost and demand functions.Hence, in general, when costs and demand are complex, the analysis of thiscondition might not be any simpler than the analysis of profit maximization.

Assume that price and AVC are constant, then (equ -- 1) may be written as

$$P.Q_{BE} = TFC + AVC.QBE$$

$$\mathbf{Q_{BE} = TFC / (P - AVC)} \quad ,,,,,,,,,,,,,,,, \text{equ} ---------------- 2$$

Where The difference (P – AVC) is often called the **average contribution margin**

Q_{BE} = output quantity

P.Q* = Target profit + TFC + AVC.Q*

Hence

$$Q_{BE} = \frac{TFC + \text{Target Profit}}{P - AVC}$$

Diagram 13.6: Effects of TFC, AVC, and P on Break-Even Output

Variable	Direction of Change	Break-even Output
Total Fixed Cost	Up	Up
Down	Down	Down
Average Variable Cost	Up	Up
Down	Down	Down
Product Price	Up	Down
Down	Up	Up

Break-Even Diagram or Break-Even Chart

The total revenueand total cost curves are straight lines because price and AVC are assumed to be constant.

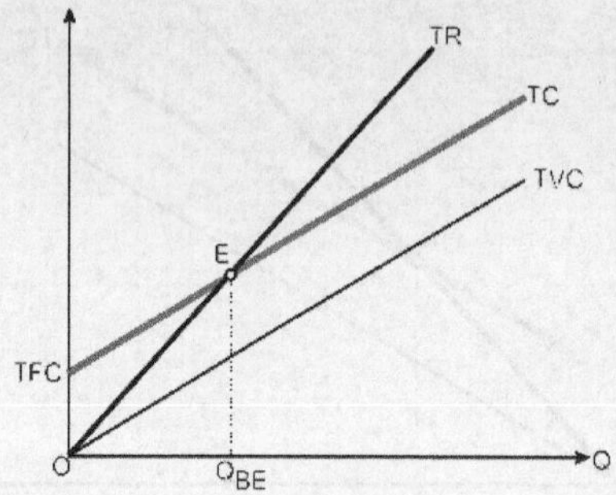

Diagram 13.7: A Break-Even Diagram with Constant Price and Constant Unit Cost

The break-even diagram can be employed to see the effects of various exogenous changes on the break-even point.The break-even diagram can also be modified to determine the output required to meet a non-zero target profit level. Effects on this output due to various changes can also be similarly studied with this diagram.

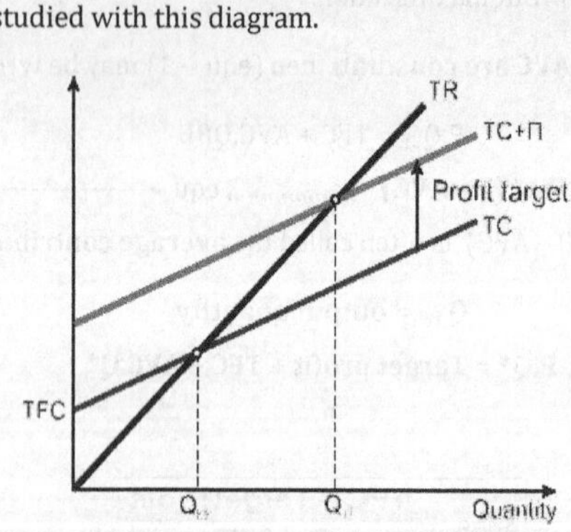

Diagram 13.8: Break-Even Diagram with a Non-Zero Target Profit

Initial change	Which curve is affected	What happens to QBE
Increase in output price	TR curve, counterclockwise	Decrease
Increase in the price of a variable input	TVC and TC curves, both counterclockwise	Increase
Higher TFC	TFC curve, parallel-shift up	Increase

(i) Degree of operating leverage :- The Degree of Operating Leverage (DOL) at a given output level Q is the percentage change in total profit that results from a one-percent change in units sold:

$$\text{Degree of Operating Leverage} = \frac{\text{Change in Profit (in percentage)}}{\text{Change in Output (in percentage)}}$$

$$DOL = \frac{TFC}{Profit} = \frac{TR - TVC}{TR - TC} = \frac{TR - TVC}{TR - TVC - TFC}$$

Possible Outcomes

- The DOL generally depends on the output level, and is equal to zero at the profit maximizing output.
- The DOL is negative (positive) below (above) the break-even level. For example if, as a result of a 1% increase in output, profit increases by 3% then the DOL is 3%/1% = 3.

(ii) Conclusion of Break-even analysis: - following conclusion are achieved.

- The main purpose of break-even analysis is to determine the output level at which TR is equal to TC. Break-even analysis is easy to do (and thus its popularity) if both the product price and the average cost are constant. Graphically, the break-even output is where the TR curve and the TC curve intersect.
- Break-even analysis is a popular tool to examine the effects of TFC and TVC on breakeven output level and on the degree of operating leverage. The degree of leverage expresses the sensitivity of total profit with respect to output .

Advantages of Break-Even Analysis

(i) It is simple to conduct and understand.

(ii) It shows profit and loss at different levels of output.

(iii) It can cope with changing circumstances. e.g. the following changes in the business environment can be shown in a break even chart.

Internal Factor Affecting Beak-Even Point (**B.E.P.).**

	Cause	Effect
1	Employing extra sales staff	Fixed costs rise, so total costs rise and beak-even point rises.
2	Price increase	Revenue rises more steeply, break-even point falls.
3	Automation replaces direct labour	Fixed costs rise while direct costs fall, effect on break-even point unclear.

External Factor Affecting Beak-Even Point (**B.E.P.)**

	Cause	Effect
1	Recession cuts demand	Break-even point unaffected, though safety margin is reduced.
2	Price war forces price cut	Revenue line rises less steeply, break-even point rises.
3	Inflation pushes up direct costs.	Direct and total cost line rise more steeply, break-even point rises.

Disadvantages of Break-Even Analysis

(i) It assumes that all output is sold at the given price (this may well be untrue) .

(ii) Although it can cope with changes in circumstances, these factors change regularly reducing its usefulness as a forecasting tool.

(iii) The model assumes that costs increase constantly and do not benefit from economies of scale. If the firm obtains purchasing economies of scale then its total cost line will no longer be straight.

(iv) Break-even analysis is only as good as the data upon which it is based. Poor quality data will lead to inaccurate conclusions being drawn.

Question for Discussion

1. Explain steps needed for ergonomic approach to design.
2. List and explain various economic factors influencing design
3. How test marketing of new product to be launched is done? Explain with examples.
4. How economic analysis is. done after evaluation of new product ideas concepts.
5. Explain and discuss generation, screening and development of new product ideas
6. What are the different kinds of cost procedures? How cost reduction is done?
7. Discuss structural unit for new product development. What are the functions of new product development unit?
8. How creative ideas are located and how right product is selected?
9. How curiosity and imagination is helpful in designing? Give examples.
10. What is systematic procedure for product innovation?
11. List and explain various economic factors influencing product design.How test marketing of New product to be launched is done? Explain with examples .
12. Discuss economic aspects of product development
13. What are the various manufacturing aspects of product development?

14. What do you understand by product development? Discuss in manufacturing aspect of product development.
15. What is the role of designer in product development ? How product promotion is carried out ?
16. Enlist the essential factors of product design.
17. Discuss the role of aesthetics in the product design.
18. Write short notes on following.
    a) What do you understand by functional aspect of design practice?
    b) Define the term 'Product development'.
    c) Discuss the importance of product promotion.
19. Discuss, the general procedure adopted by the industrial designer in connection with the product design and development.
20. What is the role of a designer in the development of a product ? Discuss in brief.
21. What do you meant by the product graphics? Discuss the basic principles of the graphic design in detail.
22. What do you understand by Manufacturing and economic aspects of product development.
23. Define Product Development. Explain the product development process in detail.
24. What is difference between creative idea and innovation? Discuss various factors affecting product design.
25. Discuss various techniques of product designing. Explain in detail the product design methodology.
26. How ergonomic considerations are useful in product design? What are various elements and concepts of visual design?
27. What are the principal requirements of good product design? What is importance of product design in industry ?
28. How analysis of functions of product is done ? also list the ways by which functions of product are evaluated ?
29. How estimation of product cost is done? Explain.
30. What do you understand by generation of idea? Give suitable example
31. How economic analysis.is done after evaluation of new product ideas/ concepts.
32. Explain standardization, simplification and specialization in relation of product design & Development
33. What is break even analyses? Explain about break even charts.

Refrences

1. Dehmand and Khosrow, "Quality control, robust design, and the Taguchi method", London, Wadsworth, 1988.
2. Edosomwan and Johnson, "People and product management in manufacturing", msterdam, Elsevier Science Publishers, 1990.
3. El Wakil and D. Sherif, "Processes and design for manufacturing", Englewood Cliffs, New Jersey, Prentice Hall, 1989.
4. Moore and S. Lewis, "How to design interventions for improved maintenance management", Japan Industrial Design Promotion Organization (JIDPO), Good Design Products 1989, Tokyo: JIDPO, 1990.
5. L.P. Sullivan, "Quality function deployment: A system to assure that customer needs drive the product design and production process", Quality Progr, Pp. 39-50, 1986.
6. A. Shugaar, "Ecodesign: New products for a greener culture", Int Herald Trib, 1990.
7. B. Shackel, "Ergonomics in design for usability", In People and Computer: Design for Usability, edited by MD Harrison and AF Monk. Cambridge: Cambridge Univ. Press, 1986.
8. Safety of Machinery: Ergonomic Design Principles, Part 1: Terminology and General Principles, Brussels: CEN.

CHAPTER 14

CAPACITY PLANNING & FORECASTING

14.0. Introduction

Design of the production system involves planning for the inputs, conversion process and outputs of production operation. The effective management of capacity is the most important responsibility of production management. The objective of capacity management (i.e., planning and control of capacity) is to match the level of operations to the level of demand. Capacity planning is to be carried out keeping in mind future growth and expansion plans, market trends, sales forecasting, etc. It is a simple task to plan the capacity in case of stable demand. But in practice the demand will be seldom stable. The fluctuation of demand creates problems regarding the procurement of resources to meet the customer demand. Capacity decisions are strategic in nature. Capacity is the rate of productive capability of a facility. Capacity is usually expressed as volume of output per period of time. Production managers are more concerned about the capacity for the following reasons: Sufficient capacity is required to meet the customers demand in time. Capacity affects the cost efficiency of operations. Capacity affects the scheduling system. Capacity creation requires an investment. Capacity planning is the first step when an organization decides to produce more or new products.

Capacity is defined as the ability to achieve, store or produce.

It is also defined as:

For an organization, capacity would be the ability of a given system to produce output within the specific time period or maximum production capacity, which can be attained within a normal working schedule.

Capacity planning is essential to be determining optimum utilization of resource and plays an important role decision-making process, for example, extension of existing operations, modification to product lines, starting new products, etc.

14.1. Measurement of Capacity Planning

The capacity of the manufacturing unit can be expressed in number of units of output per period. In some situations measuring capacity is more complicated when they manufacture multiple products. In such situations, the capacity is expressed as man-hours or machine hours. Diagram shows the relationship between the capacity planning & output .

Expressed as man-hours or machine hours .diagram shows the relationship between the capacity planning & output.

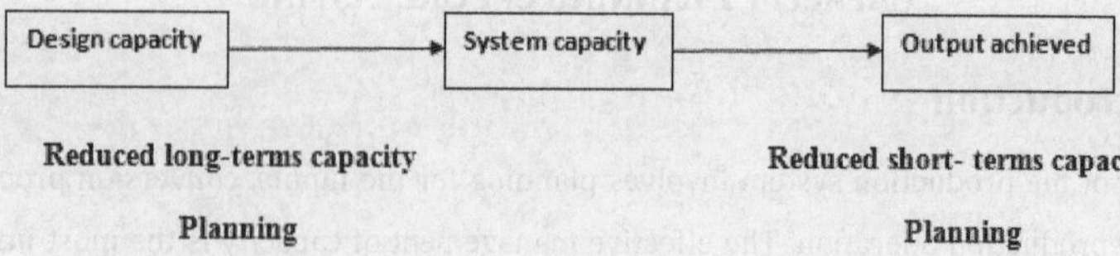

Diagram 14.1: Relationship Diagram Capacity Planning & Output

(i) **Design capacity:** Designed capacity of a facility is the planned or engineered rate of output of goods or services under normal or full scale operating conditions.

(ii) **System capacity:** System capacity is the maximum output of the specific product or product mix the system of workers and machines is capable of producing as an integrated whole. System capacity is less than design capacity or at the most equal, because of the limitation of product mix, quality specification, breakdowns.

(iii) **Licensed capacity:** Capacity licensed by the various regulatory agencies or government authorities. This is the limitation on the output exercised by the government.

(iv) **Installed capacity:** The capacity provided at the time of installation of the plant is called installed capacity.

(v) **Rated capacity:** Capacity based on the highest production rate established by actual trials is referred to as rated capacity.

14.2. Process of Capacity Planning

Capacity planning is concerned with defining the long-term and the short-term capacity needs of an organization and determining how those needs will be satisfied.

Capacity planning decisions are taken based upon the consumer demand and this is merged with the human, material and financial resources of the organization. Capacity requirements can be evaluated from two perspectives—long-term capacity strategies and short-term capacity strategies.

Long Terms or Long Range Capacity Strategies

Long-term capacity requirements are more difficult to determine because the future demand and technology are uncertain. Forecasting for five or ten years into the future is more risky and difficult. Even sometimes company's today's products may not be existing in the future. Long range capacity requirements are dependent on marketing plans, product development and lifecycle of the product.

Factors affecting Long terms or long range capacity strategies :- following are some important factors which affects the Long terms or long range capacity planning.

- **Multiple products:** Company's produce more than one product using the same facilities in order to increase the profit. The manufacturing of multiple products will reduce the risk of failure. Having more than one product helps the capacity planners to do a better job. Because products are in different stages of their life-cycles, it is easy to schedule them to get maximum capacity utilization.
- **Phasing in capacity:** In high technology industries, and in industries where technology developments are very fast, the rate of obsolescence is high. The products should be brought into the market quickly. The time to construct the facilities will be long and there is no much time as the products should be introduced into the market quickly.
- **Phasing out capacity:** The outdated manufacturing facilities cause excessive plant closures and down time. The impact of closures is not limited to only fixed costs of plant and machinery. Thus, the phasing out here is done with humanistic way without affecting the community. The phasing out options makes alternative arrangements for men like shifting them to other jobs or to other locations, compensating the employees, etc

(i) **Medium Term Capacity:** The strategic capacity planning undertaken by organization for 2 to 3 years of a time frame is referred to as medium term capacity planning.

(ii) **Short terms or short range capacity strategies:** Managers often use forecasts of product demand to estimate the short-term workload the facility must handle. Managers looking ahead up to 12 months, anticipate output requirements for different products, and services. Managers then compare requirements with existing capacity and then take decisions as to when the capacity adjustments are needed. For short-term periods of up to one year, fundamental capacity is fixed. Major facilities will not be changed. Many short-term adjustments for increasing or decreasing capacity are possible. The adjustments to be required depend upon the conversion process like whether it is capital intensive or labour intensive or whether product can be stored as inventory. Capital intensive processes depend on physical facilities, plant and equipment. Short-term capacity can be modified by operating these facilities more or less intensively than normal.

The Shorts Terms or Short -Range Capacity Strategies Includes Following Steps

- **Inventories:** Stock of finished goods during slack periods to meet the demand during peak period.
- **Backlog:** During peak periods, the willing customers are requested to wait and their orders are fulfilled after a peak demand period.
- **Employment level (hiring or firing):** Hire additional employees during peak demand period and lay off employees as demand decreases.
- **Employee training:** Develop multi-skilled employees through training so that they can be rotated among different jobs. The multi-skilling helps as an alternative to hiring employees.
- **Sub-contracting:** During peak periods, hire the capacity of other firms temporarily to make the component parts or products.
- **Process design:** Change job contents by redesigning the job.

14.3. Approach for Capacity Decisions

In general, a four step procedure is followed by the managers to make sound decisions. These steps are given below.

(i) Estimate future capacity requirements.

(ii) Find gaps by comparing the requirements with available capacity.

(iii) Build up alternate plans for filling the gaps.

(iv) Evaluate each alternative, qualitatively and quantitatively both, and make a final choice.

The Short Description of the Capacity Decisions Are Given Below:

(i) Estimate future capacity requirements: The basis for estimating long term capacity desires forecasts of demand, competition, productivity and technological changes that extend well into the future. Unfortunately, larger the time of forecast, larger is the uncertainty involved in forecast of demand. Let capacity be expressed as the number of available machines at an operation. While just one product or service is being processed, let M number of machines is required. It is computed by the following relationship: Number of machines required = Processing hours required for year's demand/ Hours available from one machine per year

$$M = DP / (N ((1-C)/100)$$

Where, D = number of units (customers) forecast per year.

P = processing time (in hours per unit or customers),

N = total number of hours per year during which the process operates.

C = desired capacity cushion.

The processing time (P) based on the process and methods decided to do the work.

N is the total number of hours that are available in the year, multiplied by a proportion that accounts for the wanted capacity cushion C. Also the capacity cushion 'C' is converted from the percent to a proportion by dividing it with 100.If multiple products and services are involved, then extra time is required to change over from one product or services to the next. Setup time is derived from process decisions, as is processing time. The total setup time is established by dividing the number of units estimated per year 'D' by the number of units made in each of lot, which provide the number of setups per year, and multiplying by the time per setup then. Number of machines required = (Processing and set hours required for year's demand summed over all products) /(Hours available from one machine per year, after deducting wanted cushion).

$$M = [(DP + (D/Q)s]_{\text{Product 1}} + [DP + (D/Q)s]_{\text{Product 2}} + \ldots\ldots\ldots\ldots + DP + (D/Q)s]_{\text{Product n}}) / N[1-(c/100)]$$

Where Q = number of units in each lot, and, s = setup time (in hours) per lot.

- **Identify Gaps:** The difference between projected demand and current capacity is called as the capacity gaps. Discovering gaps needs use of the right capacity measure. Complications can arise due to multiple operations and several resource inputs involved. The right measure of capacity was the number of departures instead of the number of seats.Overall capacity might be increased by expansions of capacity of some operations. Though, if one operation is an obstacle, the capacity may be expanded only if the capacity of the obstacle operation is extended.
- **Develop Alternatives:** In the next step, the alternative plans ought to be developed to cover the projected gaps. One of the alternatives is base case, in which firms do not have to do anything and simply lose orders from any demand that exceeds current capacity. Several timing and sizing options for adding up new capacity, by including the expansionist and wait and see strategies are also present in the other alternatives.

(i) Evaluate the Alternatives: These Alternatives are of two types..

- **Qualitative Concerns:** The managers ought to be very thoughtful at how each alternative fits the total capacity strategy and other aspects of the business not covered by the financial study. Specific concerns are about the uncertainties about demand, technological change, competitive reaction, and the cost estimates. Some of the factors may not be quantified therefore have to be accessed on the basis of judgment and experience. Others may be quantified and therefore each alternative may be analyzed using different assumptions about the future. Another type of the alternative is "what-if" analysis, which permits the managers to get an idea of each alternative's implication before making a final choice.

- **Quantitative Concerns:** The manager approximates the change in cash flows for each of the substitute over the forecast time horizon compared to the base case. The difference among the flows of funds in and out of an organization over a definite time interval, including revenues, costs, changes in properties and legal responsibilities are called as cash flow.

14.4. Capacity Planning Technique

There are four procedures for capacity planning; capacity planning using overall factors (CPOF), capacity bills, resource profiles, and capacity requirements planning (CRP). The first three are rough-cut approaches (involving analysis to identify potential bottlenecks) that can be used with or without manufacturing resource planning (MRP) systems. CRP is used in conjunction with MRP systems.

(i) Capacity using overall factors is a simple, manual approach to capacity planning that is based on the master production schedule and production standards that convert required units of finished goods into historical loads on each work center.

(ii) Bills of capacity are a procedure based on the MPS. Instead of using historical ratios, however, it utilizes the bills of material and routing sheet includes the sequence, work centers, manufacture of part, the setup and run time.

(iii) Capacity requirements can then be determined by multiplying the number of units required by the MPS by the time needed to produce each. Resource profiles are the same as bills of capacity, except lead times are included so that workloads fall into the correct periods.

(iv) Capacity requirements planning (CRP) is only applicable in firms using MRP or MRP II. CRP uses the information from one of the previous rough-cut methods, plus MRP outputs on existing inventories and lot sizing. The result is a tabular load report for each work center or a graphical load profile for helping plan-production requirements. This will indicate where capacity is inadequate or idle, allowing for imbalances to be corrected by shifts in personnel or equipment or the use of overtime or added shifts.

Goal of Capacity Planning: - The ultimate goal of capacity planning is to meet the current and future level of the requirement at a minimal wastage. The three types of capacity planning based on goal are lead capacity planning, lag strategy planning and match strategy planning.

Objective of capacity planning: -Following are very important of capacity planning.

- When you complete this section you should be able to: have defined capacity and understood what decisions have to be taken in determining capacity.
- Explored options which a company has for increasing capacity.
- Examined the approaches to capacity planning and the factors which need to be taken into.
- Have taken a brief look at forecasting demand including qualitative approaches and time series forecasting.
- Understand the application of some decision-making processes involved in capacity planning.

Factors Affecting Capacity Planning :- Effective capacity planning is dependent upon factors like production facility (layout, design, and location), product line or matrix, production technology, human capital (job design, compensation), operational structure (scheduling, quality assurance) and external structure (policy, safety regulations).

14.5. Aggregate Planning

An organization can finalize its business plans on the recommendation of demand forecast. Once business plans are ready, an organization can do backward working from the final sales unit to raw materials required. Thus annual and quarterly plans are broken down into labor, raw material, working capital, etc. requirements over a medium-range period (6 months to 18 months). This process of working out production requirements for a medium range is called aggregate planning.

Factors Affecting Aggregate Planning

Aggregate planning is an operational activity critical to the organization as it looks to balance long-term strategic planning with short term production success. Following factors affecting aggregate planning process.

- Complete information is required about available production facility and raw materials.
- A solid demand forecast covering the medium-range period.
- Financial planning surrounding the production cost which includes raw material, labor, inventory planning, etc.

- Organization policy around labor management, quality management, etc.

For Successful Aggregate Planning to be a Success, Following Steps

- An aggregate demand forecast for the relevant period.
- Evaluation of all the available means to manage capacity planning like sub-contracting, outsourcing, etc.
- Existing operational status of workforce (number, skill set, etc.), inventory level and production efficiency.

Aggregate Planning Strategies

There are three types of aggregate planning strategies available for organization to choose from. They are as follows.

(i) Level Strategy: As the name suggests, level strategy looks to maintain a steady production rate and workforce level. In this strategy, organization requires a robust forecast demand as to increase or decrease production in anticipation of lower or higher customer demand. Advantage of level strategy is steady workforce. Disadvantage of level strategy is high inventory and increase back logs.

(ii) Chase Strategy: As the name suggests, chase strategy looks to dynamically match demand with production. Advantage of chase strategy is lower inventory levels and back logs. Disadvantage is lower productivity, quality and depressed work force.

(iii) Hybrid Strategy: Hybrid strategy looks to balance between level strategy and chase strategy.

Aggregate planning helps achieve balance between operation goal, financial goal and overall strategic objective of the organization. It serves as a platform to manage capacity and demand planning.

14.6. Demand Forecasting

Demand Forecasting refers to the process of predicting the future demand for the firm's product. In other words, demand forecasting is comprised of a series of steps that involves the anticipation of demand for a product in future under both controllable and non-controllable factors.

The business world is characterized by risk and uncertainty, and most of the business decisions are taken under this scenario. An organization come across several risks, both internal or external to the business operations such as technology, attrition, unrest, employee grievances, recession, inflation, modifications in the government laws, etc .

Some Important Definitions of Demand Forecasting Are Given Below:

According to Evan J. Douglas,

Demand estimation (forecasting) may be defined as a process of finding values for demand in future time periods.

According to Cundiff and Still,

Demand forecasting is an estimate of sales during a specified future period based on proposed marketing plan and a set of uncontrollable and competitive forces.

Some others important definition,

Demand forecasting is nothing but it is a prediction, projection, or estimate of some future activity, event, or occurrence

OR

Predict the future demand for a company's products or services.

Economic forecasting is defined as,

It is defined as predict a variety of economic indicators, like money supply, inflation rates, interest rates, etc.

Technological forecasting is defined as,

Predict rates of technological progress and innovation

Demand forecasting helps in an organization to take various business decisions, such as planning the production process, purchasing raw materials, managing funds, and deciding the price of the product. An organization can forecast demand by making own estimates called guess estimate or taking the help of specialized consultants or market research agencies.

Demand forecasting Predicting the future demand for a product helps the organization in making decisions in one of the following areas:

- Planning and scheduling the production and acquiring the inputs accordingly.
- Making the provisions for finances.
- Formulating a pricing strategy.
- Planning advertisement and implementing it.

Demand forecasting holds significance in the businesses where large-scale production is involved. Since the large-scale production requires a long gestation period, a good deal of forward planning should be done. Also, the potential future demand should be estimated to avoid the conditions of overproduction and underproduction. Most often, the firms face a question of what would be the future demand for their product as they must acquire the input (labor and raw material.

The objective of demand forecasting is attained only when the forecasting is done systematically and scientifically. Thus, the following steps in demand forecasting are followed to facilitate a systematic estimation of future demand for product:

- Specifying the Objective.
- Determining the Time Perspective.
- Choice of method for Demand Forecasting.
- Collection of Data and Data Adjustment.
- Estimation and Interpretation of Results.

Thus, demand forecasting is a systematic process that assumes greater significance in large-scale producing firms.

Demand forecasting may not be a genuine issue for the small-scale firms which supply a small portion of total demand or produces the product that caters to the short demand or seasonal demand. Such firms can plan their production based on the business skills and their past experiences.

Demand: Demand is defined as the quantity of goods or services that consumers are able and willing to buy at a given price at a particular time. The law of demand provides that, if all other market factors remain constant, the demand for goods and services increases as their price decreases.

There are three types of demand.

(i) Effective demand

(ii) Latent demand and

(iii) Derived demand

- Effective demand is when the consumers' desire to purchase a particular good or service is backed up by their ability to pay for it.
- Latent demand exists when consumers are willing to buy a particular product, but they lack the purchasing power to afford the product .
- Derived demand exists when the demand for a specific product is connected to the demand for a related product.

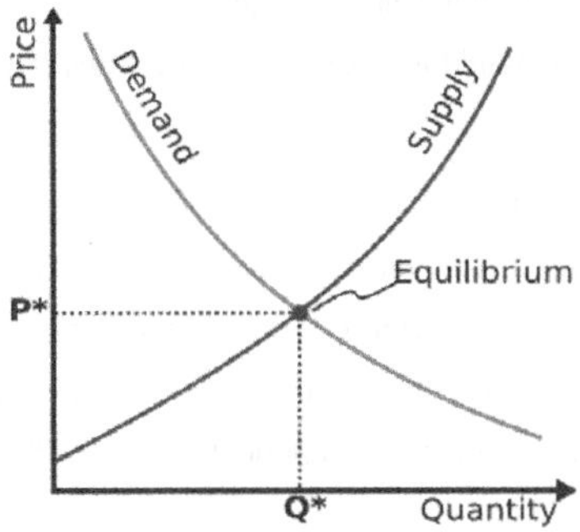

Diagram 14.2: Demand Supply Equilibrium Diagram

A demand curve is used to show the relationship between the quantity of goods or services demanded and the price of those goods over a specific period .

It Concluded that,

- When the price of a particular good falls, making the product relatively cheaper than an alternative product.
- The income effect manifests itself when the price of a specific good falls mainly because the buyer can maintain the same consumption with less expenditure.

14.7. Demand Patterns

When forecasting a particular activity, the most important requirement for generating a high quality forecasts is a fundamental understanding of the underlying pattern of the activity. Depending on its behavior over time, the underlying pattern can be classified as regular or irregular. If the pattern is regular, the future values can be predicted based on the past or historical values. Common regular patterns are a constant pattern, a trend pattern, seasonal pattern, or a combination of trend and seasonal pattern. Classical decomposition decomposes the pattern in a trend, seasonal, and random component. As long as the random variations are small compared to the underlying pattern, accurate forecasts can be obtained by popular mathematical forecasting techniques, such as regression and time series analysis. A pattern can also be irregular, when it is a singular occurrence, intermittent, or highly variable. The pattern is said to be "lumpy". Such variables are typically very difficult to forecast with any accuracy using popular mathematical models. Finally, dependent or derived demand patterns constitute a special case when an independent variable causes predictable behavior for a number of other dependent variables. In this case the value of the demand for the final product is stochastic and forecasted with the techniques described above. But once the demand for the final product is known, the demand for all the subcomponents, raw materials, and resources is known with certainty.

The derived demand is usually obtained by recursively exploding the bill of materials(BOM) of the final product to all of its individual product components. For example, the forecasted independent demand for a car model causes a dependent demand shifted back in time for the associated car engine. Most popular forecasting techniques are based on the assumptions of independent behavior with a random component . The methods are not applicable for forecasting dependent behavior.

The future logistics activities are subject to the actions of many different organizations in and outside the supply chain and to inherent uncertainty caused by truly random events Therefore, the forecasting should be a collaborative effort among many groups in the supply chain. Marketing can provide advance estimates of the impact of future promotions and product introductions. The expertise of long-term employees may judge the impact that similar conditions have had in the past. The end result of the forecasting process should a consensus-based forecast that is used for all the planning by all organizations in the supply chain.

To control variations in demand for a particular service, managers need to determine the answers to a series of important questions about the patterns of demand and their underlying causes. As you think about some of the seemingly "random" causes of fluctuations in market demand, consider how rain and cold affect the use of indoor and outdoor recreational or entertainment services.

Random fluctuations are usually caused by factors beyond management's control. Where relevant, it's useful to record weather conditions and other special factors (a strike, an accident, a big convention in town, a price change, launch of a competing service, etc.) that might have influenced demand. Detailed market analysis may sometimes reveal that a predictable **demand cycle** is concealed within a broader, seemingly random pattern.

(i) Analyses of demand as per MarketSegment: No strategy for smoothing demand is likely to succeed unless it's based on an understanding of why customers from a specific market segment choose to use the service when they do. Keeping good records of customer transactions helps when it comes to analyzing demand patterns by market segment. Computer-based services can track customer consumption patterns by date and time of day automatically and enter them into a company's database. Some services, such as telephone and electrical, even have the ability to track subscriber consumption patterns by time of day. Although this data may not always yield specific information on the purpose for which the service is being used, it is often possible to make informed judgments about the volume of sales generated by different user groups.

(ii) Multiple Influences on Demand: Periodic cycles influencing demand for a particular service can vary in length from one day to one year. In many instances, multiple cycles may operate simultaneously. For example, demand levels for public transportation may differ by time of day (highest during commute hours), day of week (less travel to work on weekends but more leisure travel), and season of year (more travel by tourists in summer). The demand for service during the peak period on a Monday in summer may be different from the level during the peak period on a Saturday in winter, reflecting day-of-week and seasonal variations jointly. Thus demand level (at a given price) and customer profiles (with resulting differences in needs and expectations) are different.

Factor Influences the Demand: Following factor generally influences the demand pattern.

(i) Taste and preference of the consumers.
(ii) Income of the people.
(iii) Change in prices of related good "s.
(iv) Advertisement expenditure.
(v) Consumers strength in markets.
(vi) Consumers' demand of products in future with respect to prices.

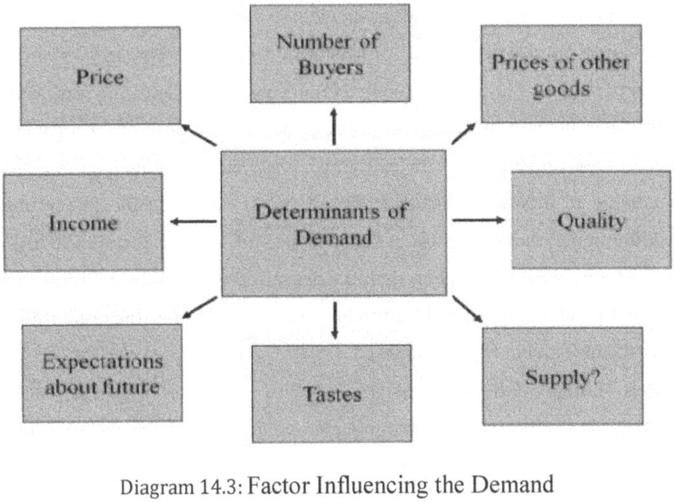

Diagram 14.3: Factor Influencing the Demand

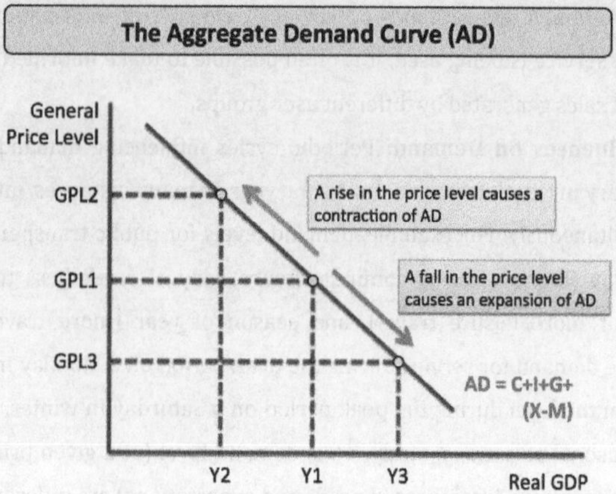

Diagram 14.4: Aggregate Demand Curve

Some important aspects of demand forecasting: These aspects are very important from subject point of view. The description is given below.

(i) **Fulfilling objectives:** It implies that every business unit starts with certain pre-decided objectives. Demand forecasting helps in fulfilling these objectives. An organization estimates the current demand for its products and services in the market and move forward to achieve the set goals.

(ii) **Preparing thebudget:** The Demand forecasting **plays** a crucial role in making budget by estimating costs andexpected revenues.

(iii) **Stabilizing employment and production:** Helps an organization to control its production and recruitment activities. Producing according to the forecasted demand of products helps in avoiding the wastage of the resources of an organization. This further helps an organization to hire human resource according to requirement.

(iv) **Expansion of business organizations:** It Implies that demand forecasting helps in deciding about the expansion of the business of the organization. If the expected demand for products is higher, then the organization may plan to expand further. On the other hand, if the demand for products is expected to fall, the organization may cut down the investment in the business.

(v) **Taking Management Decisions:** It also helps in making critical decisions, such as deciding the plant capacity,determiningthe requirement of raw material, and ensuring the availability of labor and capital.

(vi) **Evaluating Performance:** Helps in making corrections. For example, if the demand for an organization's products is less, it may take corrective actions and improve the level of demand by enhancing the quality of its products or spending more on advertisements.

(vii) **Helping Government:** Enables the government to coordinate import and export activities and plan international trade.

Factors Influencing Demand Fore-casting: Demand forecasting is a proactive process that helps in determining what products are needed where, when, and in what quantities. There are a number of factors that affect demand fore-casting.

(i) **Prevailing business conditions**: While preparing demand forecast it becomes necessary to study the general economic conditions very carefully. These include the price level changes, change in national income, per-capita income, consumption pattern, savings and investment habits, employment etc.

(ii) **Conditions within the industry:** Every business enterprise is only a unit of a particular industry. Sales of that business enterprise are only a part of the total sales of that industry. Therefore, while preparing demand forecasts for a particular business enterprise, it becomes necessary to study the changes in the demand of the whole industry, number of units within the industry, design and quality of product, price policy, competition within the industry etc.

(iii) **Conditions within the firm**: Internal factors of the firm also affect the demand forecast. These factors include plant capacity of the firm, quality of the product, price of the product, advertising and distribution policies, production policies, financial policies etc

(iv) **Factors affecting export trade**: If a firm is engaged in export trade also it should consider the factors affecting the export trade. These factors include import and export control, terms and conditions of export, exim policy, export conditions, export finance etc.

(v) **Types of Goods:** Affectthe demand forecasting process to a larger extent. Goods can be producer's goods, consumer goods, or services. Apart from this, goods can be established and new goods. Established goods are those goods which already exist in the market, whereas new goods are those which are yet to be introduced in the market. Information regarding the demand, substitutes and level of competition of goods is known only in case of established goods. On the other hand, it is difficult to forecast demand for the new goods. Therefore, forecasting is different for different types of goods.

(vi) Economic View-point: Economic point of View Play a crucial role in obtaining demand forecasts. For example, if there is a positive development in an economy, such as globalization and high level of investment, the demand forecasts of organizations would also be positive.

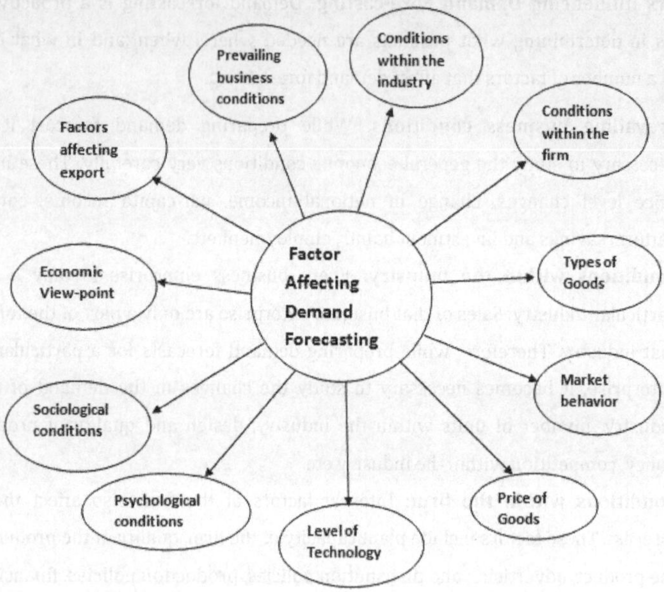

Diagram 14.5: Factors Influencing Demand Fore-casting

(vii) Level of Technology : Level of Technology Constitutes an important factor in obtaining reliable demand forecasts. If there is a rapid change in technology, the existing technology or products may become obsolete. For example, there is a high decline in the demand of floppy disks with the introduction of compact disks (CDs) and pen drives for saving data in computer. In such a case, it is difficult to forecast demand for existing products in future.

(viii) Market behavior: While preparing demand forecast, it is required to consider the market behavior which brings about changes in demand.

(ix) Sociological conditions: Sociological factors have their own impact on demand forecast of the company. These conditions relate to size of population, density, change in age groups, size of family, family life cycle, level of education, family income, social awareness etc.

(x) **Price of Goods:** Acts as a major factor that influences the demand forecasting process. The demand forecasts of organizations are highly affected by change in their pricing policies. In such a scenario, it is difficult to estimate the exact demand of products.

(xi) **Psychological conditions:** While estimating the demand for the product, it becomes necessary to take into consideration such factors as changes in consumer tastes, habits, fashions, likes and dislikes, attitudes, perception, life styles, cultural and religious bents etc.

The concept of elasticity of demand is of great importance in practical life. Its main points are given as under:

(i) **Useful for Business:** It enables the business in general and the monopolists in particular to fix the price. Studying the nature of demand the monopolist fixes higher prices for those goods which have inelastic demand and lower prices for goods which have elastic demand. In this way, this helps him to maximize his profit.

(ii) **Fixation of Prices:** It is very useful to fix the price of jointly supplied goods. In the case of joint products like paddy and straw, the cost of production of each is not known. The price of each is then fixed by its elastic and inelastic demand

(iii) **Helpful to Finance Minister**: It helps the Finance Minister to levy tax on goods. After levying taxes more and more on goods which have inelastic demand, the Government collects more revenue from the people without causing them inconvenience. Moreover, it is also useful for the planning.

(iv) **Fixation of Wages:** It guides the producers to fix wages for labourers. They fix high or low wages according to the elastic or inelastic demand for the labour.

(v) **In the Sphere of International Trade:** It is of greater significance in the sphere of international trade. It helps to calculate the terms of trade and the consequent gain from foreign trade. If the demand for home product is inelastic, the terms of trade will be profitable to the home country.

(vi) **Paradox of Poverty.** It explains the paradox of poverty in the midst of plenty. A bumper crop instead of bringing prosperity may result in disaster, if the demand for it is inelastic. This is specially so, if the products are perishable and not storable.

(vii) **Significant for Government Economic Policies.** The knowledge of elasticity of demand is very important for the government in such matters as controlling of business cycles, removing inflationary and deflationary gaps in the economy. Similarly, for price stabilization and the purchase and sale of stocks, information about elasticity of demand is most useful

(viii) Determination of Price of Public Utilities. This concept is significant in the determination of the prices of public utility services. Economic welfare of the society largely depends upon the cheap availability.

14.8. Principle of Forecasting

Following are some important principle of forecasting that are generally applied.

- Use checklists of categories of information relevant to the forecasting task.
- Establish explicit and agreed criteria for adopting a forecast method
- Keep records of forecasts and use them appropriately to obtain feedback.
- Study data in graphical rather than tabular form when making judgmental fore-casts
- Draw a best-fitting line through the data series when making judgmental forecasts from a graphical display.
- Use more than one way of judging the degree of uncertainty in time-series fore-casts
- Someone other than the person(s) responsible for developing and implementing a plan of action should estimate its probability of success.

Fore Casting Can Be of Three Types, Which Are Explained as Follows

- **Short Period Forecasts : -** it refers to the forecasts that are generally for one year and based upon the judgment of the experienced staff. Short period forecasts are important for deciding the production policy, price policy, credit policy, and distribution policy of the organization.
- **Long Period Fore-casts : -** It refers to the forecasts that are for a period of 5-10 years and based on scientific analysis and statistical methods. The forecasts help in deciding about the introduction of a new product, expansion of the business, or requirement of extra funds.
- **Very Long Period Forecasts :-** it refers to the forecasts that are for a period of more than 10 years. These forecasts are carried to determine the growth of population, development of the economy, political situation in a country, and changes in international trade in future.

Among the aforementioned forecasts, short period forecast deals with deviation in long period forecast. Therefore, short period forecasts are more accurate than long period forecasts.

Level of Fore-casts: Influences demand forecasting to a larger extent. A demand forecast can be carried at three levels, namely, **macro level, industry level, and firm level.**

- At macro level, forecasts are undertaken for general economic conditions, such as industrial production and allocation of national income.
- At the industry level, forecasts are prepared by trade associations and based on the statistical data. Moreover, at the industry level, forecasts deal with products whose sales are dependent on the specific policy of a particular industry.
- On the other hand, at the firm level, forecasts are done to estimate the demand of those products whose sales depends on the specific policy of a particular firm. A firm considers various factors, such as changes in income, consumer's tastes and preferences, technology, and competitive strategies, while forecasting demand for its products.

14.9. Nature of Fore-Casts

Constitutes an important factor that affects demand forecasting. A forecast can be specific or general. A general forecast provides a global picture of business environment, while a specific forecast provides an insight into the business environment in which an organization operates. Generally, organizations opt for both the forecasts together because over-generalization restricts accurate estimation of demand and too specific information provides an inadequate basis for planning and execution.

Steps of Demand Forecasting: The Demand forecasting process of an organization can be effective only when it is conducted systematically and scientifically. It involves a number of steps.

(i) Setting the Objective: Refers to first and foremost step of the demand forecasting process. An organization needs to clearly state the purpose of demand forecasting before initiating it. Setting objective of demand forecasting involves the following:

(a) Deciding the time period of forecasting whether an organization should opt for short-term forecasting or long-term forecasting.

(b) Deciding whether to forecast the overall demand for a product in the market or only-for the organizations own products.

(c) Deciding whether to forecast the demand for the whole market or for the segment of the market.

(d) Deciding whether to forecast the market share of the organization.

(ii) Determining Time Period: Involves deciding the time perspective for demand forecasting. Demand can be forecasted for a long period or short period. In the short run, determinants of demand may not change significantly or may remain constant, whereas in the long run, there is a significant change in the determinants of demand. Therefore, an organization determines the time period on the basis of its set objectives.

(iii) Selecting a Method for Demand Forecasting: Constitutes one of the most important steps of the demand forecasting process Demand can be forecasted by using various methods. The method of demand forecasting differs from organization to organization depending on the purpose of forecasting, time frame, and data requirement and its availability. Selecting the suitable method is necessary for saving time and cost and ensuring the reliability of the data.

(iv) Collecting Data : Requires gathering primary or secondary data. Primary' data refers to the data that is collected by researchers through observation, interviews, and questionnaires for a particular research. On the other hand, secondary data refers to the data that is collected in the past; but can be utilized in the present scenario/research work.

(v) Estimating Results: Involves making an estimate of the forecasted demand for predetermined years. The results should be easily interpreted and presented in a usable form. The results should be easy to understand by the readers or management of the organization.

14.10. Forecasting Methods

Forecasting is divided into two broad categories, techniques and routes. Techniques are further classified into quantitative techniques and qualitative techniques.

Quantitative techniques comprise of time series method, regression analysis, etc., where as qualitative methods comprise of Delphi method, expert judgment.

(i) Quantitative techniques: Based on historical information that is usually available within the company. Various techniques are:

- **Trend Analysis:** A method for forecasting sales data when a definite upward or downward pattern exists. Model includes double exponential smoothing, regression & triple smoothing.
- **Seasonal Adjustment:** Seasonal models take into account the variation of demand from season to season. Adjustments can be made to a baseline forecast to predict the impact of a seasonal demand.

- **Decomposition :** "A method of forecasting where time series data are separated into up to three components: trend, seasonal, and cyclical; where trend includes the general horizontal upward or downward movement over time; seasonal includes a recurring demand pattern such as day of the week, weekly, monthly, or quarterly; and cyclical includes any repeating, non seasonal pattern. A fourth component is random, that is, data with no pattern. The new forecast is made by projecting the patterns individually determined and then combining them".
- **Graphical Methods:** Plotting information in a graphical form. It is relatively easy to convert a spreadsheet into a graph that conveys the information in a visual manner. Trends & patterns are easier to spot & extrapolation of previous demand can be used to predict future demands.
- **Econometric Modeling:** A set of equations intended to be used simultaneously to capture the way in which dependent and independent variables are interrelated.
- **Life Cycle Modeling :** "A quantitative forecasting technique based on applying past patterns of demand data covering introduction, growth, maturity, saturation, and decline of similar products to a new product family".

(ii) Time series forecasting methods: It is defined as "A time series is simply some variable reported at regular time intervals of time" Time series forecasting methods produce forecasts completely based on historical values. Time series forecasting methods are widely used in business situations where complex statistical models is setup or problem analyses is done on the basis of simple visual perception. The time series techniques are well suited to Sales, Marketing, Finance, and Production planning. This method is simple.

- Time series methods are well suited to short-term forecasts (i.e., less than a year).
- Time series forecasting relies on sufficient past data being available and that the data is of a high quality and truly representative.
- Time series methods are best suited to relatively stable situations.

Time series forecasting method depends upon four factors, Trend , Cyclical Fluctuation , Seasonality and Residual Effect.

(i) **Trend** is the increase or decrease in the series over a long period of time. For this reason is also known as the long-term trend.

(ii) **The Cyclical Fluctuation** or (**Cyclicality**) is the wave like up and down fluctuations about the trend that is attributable to economic or business conditions. This fluctuation is also known as business cycle. During economic expansion, the cycle lies above the trend; during a downturn, beneath it.

(iii) **The Seasonality** or (**Seasonal Variation**) in a time series is the fluctuation that occurs each month, each year etc. Seasonal variations tend to be repeated from year to year.

(iv) **The Residual Effect** is what remains, having removed the Trend, Cyclical and Seasonal components of a time series. It represents the random error effect of a time series, caused by events as widespread as wars, hurricanes, strikes and randomness of human actions Time series may be easily understand by the following graph.

Graph -1 shows a hypothetical time series graph. Note that time is plotted on the horizontal axis and sales are on the vertical axis. If you look carefully, you should actually see more than one pattern occurring at the same time.

That's because many time series, particularly of sales, tend to follow somewhat predictable recurring patterns.

If retailer sales were graphed, this relatively predictable pattern would likely be easy to spot visually, assuming that monthly or even quarterly sales data were available. To better understand how to discern the patterns that may exist in time series data, it is useful to know the possible sources of variation in much time series data.

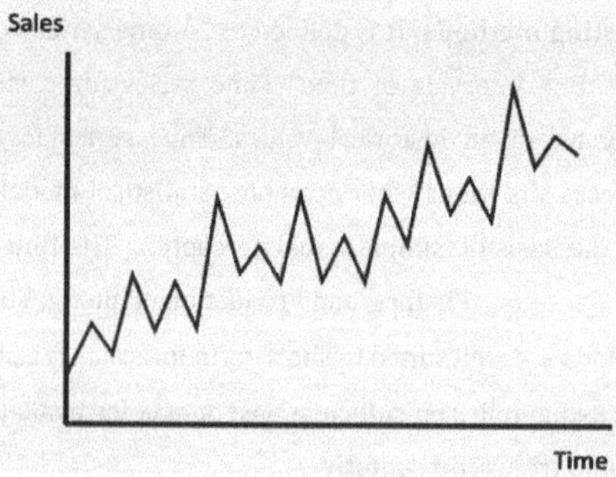

Diagram 14.6: Time Series Plot

Graph -2 shows a hypothetical time series graph, it shows the same hypothetical time series graph as above. However, in this case it also shows the overall upward trend of sales over the time period covered by the time series. Thus, the first source of variation in time series data is called the trend.

The trend shows the long term movement of the data. Trend is generally the longest term source of variation in time series data. For short term sales forecasts, trend is not especially useful.

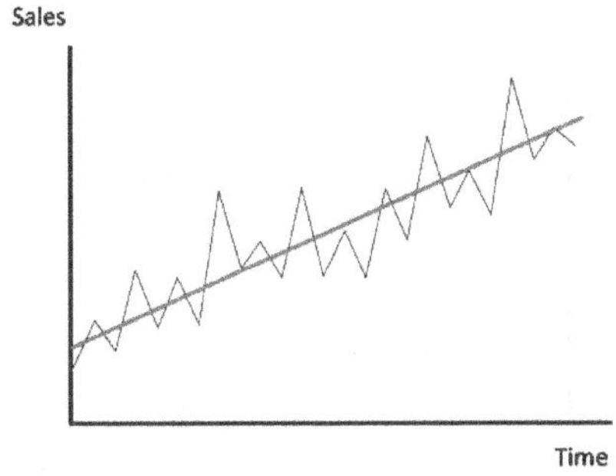

Diagram 14.7: Time Series Plot Showing Trend

Graph -3 shows a hypothetical time series graph, the red line shows the second source of time series variation, called cyclicality. When studying sales data, variation from cyclicality generally occurs with broad macroeconomic conditions. The business cycle refers to the repeating ups and downs of gross product over time. During periods of expansion, sales cyclicality increasing growth while during periods of recession, the general pattern may be one of slower growth or even contraction.

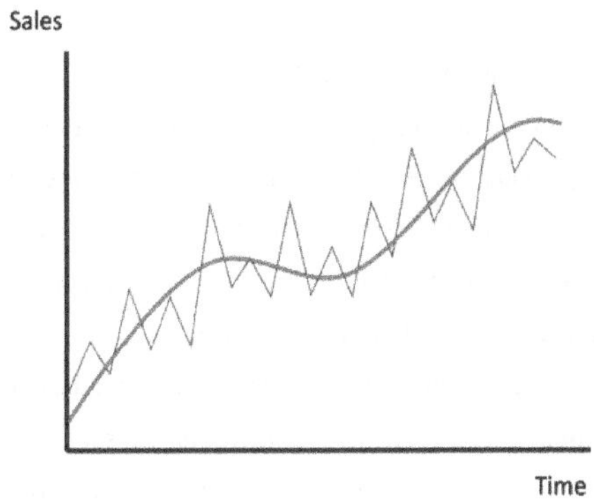

Diagram 14.8: Time Series Plot Showing Cyclicality

Graph -4 shows a hypothetical time series graph, the red line shows the third source of variation, referred to as seasonality. The term reflects the fact that much intra-year variation in sales occurs by season. For example, as noted earlier, retailers often experience spikes in sales during the Christmas holidays and declines after the first start of the year. Seasonality is the shortest term source of variation in time series.

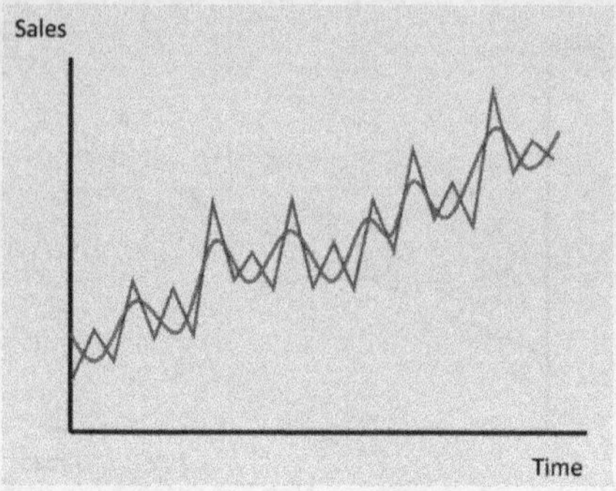

Diagram 14.9: Time Series Plot showing Seasonality

The fourth and final source of variation in time series is simply random variance that occurs in every period. This is often referred to simply as "noise." Because it is random, it cannot be predicted.

(i) Qualitative forecasting techniques: These techniques are primarily based upon judgment and intuition and especially when sufficient information and data is not available so that complex quantitative techniques cannot be used. The widely used qualitative methods are:

- **Judgmental Forecasting Techniques: The** judgmental forecasting techniques rely on the judgment of - knowledgeable people to make informed guesses about sales in the coming period or periods. Judgmental methods have the advantage of simplicity, which means they can be completed with relatively low costs and in relatively short time frames. Judgmental forecasts should rely on multiple participants, which may mean that the forecasts are somewhat more accurate. Of course, the accuracy of these forecasts depends greatly on how knowledgeable and insightful the participants are.
- **Sales-force Composite:** In organizational markets, salespeople should know the needs and plans of their customers. They should have a good sense from their customers about how much their customers expect to purchase in the coming period. As such, salespeople are often a good source of information for creating forecasts. Many companies, especially those with large sales-forces, survey their salespeople and ask them whether they believe their customers will be buying more, less, or about the same amount in the coming periods. Management should try to develop a feel for how their sales staff approach sales forecasting and make adjustments as necessary.

- **Jury of Executive Opinion:** This judgmental forecasting method asks executives to provide their best guess of what sales will be in the coming period. Their opinions are compiled in some way and a forecast produced. How the opinions are compiled is up to the manager preparing the forecast; no rules exist for combining the executives' predictions. Some executives, based on position or past predictions, may be given more weight than others. Importantly, executives need not be strictly internal to the company preparing the forecast. Companies may poll customers, outside experts, industry leaders, and others in addition to their own corporate executives. If executive opinions of expected sales are very different from one another, then the manager preparing the forecast should be somewhat concerned about the accuracy of the forecast and should seek more information from the executives about why they made the predictions that they did.
- **Delphi Method:** The Delphi method of forecasting is a somewhat more structured variation of the jury of executive opinion. The Delphi method begins with a panel or jury of knowledgeable executives; however, it utilizes an iterative approach to trying to reach a census among them. The executives, who are anonymous to one another, are asked to forecast sales and submit their opinions to the forecaster along with the reasoning behind their predictions. The forecaster collects opinions from the executives, and then summarizes them, usually with an average of the initial forecasts and some of the executives' reasons for their opinions. The summary is distributed back to the executives, who may revise their estimates and submit the revisions to the forecaster. The process continues until no more movement toward a consensus can be obtained, The Delhi method is nothing but it is an iterative method approach.
- **Consumers' expectations:** This method involves a survey of the customers as to their future needs. This method is especially useful where the industry serves a limited market. Based on the future needs of the customers a general overall forecast for the demand can be made.

(ii) Moving Average Method: moving average method of forecasting is a trend, which follows an indicator to smoothen a demand. Simple Moving Average is calculated by adding up the total demands in a fixed time period and dividing the sum total by the total number of time periods. In addition to looking for the average demand of a product, this technique also looks for a specific period of time, which could span over a week or a month.A general trend, which is followed in moving average method of demand forecasting, is to take the average of previous weeks or months depending upon the nature of the business. It can also work on Quarterly basis where fluctuations are not so volatile . The formula is illustrated as follows:

Moving Average = (n1 + n2 + n3 + ...) / n

Where n = the number of time periods in the data set.

Year	Sales Volume
2011	4.6
2012	5.3
2013	8.1
2014	7.8
2015	8.3

Moving Average n=5 = (4.6 + 5.3 + 8.1 + 7.8 + 8.3) / 5 = 6.82

The sales manager forecasts that the sales demand for 2016 will be approximately 6.82.

Example 2

Time Period	Sales	Three Period Moving Average	
2010 Q1	32,500	...	
2010 Q2	28,270	...	
2010 Q3	31,050	30,607	**Average of 2010 Q1, Q2, and Q3**
2010 Q4	30,990	30,103	Average of 2010 Q2, Q3, and Q4
2011 Q1	34,540	32,193	Average of 2010 Q3, Q4 and 2011 Q1
2011 Q2	26,780	30,770	
2011 Q3	32,460	31,260	and so on. . .
2011 Q4	31,770	30,337	
2012 Q1	34,760	32,997	
2012 Q2	30,920	32,483	

Moving Average Calculation

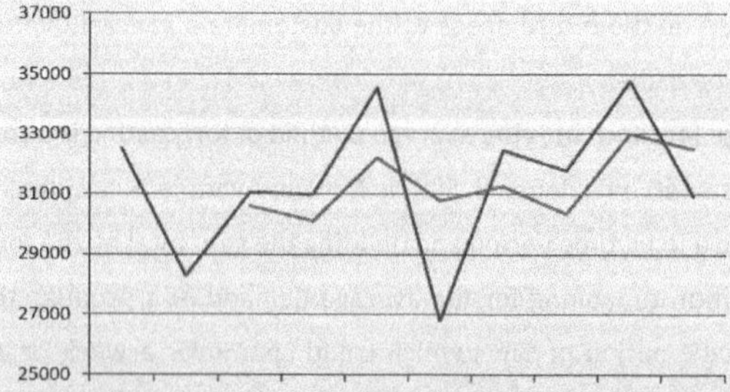

Plot of Time Series and Moving Average

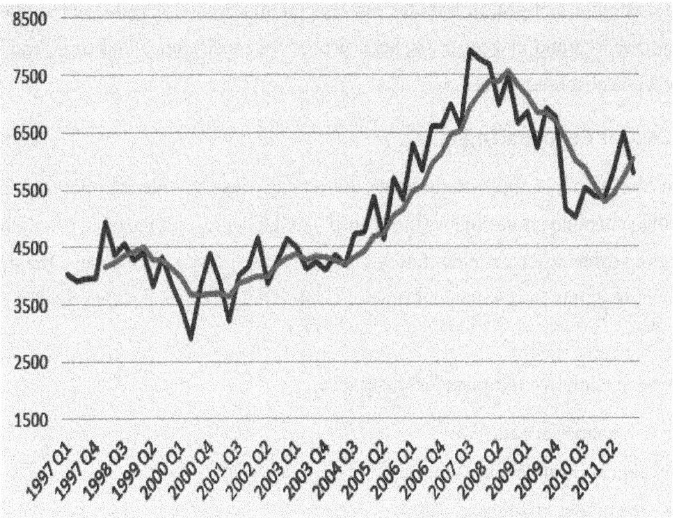

Time Series of Sales Data and Five Period Moving Average

(iii) Subjective Forecasting Methods: These are the most simple yet widely used methods of forecasting. Basically, these consist of collecting the opinions and judgments of people inside or outside the organization who are expected to have the best knowledge of current activities or future plans. These are generally employed when,

- There is not sufficient time.
- Available data is not sufficient to cover possible development in the more distant future.

No relevant data is available, e.g. while introducing a new product for which no historical data may be available.

Subjective Estimates: - For developing a demand forecast several organizations use the knowledge, experience and "sixth sense" of their own staff. In one typical case, a company asked its salesmen to submit estimates of anticipated demand in their allocated areas for a future period. These figures were then pooled at the regional level. A correction factor was then applied to take care of regional economic, demographic and other factors which were not previously considered. The adjusted regional estimates were pooled at headquarters and further adjustments were made to account for competitors, developments in international trade, and the possible government policies. The final estimate for total demand was the sum of subjective inputs from different sources.Subjective forecasts can be supplemented with the results of customer surveys, distributor surveys, or more carefully planned studies such as marketing research and market trials. Such procedures cost money and time, and should be used only when absolutely necessary.

14.11. Casual Forecasting

Causal forecasting is appropriate when there is a "cause and effect" relationship between one or more independent variables (the "cause") and a dependent variable (the "effect") such as demand or some other variable that is being forecast. Causal models have the potential to predict turning points in the demand function, something that time series models cannot do. (Why?)

The general approach to causal forecasting is:

- Collect historical data.
- Develop and validate the model.
- Use the model to forecast.

A second major category of forecasting models consists of causal models. The basic assumption for a casual model is that the future value of the forecasted variable can be expressed as a mathematical function of the known values of a set of different variables. For example, the historical sales of a product combined with the historical breakdown rates for this product allow the forecast of the number of breakdowns and required service parts during the coming year. Causal forecasting methods are often used in predicting future economic activity and future social and life science trends.

The known variables in the causal model are also called leading indicators, since they occur before or lead the forecasted variable and indicate the behavior of the forecasted variable. One leading indicator used often in predicting logistics activity is the purchase of corrugated cardboard boxes. When companies expect an increase in sales, they purchase more shipping supplies such as cardboard boxes. Causal models can be quite good in predicting major changes and trends and are often used in the medium and long-range period.The major difficulty in applying causal models is of course the development of accurate cause-and-effect relationships, since the true causal relationship is often much more complex than the causal model. Causal models based on regression and economic techniques typically have substantial forecasting errors.

The causal relationship in general is expressed as,

$$Y = f(X1, X_2, X_3 \ldots X_N)$$

Econometric models are a subclass of causal models that use only linear causal relationships, as illustrated in the following expression

$$\mathbf{Y = a_1X_1+a_2X_2+a_3X_3}$$

Multiple regression analysis: - Multiple regression analysis is one approach used to develop a causal forecasting model. It is important to note that regression implies dependence and not necessarily causation, however, causation does not have to be proven for causal forecasting model to be used effectively. The general form for a multiple linear regression equation is:

$\mathbf{Y_C = a + b_1X_1 + b_2X_2 \text{ .... } B_jX_j}$

$\mathbf{Y_C}$ = calculated (predicted) value of the dependent variable.

a = intercept (constant term).

$\mathbf{X_j}$ = jth independent (predictor) variable.

Bj = coefficient associated with the jth independent variable.

Simple Exponential Smoothing : One of the most used methods for short-term forecasting of a variable with an underlying constant pattern is exponential smoothing. To differentiate this variant of exponential smoothing from later variants for trend and seasonal patterns, this variant is also called simple exponential smoothing. Exponential smoothing has several significant advantages. It is mathematically simple, it requires a minimum amount of data, and it is self-adaptive to changes in the underlying data pattern. In exponential smoothing, the past observations are not given equal weights. The more recent observations are given more weight than earlier observations and it will be shown that the weight decreases by a constant factor when going back farther in the past.

There exist several equivalent expressions for the computation of the forecast in period based on the observed demand during period tand the forecast for time period t. The constant α is commonly called the exponentialsmoothing constantand **(1- α)** is referred to as the damping factor.

The impact of the most recent observation is dampened by the most recent forecast, which incorporates all the previous observations.

Equivalently, the new forecast is equal to the previous forecast plus a smoothing termbased on the forecast error. The smoothing constant has values strictly between 0 and 1. So the new forecast can also be interpreted as a convex combination of the most recent observation and the most recent forecast.

$$F_{t+1} = \alpha D_t + F_t(1-\alpha) = F_t + \alpha(D_t - F_t)$$

The following initial values are typically used,

$$F_2 = D_1$$

$$F_1 = D_2$$

By repeated substitution of the previous forecasts factor (F_t) with expression (2.17), we obtain the following expression for the next forecast. It shows that the weight for previous demand observations decreases geometrically with the constant factor (1- α)

$$F_{t+1} = \sum_{i=0}^{t-1} \alpha(1-\alpha)^i D_{t-1} + (1-\alpha)^t D_1$$

Where

$$t \geq 1$$

The limit for a large number of prior observations of this expression is defined for αstrictly betweenzero and one and is equal to

$$F_{t+1} = \sum_{i=0}^{\infty} \alpha(1-\alpha)^i D_{t-1}$$

Since α is strictly smaller than one, the sumofall the weights in the geometric series is defined and equal to one. The next forecast is thus a convex combination of all the previous observations.

$$\sum_{i=0}^{\infty} \alpha(1-\alpha)^i = \frac{\alpha}{1-(1-\alpha)} = 1 \qquad \text{Where } \alpha < 1$$

$$\sum_{i=0}^{\infty} a^i = \frac{1}{1-a} \qquad \text{Where } a < 1$$

Choosing the appropriate value of the exponential smoothing constant requires a degree of judgment based on the understanding of the behavior of the underlying data pattern to balance the responsiveness versus the stability of the forecasting model. A higher value of the smoothing constant assigns more weight to the most recent observations. This allowsthe forecasting model to respond more quickly to changes in the underlying data pattern. However, if the value of the smoothing constant is increased toomuch then the model will start tracking the changes inthe randomcomponent of the time series rather than in the underlying pattern.

In this case, the forecast is said to be "nervous". Many times, a more stable forecast is desirable, which implies a smaller value of the smoothing constant. Low values of the smoothing constant provide very "stable" forecasts that are not likely to be influenced by the randomness of the time series. However, the more the smoothing constant is decreased the longer it takes for the forecasting model to adjust to fundamental changes in the underlying data pattern. Higher values of the smoothing constant may be appropriate ifthe underlying data pattern is likely to change more quickly such as for introduction of a new product, discontinuing ofa product, start of a recession or economic boom, or a promotional campaign.However, if one expects that the underlying pattern is fundamentally a trend or seasonal pattern, the exponential smoothing methodsappropriate for those cases should be used rather thana larger smoothing constant for the constant data pattern.Those exponential smoothing methods are discussed later in this chapter. The typical range ofthe exponential smoothing constant is [0.2, 0.4].

The Excel spreadsheet uses 0.3 as the default value of the smoothing constant. If sufficient historical data is available, the smoothing constant can be chosen so that forecast error for the historical period is minimized.

In this case, the smoothing constant is set to the value that minimizes the sumof squared errors over the historical period.

Assume that we want start forecasting the sales of a product. The underlying sales pattern for this product is assumed to be constant.

The sales for period one were 62 units. Since this is the startup period, a smoothing constant of 0.4 was selected. The forecast for period two is 62 units, but the observed sales equaled 59 units. The forecast for the sales during the thirdperiod using exponential smoothing is then $\mathbf{F_3} = 45.9 + (1 - 0.4) X0.62 = 60.8$

The observed sales for period three are 68 and the forecast for sales during the fourth period is then,

$$\mathbf{F_4} = 4.68 + (1 - 0.4) X 60.8$$

14.12. Objectives of Demand Forecasting

The objectives of demand forecasting are divided into short and long-term objectives, which are shown in Figure 14.10.

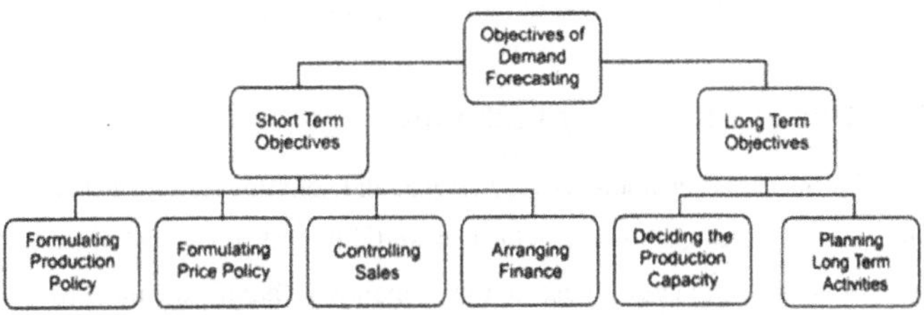

Diagram 14.10: The Objectives of Demand Forecasting

(i) Short terms objectives

(ii) Long terms objectives

The Explanation of Short Terms Objectives of Forecasting is Given Below:

- **Formulating production policy:** Helps in converting the gap between the demand and supply of the product. The demand forecasting helps in estimating the requirement of raw material in future, so that the regular supply of raw material can be maintained. It further helps in maximum utilization of resources as operations are planned according to forecasts. Similarly, human resource requirements are easily met with the help of demand forecasting.
- **Formulating price policy:** Refers to one of the most important objectives of demand forecasting. An organization sets prices of its products according to their demand. For example, if an economy enters into depression or recession phase, the demand for products fall. In such a case, the organization sets low prices of its products.
- **Controlling sales:** Helps in setting sales targets, which act as a basis for evaluating sales performance. An organization makes demand forecasts for different regions and fixes sales targets for each region accordingly.
- **Arranging finance:** Implies that the financial requirements of the enterprise are estimated with the help of demand forecasting. This helps in ensuring proper liquidity within the organization.

The Explanation of Long- Terms Objectives of Forecasting is Given Below:

- **Deciding the production capacity:** Implies that with the help of demand forecasting, an organization can determine the size of the plant required for production. The size of the plant should conform to the sales requirement of the organization.
- **Planning long-term activities:** Implies that demand forecasting helps in planning for long term. For example, if the forecasted demand for the organization's products is high, then it may plan to invest in various expansion and development projects in the long term.

14.13. Selection of Forecasting Technique

selection of forecasting technique is very critical and complex task for firm's manager, due to seasonality, sudden changes in demand levels, and different price level of similar products, the competition, strikes, and large swings of the economy. To handle the increasing variety and complexity of managerial forecasting problems, many forecasting techniques have been developed in recent years. Each has its special use, and care must be taken to select the correct technique for a particular application.

The manager as well as the forecaster has a role to play in technique selection; and the better they understand the range of forecasting possibilities.The selection of a method depends on many factors—the context of the forecast, the relevance and availability of historical data, the degree of accuracy desirable, the time period to be forecast, the cost/ benefit (or value) of the forecast to the company, and the time available for making the analysis .

When-ever a company wishes to forecast with reference to a particular product, it must consider the stage of the product's life cycle for which it is making the forecast. The availability of data and the possibility of establishing relationships between the factors depend directly on the maturity of a product, and hence the life-cycle stage is a prime determinant of the forecasting method to be used.Successful forecasting begins with collaboration between the manager and the forecaster, in which they work out answers to the following questions.

- **What is the purpose of the forecast—how is it to be used?** This determines the accuracy and power required of the techniques, and hence governs selection. Deciding whether to enter a business may require only a rather gross estimate of the size of the market, whereas a forecast made for budgeting purposes should be quite accurate. The appropriate techniques differ accordingly.Again, if the forecast is to set a "standard" against which to evaluate performance, the forecasting method should not take into account special actions, such as promotions and other marketing devices, since these are meant to change historical patterns and relationships and hence form part of the "performance" to be evaluated.
- Forecasts that simply sketch what the future will be like if a company makes no significant changes in tactics and strategy are usually not good enough for planning purposes. On the other hand, if management wants a forecast of the effect that a certain marketing strategy under debate will have on sales growth, then the technique must be sophisticated enough to take explicit account of the special actions and events the strategy entails. The manager must, decide how his or her decision will vary, depending on the range of accuracy of the forecast. This allows the forecaster to trade off cost against the value of accuracy in choosing a technique. Following diagram shows that how cost and accuracy increase with sophistication and charts this against the corresponding cost of forecasting errors.

- **What are the dynamics and components of the system for which the forecast will be made?** This clarifies the relationships of interacting variables. Generally, the manager and the forecaster must review a flow chart that shows the relative positions of the different elements of the distribution system, sales system, production system, or whatever is being studied. In the part of the system where the company has total control, management tends to be tuned in to the various cause-and-effect relationships, and hence can frequently use forecasting techniques that take causal factors explicitly into account . The flow chart has special value for the forecaster where causal prediction methods are called for because it enables him or her to conjecture about the possible variations in sales levels caused by inventories and the like, and to determine which factors must be considered by the technique to provide the executive with a forecast of acceptable accuracy.

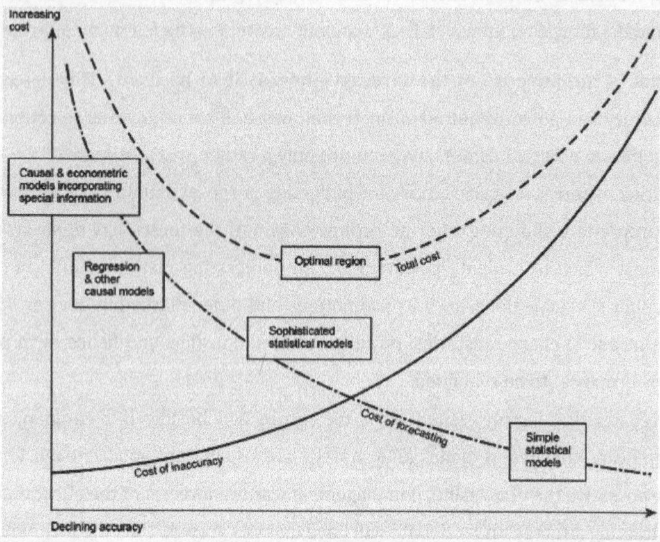

Diagram 14.11: Cost of Forecasting Versus Cost of Inaccuracy for a Medium-Range Forecast

- **How important is the past in estimating the future?** Significant changes in the system—new products, new competitive strategies, and so forth—diminish the similarity of past and future. Over the short term, recent changes are unlikely to cause overall patterns to alter, but over the long term their effects are likely to increase. The executive and the forecaster must discuss these fully.

Here we are describing some forecasting technique in tabular form,

Basic Forecasting Technique (Qualitative Method)

TECHNIQUE	DELPHI METHOD	MARKET RESEARCH	PANEL CONCENSUS
Description	A panel of experts is interrogated by a sequence of questionnaires in which the responses to one questionnaire, any set of the information available to some experts and not others is thus passed on to the others enablingall the experts to have access to all the information for forecasting. this technique eliminates the bandwagon effect of majority option .	The systematic, formal, conscious procedure for evolving and testing hypotheses about real markets .	The technique is based on the assumption thatseveral experts can arrive at a better forecast than one person there is no secrecy and communication is encouraged. the forecast is some times influenced by social factors, and may not reflect the true consensus
Accuracy Short terms time duration 0 to 3 month.	**Fair to very good**	**Excellent**	**Poor to fair**
Medium termstime duration 3 to 2 years .	**Fair to very good**	**Good**	**Poor to fair**
Long terms time duration above 2years .	**Fair to very good**	**Fair to Good**	**Poor**
Identification of turning points	**fair to good**	**Fair to very good**	**Poor to fair**
Typical application	Forecast of long range and new sales ,forecast of margin .	Forecast of long range and new sales ,forecast of margin .	Forecast of long range and new sales ,forecast of margin .
Data required	A coordinators issues the sequences of questionnaire editing and consolidating the responces	As a minimum to sets of reports over time , one need a consideration collection of markets data from questionnaire &surveys and and time series analyses of mrkets variable .	Information from panel of experts is presented openly in group meeting to arrive at a consensus forecast again a minimum of two sets of reports over times.
Cost of forecasting	Between ₹ 110 , 000 to 130, 000 .	Between ₹ 275, 000 to 325 , 000 .	Between ₹ 55,000 to 65 ,000.
Time required for developed and application & forecast	2 months or may be above	3 months or may be above	2 weeks or may be above

Basic Forecasting Technique (Qualitative Method)

TECHNIQUE	**Qualitative Method**		**Time series analyses**	
	Visionary Forecasting	**Historical Analogy**	**Moving average**	**Exponential smoothing**
DESCRIPTION	A prophecy that uses personal insight judgment and when possible facts about different scenario of the future. It is characterized by different guess work & non- scientific method are generally applied.	This is comparative analyses of introduction & growth of similar new product .	Each point of moving average of the series is the arithmetic average of consecutive point of series the. In order to eliminate the irregularity & seasonality .	This technique to moving average except yhat more data points are given ' The new forecast is equal to the old one plus previous forecasting error . There are variance in exponential smoothing some are more versatile , some are complex and some are taken more time .
Accuracy Short terms time duration 0 to 3 month.	Poor	Poor	Poor to good	Fair to very good
Medium termstime duration 3 to 2 years .	Poor	Fair to Good	Poor	Poor to good
Long terms time duration above 2years	Poor	Fair to Good	Fair to Good	Very poor
Identification of turning points	Poor	Poor to fair	Poor	Poor
Typical application	Forecast of long range And new product sales forecast of margin .	Forecast of long range And new product sales forecast of margin .	Inventory control of low volume items	Production & Inventory control forecast of margine & other financial data
Data required	A set of possible scenarios about the future prepared by the experts in light of past events .	Several year history of products .	A minimum of two years of sales history. if seasonal are presents .	The same as for a moving average method .
Cost of forecasting	Between ₹ 5500 to 6500 . or may above	Between ₹ 55000 to 65000 . or may above	Between ₹ 0.0275 to 0 .325 or may above	Between ₹ 0.0275 to 0 .325 or may above
Time required for developed and application & forecast	1 weeks or may be above	1 months or may be above	1 days or may be above	1 days or may be above

Basic Forecasting Technique (Qualitative Method)

Technique	Time series & projection		
	Box – Jenkins	**x-11**	**Trends projection**
Description	Exponential smoothing is a special case of the box Jenkins technique. The time series is finished with a mathematical model that is optimal in sense that is assign smaller error to history than only other model. The type of model must be identified and the parameters than eliminated, this is opporently most accurate mathematical model but very costly & time consuming.	Developed by Julius shiskins of the census bureau . this technique decomposes a time series into seasonal trends , cycles and irregular elements . primarily used for detailed time series analyses . but we have extended its uses to forecasting tracking and warning by in cooperating other analytical method . used with special knowledge . it is applied for medium range forecasting 3 - month to 1- year , allowing one to predict turning point and to time special events .	These techniques fit a trends line to a mathematical equation and it projects it into future by means of this equations. There are several variation, slope characteristics method, polynomials and logarithms and so on.
Accuracy Short terms time duration 0 to 3 month. **Medium terms** time duration 3 to 2 years . **Long terms** time duration above 2years.	very good to Excellent poor to good Very Poor	very good to Excellent Good Very Poor	very good Good Good
Identification of turning points	Fair	very good	poor
Typical application	Production & Inventory control forecast of large volume , forecast of cash balances	Traking & warning Forecast of company divisions , and departments sales	New product forecast particularly for long terms & intermediate forecasting
Data required	The same as that of moving average	Minimum of three year of history to starts .	Varies with the technique used .however a good rule of thumb is used to a minimum of five years , annual data to starts there after the complete history
Cost of forecasting	Between ₹ 550 to 650 . or may above	Between ₹ 550 to 650 . or may above	Depends upon application
Time required for developed and application & forecast	1- 2 days	1 days	1 days

Basic Forecasting Technique (Causal Method)

REGRESSION MODEL	ECONOMETRIC MODEL	INTENTION TO BUY & ANTCLPATIONS SURVEYS	INPUT – OUTPUT MODELS
This functionally relates to soles to other economics , competitive or internal variables and estimates on eqation using the least squares technique . relationship are primarialy analyzed statistically .although any relationship should be selected for testing on a rotational ground	An econometric model is a system of interdependent regression equation that described some sector some sectors of economic sales of the profit activity . the promoter of regration equations are usually estimated simultaneously . as a rule these models are relatively expensive to develop and can easily cost between 30000 and 600000 . depending on details . how ever due to the system of equation , inherent in such models , they will better for example press the causalities involved than on ordinary regression equation and hence will predict turning points more accurately .	These surveys of the general public . (a)determine interations to buy certainproducts (b) drive on index that measures central feeling about the present and the future and estimates how this feeling affects buying habits . these approaches to forecasting are more useful for tracking and warning than forecasting . the basic problem in using them is that a turning points may be signaled incorrectly .	The method of analyses concentrated with the inter industries of interdependent mental flow of goods or sevices in the economy or a company and its markets .it shows whats flows of inputs must occurs to obtains certain outputs .considerable effects must be expanded to use these models properly and additional details not normally available must be obtained if they are to be applied to specific businesses . corporation using input output models have expended as much as 60,00000 and more annually to developed useful applications.
good to very good good to very good Poor	good to very good Very good to excellent Good	poor to good poor to good Very poor	Not applicable good to very good good to very good
very good	excellent	Good	Fair
Forecast of sales by products classes , forecasts of margin	Forecast of sales by products classes , forecasts of margin	Forecast of sales by products classes ,	Forecasts of company sales and divisions sales for industrial sectors and subsectors
Several years quartly history to obtain good meaningful relationship . mathematically necessary to have two more observations than there are independent than there are independent variables	Similarly to regression	Several year data are usually required to relate such indexes to company sales	Ten or fifteen years history consideration amount of information on products and service flows with in e corporation for each year for which on input output analyses is desired .
Between ₹ 5500 to 6500 . or may above	Between ₹ 275,000 to 325,000 . or may above	Between ₹ 275,000 to 325,000 . or may above	Between ₹ 275,0000 to 325,0000. or may above
Depends on ability to identify relationship	2- months or above	Several weeks	6- month or above

Basic Forecasting Technique (Causal Method)

Technique	**Economic input out-put models**	**Diffusion index**	**Leading indicators**	**Life cycle analyses**
Description	Economic models and input-output models aresometimescombined for forecasting. Theinput – output models are used to provide long-terms trends .for the economics model and also stabilizes the economics models.	The percentage of a group of economic indicators that is going t up & down , it is called Diffusion index	a time series of an economic activity whose movements in a given direction precedes the movements of some other time series in the same direction is a leading indicators	this is an analyses and forecasting of new product growth rates based on s- curves . the phases of the products acceptance by the various group such as innovator , early adopter ,early majority , late majority and laggards are central to the analyses .
Accuracy **Short terms** time duration 0 to 3 month. **Medium terms** time duration 3 to 2 years. **Long terms** time duration above 2 years.	Not applicable Good to very good Good to excellent	poor to good poor to good very poor	poor to good poor to good very poor	poor poor to good poor to good
Identification of turning points	Good	Good	Good	poor to good
Typical application	company sales for industrial sectors and subsectors	Forecast of sales by products classes	Forecast of sales by products classes ,	Forecast of new product sales
Data required	The same as for a moving average and x-11	The same as an intension to buy survey	The same as an intention to buy survey + 5 to 10 years history .	As a minimum, the annual sales of the products being considered or of a similar product . it is often necessary to do market survey .
Cost of forecasting	Between ₹ 5500,000 to 6500,000 but not feasible	Between ₹ 55,000 to 65,000 . or may above	Between ₹ 55,000 to 65,000 . or may above	Between ₹ 82,500 to 97,500 . or may above
Time required for developed and application & forecast	6- month or above	1- month or above	1- month or above	1- month or above

Forecasting Models

Many models are used to predict demand, such as regression analysis, time series, the Delphi method, and market research-or some combination of these quantitative and qualitative methods. Forecast data may be broken down in an attempt to uncover the components of demand, such as trend, seasonality, and cyclical and random patterns. The base component reflects the demand for an item without applying the patterns.

- **Baseline methods**. Baseline demand is the percentage of a company's demand derived from continuing contracts and existing customers. It usually is a predictable component of demand.
- **Time series**. Time series is a technique that projects historical data patterns by looking at past forecasts and forecast errors. A time series may contain seasonal, cyclical, trend, and random components.
- **Exponential smoothing**. Exponential smoothing is a forecasting technique using a weighted moving average, where past observations are adjusted according to their age. The most recent data typically are weighted the heaviest. A smoothing constant is applied to the difference between the most recent forecast and critical sales data, avoiding the necessity of maintaining historical sales data. Alternatives to exponential smoothing include moving average and weighted moving average models.
- **Trend**. Trend is the general upward or downward movement of demand over time.
- **Seasonality**. Seasonality is a cyclical pattern of demand, where some periods of the year are higher or lower than others.
- **Regression models.** Regression models are statistical techniques used to determine the best mathematical expression that describes the relationship between a dependent variable, such as demand, and one or more independent variables. Focus forecasting. Focus forecasting is a system that allows a user to simulate the effectiveness of numerous forecasting techniques and select the best.

Error measurement in forecasting technique: Error is the difference between a forecast and the actual demand.

- **Bias**. Bias refers to consistent errors that cause a forecast to go either too high or too low. A forecast is biased if the current forecast errors are greater or less than zero.
- **Standard deviation.** Standard deviation is a measurement of the dispersion of data or of a variable. It is calculated by finding the differences between average and actual

observations, squaring each difference, adding the squared differences, dividing by n minus 1 (for a sample), and finally taking the square root of the result.

- **Mean absolute deviation (MAD).** MAD is the average of the absolute values of the deviations of observed values from forecast values. It is the arithmetic mean of past absolute errors.
- **Mean absolute percent error.** Mean absolute percent error is computed by taking the MAD, dividing by the average demand, and then multiplying by 100.
- **Tracking signal.** Tracking signal refers to the ratio of the cumulative algebraic sum of the deviation between forecasts and actual values to the mean absolute deviation. This is used to alert that the forecast model is biased.

Question for Discussion

1. Identify the benefits , reasons and objectives of forecasting.
2. Distinguish between forecasting and prediction .Discuss the concept of economic forecasting.
3. What is production planning? How it is related with capacity planning?
4. List down and brieily explain the main uses of demand and forecasting.
5. What is meant by Time Series? Discuss the basic patterns observed in different demand time series?
6. what is aggregate planning? What are the different strategies under aggregate planning? Discuss them.
7. What do you mean by qualitative and quantitative methods of demand forecasting ? When do we use each of these methods of forecasting ? Explain the Time series analysis in detail .
8. What is sales forecasting? List various methods of sales forecasting and explain exponential smoothing method.
9. What are the various factors which affect the demand forecasting? Also list the principle of forecasting.
10. Define forecasting? Classify them.
11. How can we implement the demand forecasting process of an organization? Explain the various steps associated with it
12. Explain in brief various type of forecasting method?
13. What are the various factors which must be considered for selecting the forecasting in any organization? Explains.
14. List and briefly explain about forecasting models? Any kind of error may be persists during forecasting if yes justify them.

References

1. E.E. Adams and R.J. Ebert, "Production and Operations Management", Englewood Chiffs,Prentice Hill, Inc, USA, 1978.
2. J. Dejonckleere, S.M. Disney, M. lambrecht and D.R. Towill, "The Dynamics of Aggregate Planning. Production Planning and control", Hillcon Co. Ltd, Ikeja, Lagos, Nigeria, 2003.
3. C.H Gowon, "Basic operations Management", Henry Co. Ltd, Ondo, Nigeri, 2011.
4. M. Gray John and J. Blake, "Capacity Planning Analysis", High-Hill Co. Ltd,Owerri, 1989.
5. J. Ihemeje and J.C. Ihimije, "Operations Management; Principle and practices", Onaivi, Co.Ltd, Keffi, Nassavawa, Nigeria, 2010.
6. E.U.L. Imaga, "Theory and Practice of Production and Operations Management", Rhyce Kerex, Ogui, Enugu, Nigeria, 2003.
7. M.D. Mark, N.J. Aquilano and B.C. Richard, "Fundamentals of operations Management", McGraw-Hill/Irwin, New York, U.S.A, 1991.
8. C.J. Wisdom and R. Johnson, "Planning of Capacity Requirements", Joe-vad Co. Ltd, Nnewi, Imo, Nigeria, 1990.
9. W.T. Huh, R.O. Roundy and M. Cakanyildirim, "A General Strategic Capacity Planning Model under Demand Uncertainty", International Journal of Naval Research Logistics, Vol. 53, Pp. 137-150, 2006
10. R.U. Ayres, "Future trends in factory automation", Manufacturing Review, Vol. 1, Pp. 93-103, 1988.
11. D.J. Stockton and L. Quinn, "Aggregate production planning using genetic algorithms", Proceedings of the Institution of Mechanical Engineers, 1995.
12. Monaghan, A, "Strategic Capacity Planning for Products and Services", Powerpoint presentation.
13. F.R. Jacobs and R.B. Chase, "Operations and supply management", The Core. New York, NY: The McGraw-Hill Companies, Inc., 2008.
14. G. Reiner and J. Fichtinger, "Demand forecasting for supply processes in consideration of pricing and market information", International Journal of Production Economics, Vol. 118, No. 1, Pp. 55-62, 2009.
15. J. Scott Armstrong, "Research needs in forecasting", International Journal of Forecasting, Vol. 4, Pp. 449-465, 1988.
16. Y. Barlas, B. Gunduz, "Demand forecasting and sharing strategies to reduce fluctuations and the bullwhip effect in supply chains", Journal of Operation Research Society, Vol. 62, No. 3, Pp. 458-473, 2011.

17. R. Kamalpur, "Impact of Forecast Error in CPFR Collaboration Strategy", American Journal of Industrial and Business Management, Vol. 3, Pp. 389-394, 2013.
18. A. Nasiri Pour, B. Rostami Tabar and A. Rahim Zadeh, "A Hybrid Neural Network and Traditional approach for Forecasting lumpy demand".
19. E. Babiloni, M. Cardos, J.M. Albarracin and M.E. Palmer, "Demand Categorization, Forecasting, and Inventory Control for Intermittent demand items", South African Journal of Industrial Engineering, Vol. 21, No. 2, Pp. 115-130, 2010.
20. K.B. Khan, "An exploratory investigation of new product forecasting practices", The Journal of Product Innovation Management, Vol. 19, Pp. 133-143, 2002.
21. H. Sun, H.K. You and E.K.M. Suen, "The simultaneous impact of supplier and customer involvement on new product performance", Journal of Technology Management and Innovation, Vol. 5, No. 4, 2010.
22. A.H.C. Eaves and B.G. Kingsman, "Forecasting for the ordering and stock holding of spare parts", Journal of the Operational Research Society, Vol. 50, Pp. 431-437, 2004.
23. T.F. Wallace and R.A. Stahl, "Sales and operation planning: The how to handbook", 3rd ed. S.L., T.F. Wallace and Company 2008
24. G. Zhang, B.E. Patuwo and M.Y. Hu, "Forecasting with neural networks: The state of the art", International Journal of Forecasting, Vol. 14, No. 1, Pp. 35-62, 1998.
25. A.N. Borisov and V.A. Pavlov, "Prediction of a continuous comparing mean square forecast errors", Journal of Forecasting, 1995.
26. J. Armstrong and Scott, "Research on forecasting: A quarter-century review", Interfaces, Vol. 16, Pp. 89-109, 1986.
27. J. Armstrong, Scott and E. Lusk, "Research on the accuracy of alternative extrapolation models: Analysis of a forecasting competition through open peer review", Journal of Forecasting, Vol. 2, Pp. 259-311, 1983.
28. Makridakis and C. Steven Wheelwright, "The Handbook of Forecasting: A Manager's Guide", 2nd ed.Wiley, New York, Pp. 118-134.
29. Cerullo, J. Michael and A. Avila, "Sales forecasting practices: A survey", Managerial Planning, Vol. 24, Pp. 33-79, 1975.
30. Dalrymple, Douglas and B.E. King, "Selecting parameters for short-term forecasting techniques", Decision Sciences, Vol. 12, Pp. 661-669, 1981.
31. Fischhoff and Baruch, "Judgmental aspects of forecasting: Needs and possible trends", International Journal of Forecasting, Vol. 4, No. 3, Pp. 331-339, 1988.

32. Gardner, S. Everett and E. McKenzie, "Forecasting trends in time series", Management Science, Vol. 31, Pp. 1237-1246, 1985.
33. Hanke and John, "Forecasting in business schools: A follow-up survey", International Journal of Forecasting, Vol. 5, Pp. 259-262, 1989.
34. Hogan and J. Andrew, "Combining forecasts: Some managerial experiences with extrapolation", Socio-economic Planning Sciences, Vol. 3, Pp. 205-211, 1987.
35. Kahneman Daniel, P. Slovic and A. Tversky, "JudgmentUunder Uncertainty: Heuristics and Biases", Cambridge University Press, Cambridge, 1982.
36. S. Chopra and P. Meindl, "Supply Chain Management: Strategy, Planning and Operation", Prentice Hall, 2004.
37. Gerson Lachtermacher and J. David Fuller, "Back Propagation in TimeSeries Forecasting", 2006.
38. Karin Kandananond, "Consumer Product Demand Forecasting Based on Artificial Neural Network and Support Vector Machine", World Academy of Science, Engineering and Technology, Vol. 63, Pp. 372-375, 2012.
39. B. Billah, M.L. King, R.D. Snyder and A.B. Koehler, "Exponential smoothing model selection for forecasting", International Journal of Forecasting, Vol. 22, Pp. 239–247, 2006.
40. J.D. Croston, "Forecasting and stock control for intermittent demands", Operational Research Quarterly, Vol. 23, Pp. 289-303, 1972.
41. A.H.C. Eaves and B.G. Kingsman, "Forecasting for the ordering and stock-holding of spare parts. Journal of the Operational Research Society, Vol. 55, PP. 431-437, 2004.
42. F.R. Johnston and J.E. Boylan, "Forecasting for items with intermittent demand", Journal of the Operational Research Society, Vol. 47, Pp. 113-121, 1996a.
43. J.K, Snyder, R.D. and A. Beaumont, "Forecasting for Intermittent Demand for Slow-Moving Items", Department of Econometrics and Business Statistics, Monash University Working, 2010.
44. A.V. Rao, "A comment on: Forecasting and stock control for intermittent demands", Operational Research Quarterly, Vol. 24, Pp. 639-640, 1973.
45. B. Sani and B.G. Kingsman, "Selecting the best periodic inventory control and demand forecasting methods for low demand items", Journal of the Operational Research Society, Vol. 48, Pp. 700-713, 1997.
46. R.D. Snyder, "Forecasting sales of slow and fast moving inventories", European Journal of Operational Research, Vol. 140, Pp. 684-699, 2002.

47. A.A. Syntetos, K. Nikolopoulos and J.E. Boylan, "Judging the judges through accuracy implication metrics: The case of inventory forecasting", International Journal of Forecasting, Vol. 26, Pp. 134-143, 2010.
48. R.H. Teunter and L. Duncan, "Forecasting intermittent demand: a comparative study", Journal of the Operational Research Society, Vol. 60, Pp. 321-329, 2009.
49. T.R. Willemain, C.N. Smart and H.F. Schwartz, "A new approach to intermittent forecasting demand for service parts inventories", International Journal of Forecasting, Vol. 20, Pp. 375-387, 2004.
50. Syntetos, K. Nikolopoulos and J.E. Boylan, "Judging the judges through accuracy-implication metrics: The case of inventory forecasting", International Journal of Forecasting, Vol. 26, No. 1, Pp. 134-143, 2010.
51. M. Nakano, "Collaborative forecasting and planning in supply chains: The impact on performance in Japanese manufacturers", International Journal of Physical Distribution and Logistics Management, Vol. 39, No. 2, Pp. 84-105, 2009.
52. F.R. Johnston and J.E Boylan, "Forecasting for Items with Intermittent Demand", The Journal of the Operational Research Society, Vol. 47, No. 1, Pp. 113-121, 1996.
53. A.A. Syntetos and J.E. Boylan, "The accuracy of intermittent demand estimates", Internationa journal production economics, Vol. 21, Pp. 303-314, 2005.
54. J.D. Croston, "Forecasting and stock control for intermittent demands", Operational Research Quarterly, Vol. 23, No. 3, Pp. 289–303, 1972.
55. R. Snyder, "Forecasting sales of slow and fast moving inventories", European Journal of Operational Research, Vol. 140, Pp. 684–699, 2002.
56. North and pyke, "Probes of the technological future", HBR, 1969.
57. Bass king and passimier, "Application of the science and marketing management", new York, john wiley & sons inc., 1968.
58. R.D. Irwin, "Spener clark and hoguet, Business & economic fore casting", home wood, Illinois , inc ., 1961.
59. Hadley, Introduction to business statistics, san Francisco , holden day .inc 1968 Brown, less risk in inventory estimation, HBR, 1959.
60. Box-jenkins, Time series analyses, forecasting & control, san francisco , holden-day, inc, 1970.
61. Mcloughin and boyle, "Time series forecasting", American marketing association booklet, marketing research technique, 1962.

62. Hadley, "Introduction to business statistics (san Francisco, holden day .inc 1968) Oliver & boyd", Technique of production control ; imperial chemical industries, 1964.
63. Clelland, De coni, Brown. bush & murrary, "Basic statistics with business applications", New york, john wile & sons, inc, 1966.
64. Evons, "Macro- economics activity : theory", Forecasting & control, new York, harper & row publishers, inc., 1969.
65. Publication of survey research center, institute for social research, university of michigon and of bureau of the census.
66. Leontief , input output economics, new York , oxford university press, 1966.
67. Evans & presion, "Discussion papers 138 .' Wharton school of finance & commerce", the university of Pennsylvania.
68. Evans, "Macro-economics activity", Theory , forecasting & control, new York, harper & row publishers , inc., 1969.
69. Bass, "A new products growth models for consumers durables", Management Science, 1969.

CHAPTER 15

OPERATIONS RESEARCH

15.0. Introduction

The term Operations Research (OR) describes the discipline that is focused on the application of information technology for informed decision-making. In other words, OR represents the study of optimal resource allocation. The goal of OR is to provide rational bases for decision making by seeking to understand and structure complex situations, and to utilize this understanding to predict system behavior and improve system performance. Much of the actual work is conducted by using analytical and numerical techniques to develop and manipulate mathematical models of organizational systems that are composed of people, machines, and procedures. This article introduces some of the methods and application that are affiliated with OR, and elaborates on some of the benefits that may be gained by incorporating OR into the actual business framework. The tools of operations research are not from any one discipline; rather Mathematics, Statistics, Economics, Engineering, Psychology, etc. have contributed to this newer discipline of knowledge. Today, it has become a professional discipline that deals with the application of scientific methods for decision-making, and especially to the allocation of scare resources. In India first unit of OR started in the year 1957 with its base at RRL Hyderabad. The other group was set up in Defence Science Laboratory which was followed by similar units at different parts of the country. The popular journal of OPSEARCH was established in 1963, to promote research in this field. Keeping in view the critical economic situation which required drastic increase in production efficiency, OR activities were directed, in all areas of business activities. In the late 50's OR was introduced at university level. With the development of PC's the use of OR techniques became prominent and effective tool as large amount of computation is required to handle complex problems. Inrecent years application of OR techniques have achieved significance in all walk of life, may it be industry or office work for making strategically decisions more scientifically

Today, almost all organizations make use of OR techniques for decision-making at all levels. This general acceptance to OR has come as managers have learned the advantage of the scientific approach to all industrial problems. Some of the Indian organizations using operations research techniques to solve their varied complex problems are: Railways, Defence, Indian Airlines, Fertilizer Corporation of India, Delhi Cloth Mills, Tata Iron and Steel Co. etc. A purpose of OR is to provide a rational basis for making decisions in the absence of complete information. OR can also be treated as science devoted to describing, understanding and predicting the behaviour of systems, particularly man-machine systems

History of Operations Research: Operation Research is a relatively new discipline. Whereas 70 years ago it would have been possible to study mathematics, physics or engineering (for example) at university it would not have been possible to study Operation Research, indeed the term O.R. did not exist then. It was really only in the late 1930's that operational research began in a systematic fashion, and it started in the UK. As such it would be interesting to give a short history of operation research.

In year 1936 the British Air Ministry established Bawdsey Research Station, on the east coast, near Felixstowe, Suffolk, as the centre where all pre-war radar experiments for both the Air Force and the Army would be carried out. Experimental radar equipment was brought up to a high state of reliability and ranges of over 100 miles on aircraft were obtained.

In year 1937 the experimental radar station at Bawdsey Research Station was brought into operation and the information derived from it was fed into the general air-defense warning and control system. From the early warning point of view this exercise was encouraging, but the tracking information obtained from radar, after filtering and transmission through the control and display network, was not very satisfactory.

In July 1938 a second major air-defense exercise was carried out. Four additional radar stations had been installed along the coast and it was hoped that Britain now had an aircraft location and control system greatly improved both in coverage and effectiveness. Not so! The exercise revealed, rather, that a new and serious problem had arisen. This was the need to coordinate and correlate the additional, and often conflicting, information received from the additional radar stations. With the outbreak of war apparently imminent, it was obvious that something new - drastic if necessary - had to be attempted. Some new approach was needed.

In the summer of 1939 Britain held what was to be its last pre-war air defence exercise. It involved some 33,000 men, 1,300 aircraft, 110 antiaircraft guns, 700 searchlights, and 100 barrage balloons. This exercise showed a great improvement in the operation of the air defence warning and control system. The contribution made by the OR team was so apparent that the Air Officer Commander-in-Chief RAF Fighter Command (Air Chief Marshal Sir Hugh Dowding) requested that, on the outbreak of war, they should be attached to his headquarters at Stanmore in north London. Initially, they were designated the "Stanmore Research Section". In 1941 they were re-designated the "Operational Research Section" when the term was formalized and officially accepted, and similar sections set up at other RAF commands.

On May 15th 1940, with German forces advancing rapidly in France, Stanmore Research Section was asked to analyses a French request for ten additional fighter squadrons (12 aircraft a squadron - so 120 aircraft in all) when losses were running at some three squadrons every two days (i.e. 36 aircraft every 2 days). They prepared graphs for Winston Churchill (the British Prime Minister of the time), based upon a study of current daily losses and replacement rates, indicating how rapidly such a move would deplete fighter strength. No aircraft were sent and most of those currently in France were recalled.

In 1941, an Operational Research Section (ORS) was established in Coastal Command which was to carry out some of the most well-known OR work in World War II. The responsibility of Coastal Command was, to a large extent, the flying of long-range sorties by single aircraft with the object of sighting and attacking surfaced U-boats (German submarines). The technology of the time meant that (unlike modern day submarines) surfacing was necessary to recharge batteries, vent the boat of fumes and recharge air tanks. Moreover U-boats were much faster on the surface than underwater as well as being less easily detected by sonar. Thus the Operation Research started just before World War II in Britain with the establishment of teams of scientists to study the strategic and tactical problems involved in military operations. The objective was to find the most effective utilization of limited military resources by the use of quantitative techniques. Following the end of the war OR spread, although it spread in different ways in the UK and USA.

In 1951 a committee on Operations Research formed by the National Research Council of USA, and the first book on "Methods of Operations Research", by Morse and Kimball, was published.

In 1952 the Operations Research Society of America came into being. Success of Operations Research in army attracted the attention of the industrial mangers who were seeking solutions to their complex business problems. Now a days, almost every organization in all countries has staff applying operations research, and the use of operations research in government has spread from military to wide variety of departments at all levels.India was one the few first countries who started using operations research.

In India, Regional Research Laboratory located at Hyderabad was the first Operations Research unit established during 1949. At the same time another unit was set up in Defense Science Laboratory to solve the Stores, Purchase and Planning Problems. In 1953, Operations Research unit was established in Indian Statistical Institute, Calcutta, with the objective of using Operations Research methods in National Planning and Survey. In 1955, Operations

Research Society of India was formed, which is one of the first members of International Federation of Operations Research societies. **Today Operations Research is a popular subject in Management Institutes, Engineering Institutes and Schools of Mathematics.**

15.1. Contribution of Operation Research In Manufacturing Production

(i) As OR has made (over the years) significant contributions in virtually all industries, in almost all managerial and decision-making functions, and at most organizational levels, the list of OR applications is prodigious. Hence, this article focuses in the next few paragraphs on the manufacturing industry, and introduces some of the application where OR is being used. The term operations in OR may suggests that the manufacturing application category represents the original home of OR. That is not quite accurate, as the name originated from military operations, not business operations. Nevertheless, it is a true statement that OR's successes in contemporary business pervade manufacturing and service operations, logistics, distribution, transportation, and telecommunication. The myriad applications include scheduling, routing, workflow improvements, elimination of bottlenecks, inventory control, business process re-engineering, site selection, or facility and general operational planning. Revenue and supply chain management reflect two growing applications that are distinguished by their use of several OR methods to cover several functions.

Revenue management entails first to accurately forecasting the demand, and secondly to adjust the price structure over time to more profitably allocate fixed capacity. Supply chain decisions describe the who, what, when, and where abstractions from purchasing and transporting raw materials and parts, through manufacturing actual products and goods, and finally distributing and delivering the items to the customers. The prime management goal here may be to reduce overall cost while processing customer orders more efficiently than before. The power of utilizing OR methods allows examining this rather complex and convoluted chain in a comprehensive manner, and to search among a vast number of combinations for the resource optimization and allocation strategy that seem most effective, and hence beneficial to the operation.

(ii) Businesses and organizations frequently face challenging operational problems whose successful solution requires certain expertise in applied statistics, optimization, stochastic modeling, or a combination of these areas. To illustrate, a company may need to design a sampling plan in order to meet specific quality control objectives. In a manufacturing environment, operations that compete for the same resources must be scheduled in a way that deadlines are not violated. The manager of a supermarket must determine how many checkout lines to keep open at various times during the day and evening so that shoppers are not unnecessarily delayed. Or as a final example, the size of the areas reserved for storing work in process at anumber of bottleneck stations has to be determined so that a smooth flow of work results, even at the busiest (peak) production times.

The area of operations research that concentrates on real-world operational problems is called production systems. Production systems problems may arise in settings that include, but are not limited to, manufacturing, telecommunications, health-care delivery, facilitylocation and layout, and staffing.

The area of production systems presents special challenges for operations researchers. Production problems are operations research problems, hence solving them requires a solid foundation in operations research fundamentals. Additionally, the solution of production systems problems frequently draws on expertise in more than one of the primary areas of operations research,implying that the successful production researcher cannot be one-dimensional. Furthermore, production systems problems cannot be solved without an in-depth understanding of the real problem, since invoking assumptions that simplify the mathematical structure of the problem may lead to an elegant solution for the wrong problem. Common sense and practical insight are common attributes of successful production planners. At the current time, the field of OR is extremely dynamic and ever evolving. To name a few of the contemporary (primary) research projects, current work in OR seeks to develop software for material flow analysis and design of flexible manufacturing facilities using pattern recognition and graph theory algorithms.

Further, approaches for the design of re-configurable manufacturing systems and progressive automation of discrete manufacturing systems are under development. Additional OR projects focus on the industrial deployment of computer-based methods for assembly line balancing, business process reengineering, capacity planning, pull scheduling, and setup reduction, primarily through the integration of the philosophies of the Theory of Constraints and Lean Manufacturing.

Please see Appendix A for a short introduction to the objectives and key principles of the Lean Manufacturing principles).

15.2. Definition of Operation Research

There are so many definitions are given in previous literature by the various authors. Some of the definitions regarding the subject point of view some are given below.

(i) According to Pocock , OR is defined as .... It is an applied science.

OR is scientific methodology-analytical, experimental, quantitative—which by assessing the overall implication of various alternative courses of action in a management system, provides an improved basis for management decisions.

(ii) According to Morse and Kimball

A scientific method of providing executive departments with a quantitative basis for decisions regarding the operations under their control.

(iii) According to Miller and Starr

OR is applied decision theory. It uses any scientific, mathematical or logical means to attempt to cope with the problems that confront the executive, when he tries to achieve a thorough going rationality in dealing with his decision problem.

(iv) According to Saaty

OR is the art of giving bad answers to problems which otherwise have worse answers.

(v) According to Operations Research Society, America

OR is concerned with scientifically deciding how to best design and operate man-machine system usually requiring the allocation of scare resources.

(vi) According to Daellenbach and George

OR is essentially a collection of mathematical techniques and tools which in conjunction with system approach, are applied to solve practical decision problems of an economic or engineering nature.

(vii) According toThieraub and Klekamp

OR utilizes the planned approach (updated scientific method) and an interdisciplinary team in order to represent complex functional relationships as mathematical models for the purpose of providing a quantitative analysis.

(viii) According toH.A. Taha

OR is a scientific knowledge through interdisciplinary team effort for the purpose of determining the best utilization of limited resources.

(ix) According toH.M. Wagner

OR is a scientific approach to problem solving for executive management.

Objectives of Operation Research

(i) Decrease Cost or Investment.

(ii) Increase Revenue or Return on Investment.

(iii) Increase Market Share.

(iv) Manage and Reduce Risk.

(v) Improve Quality.

(vi) Increase Throughput while Decreasing Delays.

(vii) Achieve Improved Utilization form Limited Resources.

(viii) Demonstrate Feasibility and Workability.

15.3. Important Aspect of Operation Research

There are so many aspects are covered by the operation research. Some of them are given below.

(i) Decision-making: Every industrial organization faces multifaceted problems to identify best possible solution to their problems. OR aims to help the executives to obtain optimal solution with the use of OR techniques. It also helps the decision maker to improve his creative and judicious capabilities, analyze and understand the problem situation leading to better control, better co-ordination, better systems and finally better decisions.

(ii) Scientific Approach: OR applies scientific methods, techniques and tools for the purpose of analysis and solution of the complex problems. In this approach there is no place for guess work and the person bias of the decision maker.

(iii) Inter-disciplinary: Team Approach. Basically the industrial problems are of complex nature and therefore require a team effort to handle it. This team comprises of scientist/mathematician and technocrats. Who jointly use the OR tools to obtain a optimal solution of the problem. The tries to analyze the cause and effect relationship between various parameters of the problem and evaluates the outcome of various alternative strategies.

(iv) System Approach: The main aim of the system approach is to trace for each proposal all significant and indirect effects on all sub-system on a system and to evaluate each action in terms of effects for the system as a whole. The interrelationship and interaction of each sub-system can be handled with the help of mathematical/analytical models of OR to obtain acceptable solution.

(v) **Use of Computers:** The models of OR need lot of computation and therefore, the use of computers becomes necessary. With the use of computers it is possible to handle complex problems requiring large amount of calculations. The objective of the operations research models is to attempt and to locate best or optimal solution under the specified conditions. For the above purpose, it is necessary that a measure of effectiveness has to be defined which must be based on the goals of the organization. These measures can be used to compare the alternative courses of action taken during the analysis.

15.4. Important Steps in Operation Research

OR is a logical and systematic approach to provide a rational basis for decision-making , The various steps required for the analysis of a problem under , there are six steps are generally applied in operation research . These steps are given below.

- **Step I:** Observe the problem environment.
- **Step II:** Analyze and define the problem.
- **Step III:** Develop a model.
- **Step IV:** Select appropriate data input.
- **Step V:** Provide a solution and test its reasonableness.
- **Step VI:** Implement the solution.

(i) **Step I: Observe the problem environment: The** first step in the process of O.R. development is the problem environment observation. This step includes different activities; they are conferences, site visit, research, observations etc. These activities provide sufficient information to the O.R. specialists to formulate the problem.

(ii) **Step II: Analyze and define the problem:** This step is analyzing and defining the problem. In this step in addition to the problem definition the objectives, uses and limitations of O.R. study of the problem also defined. The outputs of this step are clear grasp of need for a solution and its nature understanding.

(iii) **Step III: Develop a model:** This step develops a model; a model is a representation of some abstract or real situation. The models are basically mathematical models, which describes systems, processes in the form of equations, formula/relationships. The different activities in this step are variables definition, formulating equations etc. The model is tested in the field under different environmental constraints and modified in order to work. Sometimes the model is modified to satisfy the management with the results.

(iv) **Step IV: Select appropriate data Input:** A model works appropriately when there is appropriate data input. Hence, selecting appropriate input data is important step in the O.R. development stage or process. The activities in this step include internal/external data analysis, fact analysis, and collection of opinions and use of computer data banks. The objective of this step is to provide sufficient data input to operate and test the model developed in

(v) **Step V: Provide a solution and test its reasonableness:** This step is to get a solution with the help of model and input data. This solution is not implemented immediately, instead the solution is used to test the model and to find there is any limitations. Suppose if the solution is not reasonable or the behaviour of the model is not proper, the model is updated and modified at this stage. The output of this stage is the solution(s) that supports the current organizational objectives.

(vi) **Step VI: Implement the solution:** At this step the solution obtained from the previous step is implemented. The implementation of the solution involves mo many behavioural issues. Therefore, before implementation the implementation authority has to resolve the issues. A properly implemented solution results in quality of work and gains the support from the management.

15.5. Applications of Operations Research

Today, almost all fields of business and government utilizing the benefits of Operations Research. There are voluminous of applications of Operations Research. Although it is not feasible to cover **all** applications of O.R. in brief. Or is applied in various field like Accounting , Construction, Facilities Planning, Finance, Manufacturing, Marketing, Research and Development .

Accounting

- Assigning audit teams effectively.
- Credit policy analysis.
- Cash flow planning.
- Developing standard costs.
- Establishing costs for byproducts.
- Planning of delinquent account strategy.

Construction

- Project scheduling, monitoring and control.
- Determination of proper work force.

- Deployment of work force.
- Allocation of resources to projects.

Facilities Planning

- Factory location and size decision.
- Estimation of number of facilities required.
- Hospital planning.
- International logistic system design.
- Transportation loading and unloading.
- Warehouse location decision.

Finance

- Building cash management models.
- Allocating capital among various alternatives.
- Building financial planning models.
- Investment analysis.
- Portfolio analysis.
- Dividend policy making.

Manufacturing

- Inventory control.
- Marketing balance projection.
- Production scheduling.
- Production smoothing.

Marketing

- Advertising budget allocation.
- Product introduction timing.
- Selection of Product mix.
- Deciding most effective packaging alternative.

Organizational Behavior / Human Resources

- Personnel planning.
- Recruitment of employees.
- Skill balancing.
- Training program scheduling.
- Designing organizational structure more effectively.

Purchasing

- Optimal buying.
- Optimal reordering.
- Materials transfer.

Research and Development

- R & D Projects control.
- R & D Budget allocation.
- Planning of Product introduction.

15.6. Introduction to Linear Programming

Linear Programming is a special and versatile technique which can be applied to a variety of management problems viz. Advertising, Distribution, Investment, Production, Refinery Operations, and Transportation analysis. The linear programming is useful not only in industry and business but also in non-profit sectors such as Education, Government, Hospital, and Libraries. The linear programming method is applicable in problems characterized by the presence of decision variables. The objective function and the constraints can be expressed as linear functions of the decision variables. The decision variables represent quantities that are, in some sense, controllable inputs to the system being modeled. An objective function represents some principal objective criterion or goal that measures the effectiveness of the system such as maximizing profits or productivity, or minimizing cost or consumption. There is always some practical limitation on the availability of resources viz. man, material, machine, or time for the system. These constraints are expressed as linear equations involving the decision variables. Solving a linear programming problem means determining actual values of the decision variables that optimize the objective function subject to the limitation imposed by the constraints. The main important feature of linear programming model is the presence of linearity in the problem. The use of linear programming model arises in a wide variety of applications. Some model may not be strictly linear, but can be made linear by applying appropriate mathematical transformations. Still some applications are not at all linear, but can be effectively approximated by linear models. The ease with which linear programming models can usually be solved makes an attractive means of dealing with otherwise intractable nonlinear models.

Formulating a Linear Programming

(i) Choose decision variables.

(ii) Choose an objective and an objective function – linear function in variables.

(iii) Choose constraints – linear inequalities.

(iv) Choose sign restrictions.

Linear Program (LP) in a Standard Form (Maximization)

Max $c_1x_1 + c_2x_2 + ... + c_nx_n$... Objective function

Subjected to $a_{11}x_1 + a_{12}x_2 + ... + a_{1n}x_n \leq b1$ .... (1)

$a_{21}x_1 + a_{22}x_2 + ... + a_{2n}x \leq b$ .... (2)

$a_{m1}x_1 + a_{m2}x_2 + ... + a_{mn}x_n \leq b_m$ ....(3)

Where -- $x_1, x_2, ..., x_n \geq 0$

Equation 1, 2, 3 .... are constraints of the problem.

Feasible solution (point) P = $(p_1, p_2, ..., p_n)$ is an assignment of values to the $p_1, ..., p_n$ to variables $x_1, ..., x_n$ that satisfied all constraints and all sign restrictions.

Feasible region = the set of all feasible points.

Optimal solution = A feasible solution with maximum value of the objective function.

Problem 15.61 --Suppose an industry is manufacturing two types of products P1 and P2. The profits per Kg of the two products are Rs.30 and Rs.40 respectively. These two products require processing in three types of machines. The following table shows the available machine hours per day and the time required on eachmachine to produce one Kg of P1 and P2. Formulate the problem in the form of linear programming model.

Profit/Kg	P_1(Rs.30)	P_2(Rs.40)	Total available machine hour / day
Machine 1	3	2	600
Machine 2	3	5	800
Machine 3	5	6	1100

Solution

The procedure for linear programming problem formulation is as follows:

Introduce the decision variable as follows:

Let x_1 = amount of P_1

x_2 = amount of P_2

In order to maximize profits, the objective function will be,

Objective function = Maximize $(30x_1 + 40x_2)$

Since one Kg of P_1 requires 3 hours of processing time in machine 1. On the other handcorresponding requirement of P_2 is 2 hours. So therefore the constraints will be expressed as:

$3x1 + 2x2 \leq 600$ .... Constraints for machine - **1**

$3x1 + 5x2 \leq 800$ .... Constraints for machine -**2**

$5x1 + 6x2 \leq 1100$ ....Constraints for machine -**3**

Where X_1 and $X_2$is always greater than or equal to zero.

Problem 15.62 -- a person want to decide the constituent of a diet which will fulfill his daily requirement of vitamins, proteins and carbohydrates at the minimum cost. The tables show the requirements of the foods elements.

Food type	Yields per units			Cost per units **(Rs.)**
	Vitamins	Proteins	carbohydrates	
1	4	3	6	42
2	5	2	5	43
3	8	6	5	84
4	7	4	6	68
Minimum requirement	800	200	700	

Formulate the linear programming model for the problem.

Let x_1 , x_2 , x_3 and x_4 is representing the no of units of foods of type 1 , 2, 3, 4 .

Here our objective function is to minimizing the cost per elements.

Minimize (Z) = Mini($42x_1 + 43\ x_2 + 84x_3 + 68x_4$)

Constraints related to problem are

For Vitamins .... $4x_1 + 5x_2 + 8x_3 + 7x_4 \geq 800$

For Proteins .... $3x_1 + 2x_2 + 6x_3 + 4x_4 \geq 200$

For carbohydrates .... $6x_1 + 5x_2 + 5x_3 + 6x_4 \geq 700$

Where $x_1, x_2, x_3, x_4 \geq 0$

Graphical Solution of Linear Programming

Problem 15.63 : Suppose an industry is manufacturing two types of products P1 and P2. The profits per Kg of the two products are Rs.30 and Rs.40 respectively. These two products require processing in three types of machines. The following table shows the available machine hours per day and the time required on eachmachine to produce one Kg of P1 and P2. Formulate the problem in the form of linear programming model.

Profit/Kg	**P_1**(Rs.30)	**P_2**(Rs.40)	**Total available machine hour / day**
Machine 1	3	2	600
Machine 2	3	5	800
Machine 3	5	6	1100

Objective function is to maximize the profit.

Objective function = Maximize (30x1 + 40x2)

Subject to constraints :-

$$3x_1 + 2x_2 \leq 600 \quad \text{.......... (1)}$$

$$3x_1 + 5x_2 \leq 800 \quad \text{.......... (2)}$$

$$5x_1 + 6x_2 \leq 1100 \quad \text{..........(3)}$$

Where $x1 \geq 0, x2 \geq 0$

- From the first constraints $3x1 + 2x2 \leq 600$, we have a line $3x1 + 2x2 = 600$ which passes through the point (200, 0) and (0, 300)
- From the second constraints $3x1 + 5x2 \leq 800$, we have a line $3x_1 + 5x_2 = 800$ which passes through the point (266, 0) and (0, 160)
- From the third constraints $5x1 + 6x2 \leq 1100$, we have a line $5x_1 + 6x_2 = 1100$ which passes through the point (220, 0) and (0, 183)

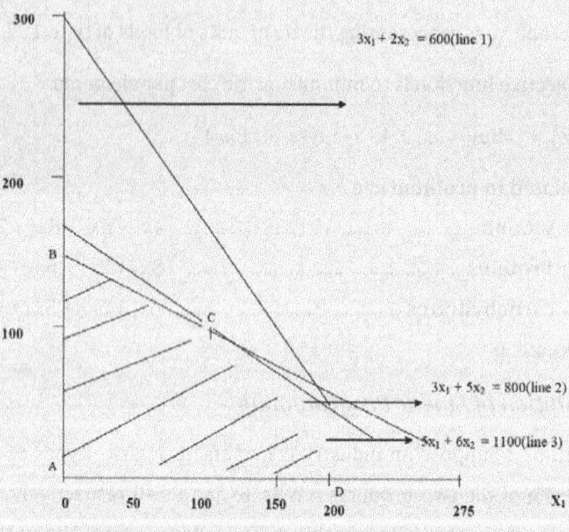

Graphical Representation of the Constraints

In this case we must decide in which side of the line $3x1 + 2x2 = 600$ the half plane is located. The easiest way to solve the inequality for x2 is $3x1 \leq 600 - 2x2$ And for the fixed x1, the coordinates satisfy this inequality are smaller than the corresponding ordinate on the line and thus the inequality is satisfied for all the points below the line 1. Similarly, we have to determine the closed half planes for the inequalities $3x1 + 5x2 \leq 800$ and $5x1 + 6x2 \leq 1100$

(line 2 and line 3 in the graph). Since all the three constraints must be satisfied simultaneously we have consider the intersection of these three closed half planes. The complete intersection of these three closed half planes is shown in the above graph as ABCD. The region ABCD is called the feasible region, which is shaded in the graph.

Feasible Solution: Any non-negative value of x1, x2 that is x1 ≥ 0 and x2 ≥ 0 is known as feasible solution of the linear programming problem if it satisfies all the existing constraints.

Feasible Region: The collection of all the feasible solution is called as the feasible region.

Problem 15.64: A company purchasing scrap material has two types of scarp materials available. The first type has 30% of material X, 20% of material Y and 50% of material Z by weight. The second type has 40% of material X, 10% of material Y and 30% of material Z. The costs of the two scraps are Rs.120 and Rs.160 per kg respectively. The company requires at least 240 kg of material X, 100 kg of material Y and 290 kg of material Z. Find the optimum quantities of the two scraps to be purchased so that the company requirements of the three materials are satisfied at a minimum cost.

Solution

First we have to formulate the linear programming model. Let us introduce the decision variables x_1 and x_2 denoting the amount of scrap material to be purchased.

Here the objective is to minimize the purchasing cost. So, the objective function here is

Objective function = Minimize $(120x_1 + 160x_2)$

Subject toconstraints :- $0.3x_1 + 0.4x_2 \geq 240$ .... (1)

$0.2x_1 + 0.1x_2 \geq 100$ ....(2)

$0.5x_1 + 0.3x_2 \geq 290$ ....(3)

Where x1 ≥ 0; x2 ≥ 0

Multiply by 10 both sides of the inequalities, then the problem becomes,

Minimize $(120x_1 + 160x_2)$ ....(4)

Subject to: $3x_1 + 4x_2 \geq$ **2400** .... **(5)**

$2x_1 + x_2 \geq$ **1000** .... **(6)**

$5x_1 + 3x_2 \geq$ **2900** .... **(7)**

Where x1 ≥ 0; x2 ≥ 0 (always)

In next steps we will calculate the minimum value of objective function.

The various values of the are tabulated below

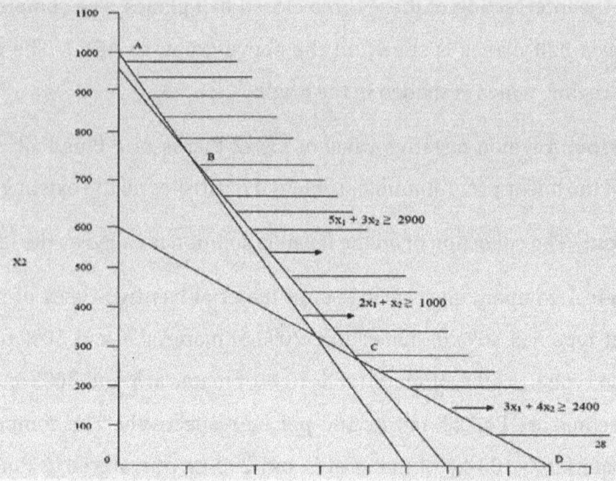

Extreme Point	Coordinate		Value of objective function
A	$X_1 = 0$	$X_2 = 1000$	160000
B	$X_1 = 150$	$X_2 = 740$	136400
C	$X_1 = 400$	$X_2 = 300$	96000
D	$X_1=800$	$X_2=0$	96000

Conclusion :- The value of objective function is at point "c" = **96000** and Optimum value variables are **x1 = 400, and x2 = 300**. Similarly. And at the point D the value of objective function is =**96000**, and the optimum value variables are **x1 = 800, and x2 = 0.**

From the conclusion, every point on the line CD minimizes objective function value and the problem contains multiple optimal solutions.

15.7. Transportation Problem

A key problem manager's face is how to allocate scarce resources among various activities or projects. Linear programming, or LP, is a method of allocating resources in an optimal way. It is one of the most widely used operations research tools and has been a decision-making aid in almost all manufacturing industries and in financial and service organizations. One of the most important and successful applications of quantitative analysis to solving business problems has been in the physical distribution of products, commonly referred to as transportation problems. Basically, the purpose is to minimize the cost of shipping goods from one location to another so that the needs of each arrival area are met and every shipping location operates within its capacity. However, quantitative analysis has been used for many problems other than the physical distribution of goods.

Mathematical Formulation

Let us consider the m-plant locations (origins) as O_1 , O_2 , ..., O_m and the n-retail shops (destination) as D_1 , D_2 , ..., $D_n$respectively. Let $a_i \geq 0$, $i= 1,2, ...m$, be the amount available at the ith plant O_i.

Let the amount required at the jth shop D_j be $b_j \geq 0$, $j= 1,2,....n$.Let the cost of transporting one unit of soft drink form ith origin to *jth* destination be C_{ij}, $i= 1,2, ...m$, $j=1,2,....n$. If $x_{ij} \geq 0$ be the amount of soft drink to be transported from i^{th} origin to *jth* destination , then the problem is to determine x_{ij} so as to Minimize.

$$z = \sum_{i=1}^{m} \sum_{j=1}^{n} x_{ij} c_{ij}$$

Subjected to Constraints

and $x_{ij} \geq 0$, for all i and j.

$$\sum_{j=1}^{n} x_{ij} = a_i, i = 1,2,..m \qquad \sum_{i=1}^{m} x_{ij} = b_j, j = 1,2,...n.$$

This linear programming problem is called a ***Transportation Problem.***

A necessary and sufficient condition for the existence of a feasible solution to the transportation problem is that

$$\sum_{i=1}^{m} a_i = \sum_{j=1}^{n} b_j$$

$$\sum_{i=1}^{m} x_{ij} = b_j \text{ and } \sum_{j=1}^{n} x_{ij} = a_i$$

Remark. The set of constraints, Represents *(m+n) equations* in (mn) non -negative variables.

Each variable x_{ij} appears in exactly two constraints, one is associated with the origin and the other is associated with the destination.

The Transportatyion Table

		D_1	D_2	......	......	D_n	Supply
	O_1	C_{11}	C_{12}	......	C_{1n}	a_1	
	O_2	C_{21}	C_{22}	......	C_{2n}	a_2	
	O_m	C_{m1}	C_{m2}	......	C_{mn}	a_m	
Requirement		b_1	b_2	.....	....	b_n	

Definition. (Loop). In a transportation table, an ordered set of four or more cells is said to form a loop if:

(i) Any two adjacent cells in the ordered set lie in the same row or in the same column.

(ii) Any three or more adjacent cells in the ordered set do not lie in the same row or in the same column.

Result

A feasible solution to a transportation problem is basic if and only if the corresponding cells in the transportation table do not contain a loop. To find an initial basic feasible solution we apply:

(i) The North-West corner rule.

(ii) Vogel`s Approximation method.

The North-West Corner Rule

Step (1)- The first assignment is made in the cell occupying the upper left-hand (North West) corner of the transportation table. The maximum feasible amount is allocated there,

$$\textbf{i.e. ; } x_{11} = \min(a_1, b_1).$$

Step (2)- If $b_1 > a_1$, the capacity of origin O_1 is exhausted but the requirement at D_1 is not satisfied. So move downs to the second row, and make the second allocation : -

$x_{21} = \min(a_2, b_1 - x_{11})$ in the cell (2,1).If $a_1 > b_1$, allocate $x_{12} = \min(a_1 - x_{11}, b_2)$ in the cell (1,2) .Continue this until all the requirements and supplies are satisfied.

15.8. Vogel's Approximation Method (VAM)

Step 1- For each row of the transportation table, identify the smallest and the next to-smallest costs. Determine the difference between them for each row. Display them alongside the transportation table by enclosing them in parenthesis against the respective rows. Similarly compute the differences for each column.

Step 2- Identify the row or column with the largest difference among all the rows and columns. If a tie occurs, use any arbitrary tie breaking choice. Let the greatest difference correspond to i^{th} row and the minimum cost be C_{ij} . Allocate a maximum feasible amount

$$x_{ij} = \min (a_i , b_j) \text{ in the } (i, j)^{th} \text{ cell, and cross off the } i^{th} \text{ row or } j^{th} \text{ column.}$$

Step 3- Re compute the column and row differences for the reduced transportation table and go to next step .

step 2 - Repeat the procedure until all the rim requirements are satisfied. VAM calculate an initial basic feasible solution, which is very close to the optimum solution.

Problem 15.65 : Solve the following transportation problem by North-West corner rule, Row Minima, Column Minima, Matrix Minima and VAM Method.

Factories	W1	W2	W3	W4	Supply
F1	6	4	1	5	14
F2	8	9	2	7	16
F3	4	3	6	2	05
Demand	6	10	15	4	35

Solution : Since is a balanced transportation problem, since supply is equal to demand .

(i) North-West Corner Rule Method

Factories	W1	W2	W3	W4	Supply
F1	6(6)	4(8)	1	5	14
F2	8	9(2)	2(14)	7	16
F3	4	3	6(1)	2(4)	05
Demand	06	10	15	04	35

The Total feasible transportation cost = **6(6) + 4(8) + 9(2) + 2(14) + 6(1) + 2(4) = Rs. 128/-**

(ii) Row Minima Method

Factories	W1	W2	W3	W4	Supply
F1	6	4	1(14)	5	14
F2	8(6)	9(5)	2(1)	7(4)	16
F3	4	3(5)	6	2	05
Demand	06	10	15	04	35

The Total feasible transportation cost = **1 (14) + 8(6) + 9(5) + 2(1) + 7(4) + 3(5) = Rs.152/-**

(iii) Column-Minima Method

Factories	W1	W2	W3	W4	Supply
F1	6(1)	4(10)	1(3)	5	14
F2	8	9	2(12)	7(4)	16
F3	4(5)	3	6	2	05
Demand	06	10	15	04	35

The Total feasible transportation cost = **6 (1) + 4 (10) + 1(3) + 2(12) + 7(4) + 4(5) = Rs. 121/-**

(iv) Matrix-Minima Method or Least Cost Method

Factories	W1	W2	W3	W4	Supply
F1	6	4	1(14)	5	14
F2	8(6)	9(9)	2(1)	7	16
F3	4	3(1)	6	2(4)	05
Demand	06	10	15	04	35

The Total feasible transportation cost = **1(14) + 8(6) + 9(9) + 2(1) + 3(1) + 2(4) = 156**

Problem 15.66 : - There are 3 Parties who supply the following quantity of coal P1= 14t, P2=12t, P3 = 5t.There are 3 consumers who require the coal as follows C1 = 6t, C2 = 10t, C3 = 15t. The cost matrix in Rs. Per ton is as follows. Find the schedule of transportation policy which minimizes the cost:

6	8	4
4	9	3
1	2	6

Solution:

Factories	W1	W2	W3	Supply
F1	6	8	4	14
F2	4	9	3	12
F3	1	2	6	5
Demand	6	10	15	31

Factories	W1	W2	W3	Supply
F1	6	8(5)	4(9)	14
F2	4(6)	9	3(6)	12
F3	1	2(5)	6	05
Demand	6	10	15	31

Therefore the total feasible transportation cost

= 8 (5) + 4(9) + 4(6) + 3(6) + 2(5) = Rs. 128/-

I. Check for Degeneracy :- (m+n-1) = (3+3-1) = 5 = Number of allocations, Hence there is no degeneracy.

II. Checking optimality using MODI method.

- For allocated cells Cij – (U_i+V_j) = 0
- For unallocated cells C_{ij} – (U_{ij}) +V_j) = 0

V1=5	V2=8	V3=4	
6	8(5)	4(9)	U1= 0
4(6)	9	3(6)	U2=-1
1	2(5)	6	U3=-6

Since all the marginal costs for the unallocated cells are positive, it gives an optimal solution and the total minimum transportation cost = Rs. 128/

Problem 15.67 : A company has three factories F_1, F_2, F_3. from which transport to 4 ware houses W_1, W_2, W_3, W_4. A unit cost of production at the three factories is RS – 4, 3, 5. Given the following data unit cost of transportation. Capacity and requirement are given below. find the optimal allocation.

Factories	W1	W2	W3	W4	SUPPLY
F1	5	7	3	8	300
F2	4	6	9	5	500
F3	2	6	4	5	200
	200	300	400	100	1000

Since the unit cost of production is Rs. 4, 3 and 5 at the three factories, therefore

Total cost = Production cost + cost

Factories	W1	W2	W3	W4	Supply
F1	9	11	7(300)	12	300
F2	7(100)	9(300)	12	8(100)	500
F3	7(100)	11	9(100)	10	200
Demand	200	300	400	100	1000

Therefore the total feasible solution

= 7 (300) + 7 (100) +9 (300) + 8 (100) +7 (100) + 9 (100) = Rs. 7900 /-

Factories	W1	W2	W3	W4	Supply
F1	9	11	7(300)	12	300
F2	7(100)	9(300)	12	8(100)	500
F3	7(100)	11	9(100)	10	200
Demand	200	300	400	100	1000

Therefore the total feasible transportation solution

= 7(300) + 7(100) + 9(300) + 8(100) + 7(100) + 9(100)

= Rs.7900/-

I. Check for Degeneracy :- (m+n-1) = (3+4-1) = 6 = Number of allocations , Hence there is no degeneracy.

II. Checking optimality using MODI method,

- For allocated cells Cij – $(U_i+V_j) = 0$
- For unallocated cells $C_{ij} - (U_{ij}) + V_j) \geq 0$

V_1	V_2	V_3	V_4	
9	11	7 (300}	12	U1 = -2
7(100)	9(300)	12	8(100)	U2 = 0
7(100)	11	9(100)	10	U3 = 0

Problem 15.68 Four factories manufacturing the same products supplies to three sales department s_1, s_2, s_3. Determine the most profitable production and distribution schedule and corresponding profits.

Departmentas					Supply	Selling price
S1	3	9	5	4	80	34
S2	1	7	4	5	120	32
S3	5	8	3	6	50	31
Demand	10	150	150	100		
Production-cost /units	15	18	14	13		
Material cost / units	10	9	12	9		

Solution

Departmentas	**F1**	**F2**	**F3**	**F4**	**Supply**	
S1	**6**	**-2**	**3**	**8**		**80**
S2	**6**	**-2**	**2**	**5**		**120**
S3	**1**	**-4**	**2**	**3**		**150**
Demand	**10**	**150**	**150**	**100**	**410**	**350**

Departmentas	**F1**	**F2**	**F3**	**F4**	**Supply**
S1	**2**	**10**	**5**	**0**	**80**
S2	**3**	**10**	**6**	**3**	**120**
S3	**7**	**12**	**6 (150)**	**5**	**150**
D	**0**	**0**	**0**	**0 (60)**	**60**
Demand	**10**	**150**	**150**	**100**	**410**

Departmentas	F_1	F_2	F_3	F_4	**Supply**
S_1	**2**	**10 (40)**	**5**	**0 (40)**	**80**
S_2	**3 (10)**	**10(10)**	**6**	**3**	**120**
S_3	**7**	**12**	**6 (150)**	**5**	**150**
D	**0**	**0**	**0**	**0 (60)**	**60**
Demand	**10**	**150**	**150**	**100**	**410**

Initial Feasible Solution

Z = (-2 x 40) + (8 x 40) + (6 x 10) + (-2 x 110) + (2 x 150).

Check for Degeneracy

(m + n -1) = 7 hence degeneracy exists.

$V_1 = 2$	$V_2 = 10$	$V_3 = 1$	$V_4 = 0$	
2	10(40)	5	0(40)	U1 = 0
2(10)	10(110)	6	3	U2 = 0
7	12	6(150)	5	U3 = 5
0 (-2)	0(-10)	0	0(60)	U4 = 0

V1 = 2	V2 = 10	V3 = 1	V4 = 0	
2	10(40 - x)	5	0(40 + x)	U1 = 0
2(10)	10(110)	6	3	U2 = 0
7	12	6(150)	5	U3 = 5
0 (-2)	0(-10)	0	0(60 - x)	U4 = 0

V1 = 2	V2 = 10	V3 = 1	V4 = 0	
2	10(40 - x)	5	0(40 + x)	U1 = 0
2(10)	10(110)	6	3	U2 = 0
7	12	6(150)	5	U3 = 5
0 (-2)	0(x)	0	0(60 - x)	U4 = 0

V1 = -8	V2 = 0	V3 = 1	V4 = 0	
2	10(40 - x)	5	0(40 + x)	U1 = 0
2(10)	10(110)	6	3	U2 = 0
7	12	6(150)	5	U3 = 5
0	0(x)	0	0(60 - x)	U4 = 0

V1 = -8	V2 = 0	V3 = 1	V4 = 0	
2	10	5	0(80)	U1 = 0
2(10)	10(110)	6 (-5)	3(-7)	U2 = 10
7	12	6(150)	5	U3 = 5
0	0(40)	0	0(20)	U4 = 0

V1 = -8	V2 = 0	V3 = 1	V4 = 0	
2	10	5	0(80)	U1 = 0
2(10)	10(110 - x)	6 (-5)	3(x)	U2 = 10
7	12	6(150)	5	U3 = 5
0	0(40 + x)	0	0(20 - x)	U4 = 0

V1 = -1	V2 = 7	V3 = 1	V4 = 0	
2	10	5	0(80)	U1 = 0
2(10)	10(110 - x)	6 (-5)	3(x)	U2 = 3
7	12	6(150)	5	U3 = 5
0	0(40 + x)	0	0(20 - x)	U4 = - 7

Thus Z = 8x80 + 6x10 + (-2x90) + 5x20 + 2x 150 = 920 /-

Question for Discussion

1. Discuss the importance of operation research in the decision-making process.
2. What are the steps involved in operation research problems ?
3. What is linear programming? What are its major assumptions and characteristics ?
4. Old hens can be brought at Rs 20 each and young ones at Rs 50 each. The old hens lay 3 eggs per week"and the young ones 5 eggs per week, each egg being worth Rs 3 a hen costs Rs 10 per week to be fed. If there are only Rs 800 available to be spent on purchasing the hens and at the most 20 hens can be accommodated in the space. Formulate this problem as LP model to determine each kind of hen that should be brought to have a maximum profit per week.
5. Solve the following LP problem using graphical method: Maximize $z = 10x_1 + 15x_2$ subject to the constraints $2x_1 + x_2 \leq 26$, $2x_1 + 4X_2 \leq 56$, $-x_1 + x_2 \leq 5$, X_1 & $X_2 \geq 0$.
6. What is linear programming and transportation model.

References

1. R.E. Walpole, "Introduction to Statistics", 3rd edn, Macmillan, New York, 1982.
2. D.J. White, "Operational Research", John Wiley & Sons, New York, 1985.
3. E.F. Wolstenholme and D.A. Corben, "A hypermedia-based Delphi tool for knowledge acquistion in model building", Journal of the Operational Research Society, Vol. 45, No. 6, Pp. 659–672, 1994.
4. F. Song, T. Rathwell, A. Long and D. Clayden, "Three value estimates of input variables and health manpower projections", Journal of the Operational Research Society, Vol. 45, No. 5, 1994.
5. A.K. Shahani, N. Korve, K.P. Jones and D.J. Paynton, Journal of the Operational Research Society, Vol. 45, No. 8, Pp. 916–926, 1994.
6. M. Pidd, "The operational research method", Operational Research for Managers, S.C. Littlechild (ed.), Philip Allan, Oxford, 1977.
7. E.M. Nascimento and J.E. Beasley, "Locating benefit posts in Brazil", Journal of the Operational Research Society, Vol. 44, No. 11, Pp. 1063–1066, 1993.
8. S. Datta and R. Bandyopadhyay, "An application of O.R. in micro-level planning in India", Computers & Operations Research, Vol. 20, No. 2, 1993.
9. T.W. Butler, "Management science/operations research projects in health care: The administrator's perspective", Health Care Management Review, Vol. 20, No. 1, Pp. 1-19, 1995.

10. R.L. Ackoff and P. Rivett, "A Manager's Guide to Operations Research", John Wiley & Sons, New York, 1963.
11. A. Bardossy, I. Bogardi and L. Duckstein, "Fuzzy nonlinear regression analysis of dose-response relationships", European Journal of Operational Research, Vol. 66, No. 1, Pp. 36–51, 1993.
12. D. Del Mar, "Operations and industrial management: Designing and managing for productivity", New York, McGraw-Hill, 1985.